OECD-FAO Agricultural Outlook 2010-2019

ORGANISATION FOR ECONOMIC CO-OPERATION AND DEVELOPMENT

FOOD AND AGRICULTURE ORGANIZATION OF THE UNITED NATIONS

ORGANISATION FOR ECONOMIC CO-OPERATION AND DEVELOPMENT

The OECD is a unique forum where governments work together to address the economic, social and environmental challenges of globalisation. The OECD is also at the forefront of efforts to understand and to help governments respond to new developments and concerns, such as corporate governance, the information economy and the challenges of an ageing population. The Organisation provides a setting where governments can compare policy experiences, seek answers to common problems, identify good practice and work to co-ordinate domestic and international policies.

The OECD member countries are: Australia, Austria, Belgium, Canada, Chile, the Czech Republic, Denmark, Finland, France, Germany, Greece, Hungary, Iceland, Ireland, Italy, Japan, Korea, Luxembourg, Mexico, the Netherlands, New Zealand, Norway, Poland, Portugal, the Slovak Republic, Spain, Sweden, Switzerland, Turkey, the United Kingdom and the United States. The Commission of the European Communities takes part in the work of the OECD.

OECD Publishing disseminates widely the results of the Organisation's statistics gathering and research on economic, social and environmental issues, as well as the conventions, guidelines and standards agreed by its members.

THE FOOD AND AGRICULTURE ORGANIZATION (FAO) OF THE UNITED NATIONS

The Food and Agriculture Organization of the United Nations leads international efforts to defeat hunger. FAO's mandate is to raise levels of nutrition, improve agricultural productivity, better the lives of rural populations and contribute to the growth of the world economy. Serving both developed and developing countries, FAO acts as a neutral forum where all nations meet as equals to negotiate agreements and debate policy. FAO is also a source of knowledge providing access to information in print and electronic format. We help developing countries and countries in transition modernize and improve agriculture, forestry and fisheries practices and ensure good nutrition for all. Since our founding in 1945, we have focused special attention on developing rural areas, home to 70 per cent of the world's poor and hungry people. FAO's activities comprise four main areas: putting information within reach; sharing policy expertise; providing a meeting place for nations; bringing knowledge to the field.

ISBN 978-92-64-08375-2 (print)
ISBN 978-92-64-08376-9 (PDF)

Also available in French: *Perspectives agricoles de l'OCDE et de la FAO 2010-2019*

Photo credits: Cover © iStockphoto.com/Robert Churchill.

Corrigenda to OECD publications may be found on line at: *www.oecd.org/publishing/corrigenda*.

Foreword

The annual Agricultural Outlook *is prepared jointly by the Organisation for Economic Co-operation and Development (OECD) and the Food and Agriculture Organization (FAO) of the United Nations. The projections and assessments provided in this report are the result of close co-operation with national experts in OECD countries, but also in several countries that are not members of the OECD, reflecting the combined knowledge and expertise of this wide group of collaborators. A jointly developed modelling system, based on the OECD's Aglink and FAO's Cosimo models, facilitates consistency in the projections. The fully documented* Outlook Database*, including historical data and projections, is available through the OECD-FAO joint website* www.agri-outlook.org.

This report covers biofuels, cereals, oilseeds, sugar, meats, and dairy products over the 2010-19 period. The market assessments are based on a set of underlying assumptions regarding macroeconomic factors, agricultural and trade policies and production technologies. They also assume normal average weather conditions and long-term productivity trends. The Outlook's *relatively stable price projections are highly conditional on these assumptions, and on the continuation of domestic policies and policy settings. For instance, an agreement of the Doha round of multilateral trade negotiations would likely have a considerable impact on the prospects for agricultural markets as contained in this assessment. Nevertheless, the* Agricultural Outlook *presents a consistent view on the evolution of global agricultural markets over the next decade and provides a baseline for further analysis of alternative economic or policy assumptions.*

Agriculture has experienced significant shocks in recent years due to high fluctuations in oil prices, commodity price spikes, food security fears and resultant trade restrictions, not to mention the most serious global economic recession since the 1930s. After a run up in prices, a rapid fall followed as supply and demand reacted to high prices and the beginning of the global economic crisis. In 2010, the economic turmoil eased and commodity prices increased from 2009 levels. Looking forward, the macroeconomic assumptions conditioning the commodity projections are more positive as compared to last year's report. The anticipated return to global economic growth, rising population, the emerging biofuel markets, but also a higher cost structure are expected to underpin the international commodity markets and prices over the Outlook *period. Developing countries are expected to be the driving force behind the expected growth in agricultural production, consumption and trade. However, the projections for the Least Developed Countries imply increased reliance on international markets, and growing exposure to commodity price changes and fluctuations in import bills.*

As high price volatility threatens farm viability, food security and needed investment, the fluctuation of commodity prices is an issue for many governments concerned about its impact on domestic producers and consumers. A chapter of this report discusses the issue of price volatility and price transmission from the international to domestic markets, and analyses certain policy options that address volatility at both the domestic and international levels. The report also identifies other trends and issues that will have an important impact on the agricultural sector and commodity markets in the future, such as developments in the fisheries sector, food security concerns, climate change and future policy directions as discussed at the OECD Agriculture Ministerial Meeting held in Paris in February 2010.

Acknowledgements

This *Agricultural Outlook* is jointly prepared by the OECD and FAO Secretariats.

At the OECD, the Outlook report was authored by the Agro-Food Trade and Markets Division of the Trade and Agriculture Directorate: Wayne Jones (Division Head), Pavel Vavra (Outlook and baseline co-ordinator), Martin von Lampe (baseline co-ordinator), Alexis Fournier, Linda Fulponi, Céline Giner, Pete Liapis, Garry Smith (Overview chapter), Gregoire Tallard and Shinichi Taya. Additional Directorate contributions were provided by Hsin Huang (Agricultural Policies and Environment Division) and Carl-Christian Schmidt (Fisheries Policies Division). The OECD Secretariat is grateful for the contributions provided by Hubertus Gay of the European Commission JRC-IPTS and Scott Pellow of the Agriculture and Agri-Food Canada. Research and statistical assistance were provided by Armelle Elasri, Alexis Fournier, Gaelle Gouarin and Claude Nenert. Meetings organisation and document preparation was provided by Christine Cameron. Technical assistance in the preparation of the Outlook database was provided by Frano Ilicic. Many other colleagues in the OECD Secretariat and member country delegations furnished useful comments on earlier drafts of the report.

At the FAO, the team of economists and commodity officers from the Trade and Markets Division contributing to this edition consisted of Alexander Sarris (Division Head), Merritt Cluff (Team Leader), Holger Matthey (baseline co-ordinator), Abdolreza Abbasssian, El Mamoun Amrouk, Pedro Arias, Concepcion Calpe, Joshua Dewbre, Cheng Fang, Adam Prakash, Matthieu Stigler, Peter Thoenes, and Doussou Traoré. Hansdeep Khaira contributed from the Statistics Division. Stefania Vannuccini and Audun Lem contributed from the Fisheries and Aquaculture Department. Research assistance and database preparation was provided by Claudio Cerquilini, Berardina Forzinetti, John Heine, Marco Milo, and Barbara Senfter. Secretarial services were provided by Rita Ashton and Valentina Banti.

Finally, the assistance and cooperation of the Executive Director, Peter Baron, and staff of the International Sugar Organisation (ISO) in London, in reviewing the country level projections and providing information on the market outlook for sugar and key emerging issues is gratefully acknowledged.

Table of Contents

Tables

Figures

This book has...

StatLinks

A service that delivers Excel® files from the printed page!

Look for the *StatLinks* at the bottom right-hand corner of the tables or graphs in this book. To download the matching Excel® spreadsheet, just type the link into your Internet browser, starting with the ***http://dx.doi.org*** prefix.

If you're reading the PDF e-book edition, and your PC is connected to the Internet, simply click on the link. You'll find *StatLinks* appearing in more OECD books.

Acronyms and Abbreviations

ACP	African, Caribbean and Pacific countries
AI	Avian influenza
AMAD	*Agricultural Market Access Database*
ARS	Argentian peso
AUD	Australian dollar
AUSFTA	Australia and United States Free Trade Agreement
bn	Billion
BNGY	Billion gallons per year
bnl	Billion litres
BNLY	Billion litres per year
BRIIC	Emerging economies of Brazil, the Russian Federation, India, Indonesia and China
BRL	Real (Brazil)
BSE	Bovine Spongiform Encephalopathy
bt	Billion tonnes
BTL	Biomass to liquid
CAD	Canadian dollar
CAFTA	Central American Free Trade Agreement
CAP	Common Agricultural Policy (EU)
CCC	Commodity Credit Corporation
CET	Common External Tariff
CIS	Commonwealth of Independent States
CMO	Common Market Organisation for sugar (EU)
CNY	Yuan (China)
COOL	Country of Origin Labelling
CO_2	Carbon dioxide
CPI	Consumer Price Index
CRP	Conservation Reserve Program of the United States
cts/lb	Cents per pound
cwe	Carcass weight equivalent
DBES	Date-based export scheme
DDA	Doha Development Agenda
DDG	Dried Distiller's Grains
dw	Dressed weight
EBA	Everything-But-Arms Initiative (EU)
ECOWAP	West Africa Regional Agricultural Policy
ECOWAS	Economic Community of West African States
EISA Act	Energy Independence and Security Act of 2007 (US)
EPAs	Economic Partnership Agreements (between EU and ACP countries)

ERS	Economic Research Service of the US Department for Agriculture
est.	Estimate
E85	Blends of biofuel in transport fuel that represent 85% of the fuel volume
EU	European Union
EU15	Fifteen member states of the European Union
EU10	Ten new member states of the European Union from May 2004
EU27	Twenty seven member states of the European Union (including Bulgaria and Romania from 2007)
EUR	Euro (Europe)
FAO	Food and Agriculture Organization of the United Nations
FCE Act	Food, Conservation and Energy Act of 2008 US Farm Bill
FMD	Foot and Mouth Disease
f.o.b.	Free on board (export price)
FR	Federal Reserve (US central bank)
FSRI ACT	Farm Security and Rural Investment Act (US) of 2002
FTA	Free Trade Agreement
G10	Group of ten countries (see glossary)
G20	Group of twenty developing countries (see glossary)
GAL	Gallons
GDP	Gross domestic product
GDPD	Gross domestic product deflator
GHG	Green House Gases
GLB	Gross land balances
GMO	Genetically modified organism
ha	Hectares
HFCS	High fructose corn syrup
hl	Hectolitre
HS	Harmonised commodity description and coding system
IEA	International Energy Agency
iLUC	Indirect land-use change
INR	Indian rupees
IPCC	Intergovermental Panel on Climate Change
kg	Kilogrammes
KRW	Korean won
kt	Thousand tonnes
La Niña	Climatic condition associated with temperature of major sea currents
LAC	Latin America and the Caribbean
lb	Pound
LDCs	Least Developed Countries
LICONSA	Leche Industralizada
lw	Live weight
MCI	Multiple cropping index
MERCOSUR	Common Market of the South America
MFN	Most Favoured Nation
Mha	Million hectares
Mn	Million
MPS	Market Price Support
Mt	Million tonnes

MTBE	Methyl tertiary butyl ether
MXN	Mexican peso
NAFTA	North American Free Trade Agreement
NLB	Net land balances
NZD	New Zealand dollar
OECD	Organisation for Economic Co-operation and Development
OIE	World Organisation for Animal Health
p.a.	*Per annum*
PCE	Private Consumption Expenditure
PIK	Payment in kind programme (US)
PPP	Purchasing power parity
PROCAMPO	Mexican Farmers Direct Support Programme
PRRS	Porcine reproductive and respiratory syndrome
PSE	Producer Support Estimate
pw	Product weight
R&D	Research and Development
RED	Renewable Energy Directive in the EU
RFS	Renewable Fuels Standard in the US, which is part of the Energy Policy Act of 2005
rse	Raw sugar equivalent
rtc	Ready to cook
RUB	Russian ruble
rwt	Retail weight
SFP	Single Farm Payment scheme (EU)
SMP	Skim milk powder
SPS	Sanitary and phytosanitary measures
STRV	Short tons raw value
t	Tonnes
t/ha	Tonnes/hectare
TFP	Total factor productivity
THB	Thai baht
TRQ	Tariff rate quota
UHT	Ultra-heat treatment is the partial sterilisation of food by heating it for a short time
UK	United Kingdom
UN	The United Nations
UNCTAD	United Nations Conference on Trade and Development
URAA	Uruguay Round Agreement on Agriculture
US	United States
USD	United States dollar
USDA	United States Department of Agriculture
v-CJD	New Creutzfeld-Jakob-Disease
VAT	Value added tax
WAEMU	West African Economic and Monetary Union
WMP	Whole milk powder
wse	White sugar equivalent
WTO	World Trade Organisation
ZAR	South African rand

Outlook in Brief

Agriculture has experienced a number of severe shocks in recent years with record high oil prices, commodity price spikes, food security fears and resultant trade restrictions, not to mention the most serious global economic recession since the 1930s. The greatest impact has been on the poor, especially in developing countries, with the world's hungry now estimated at over 1 billion people. Agriculture has shown remarkable resilience, particularly in the OECD area, with strong supply response to high prices and with continuing, albeit dampened, demand growth during the crisis. In 2010, a degree of normalcy has returned to many markets with production closer to historical levels and demand recovering. Still, many governments remain concerned about the potential for a repetition of significant shocks to such key factors as energy prices, exchange rates, and/or the macroeconomic performance of key countries and regions, and about the consequences that such shocks have on market volatility.

- The macroeconomic environment underlying the commodity projections is more positive than in the 2009 Outlook. It reflects the start of global economic recovery in late 2009 and a slow transition towards higher sustainable and non inflationary growth beyond the near term. A two-speed recovery appears to be underway characterised by weak and hesitant growth with high unemployment in many OECD countries and by stronger growth and faster recovery in the large developing countries which is slowly spreading to the rest of the developing world and helping to fuel world income growth. High energy prices have returned and are assumed to remain a feature of the period covered by this Outlook. A further increase in oil prices could be expected to increase input and production costs, having an impact on crop supplies, prices and trade flows, and reinforce feedstock demand for biofuels.
- Underpinning agricultural prices is increasingly a higher cost structure particularly in regions where energy inputs are used intensively. Global agricultural production is anticipated to grow more slowly in the next decade than in the past one, but in the absence of unexpected shocks, growth remains on track with estimated longer term requirements of a 70% increase in global food production by 2050. On a per capita basis, production growth in least developed countries is struggling to keep up with rapid population growth. Global sectoral growth will be led by the regions of Latin America and Eastern Europe and, to a lesser extent, by certain countries in Asia.
- Average crop prices over the next ten years for the commodities covered in this Outlook are projected to be above the levels of the decade prior to the 2007/08 peaks, in both nominal and real terms (adjusted for inflation). Average wheat and coarse grain prices are projected to be nearly 15-40% higher in real terms relative to 1997-2006, while for vegetable oils real prices are expected to be more than 40% higher. World sugar prices to 2019 will also be above the average of the previous decade but well below the 29-year highs experienced at the end of 2009.
- For livestock products, average meat prices in real terms, other than for pig meat, are expected to surpass the 1997-2006 average over the coming decade initially due to lower supplies, higher feed costs and rising demand. Pig meat real prices should stay relatively subdued due to an anticipated increase in supply from Brazil and China. Economic recovery will strengthen consumption of meats relative to cereals, particularly in developing countries, with most growth favouring cheaper meat – poultry and pig meat – relative to beef. Average dairy prices in real terms are expected to be 16-45% higher in 2010-19 relative to 1997-2006, with butter prices showing most gains, supported by higher energy and vegetable oil prices.
- Biofuel markets depend heavily on government incentives and mandates, but prospects remain uncertain, due to unpredictable factors such as the future trend in crude oil prices, changes in policy interventions and developments in second-generation technologies. Continued expansion of biofuel production to meet mandated use will create additional demand for wheat, coarse grains, vegetable oils and sugar used as feedstocks.

- Developing countries will provide the main source of growth for world agricultural production, consumption and trade. Demand from developing countries is driven by rising per capita incomes and urbanisation, reinforced by population growth, which remains nearly twice that of the OECD area. As incomes rise, diets are expected to slowly diversify away from staple foods towards increased meats and processed foods that will favour livestock and dairy products. Also, with increasing affluence and an expanding middle class, food consumption in these countries should become less responsive to price and income changes, as is currently the case in OECD countries. This implies that larger changes in price and incomes will be required for consumption to adjust to any unforeseen shocks.
- For virtually all commodities the projected growth in imports and exports of developing countries exceeds that of the OECD area. Only exports of protein meals increase faster in the OECD area by 2019. The higher share of developing countries in trade is reflected in expanding South-South trade in addition to North-South trade. Nevertheless, OECD countries will continue to dominate exports in 2019 (shares in brackets) of wheat (52%), coarse grains (59%), pig meat (80%), butter (80%), cheese (63%), whole milk powder (66%) and skim milk powder (74%). Developing countries will hold dominate shares in 2019 for: rice (88% share), oilseeds (56%), protein meals (80%), vegetable oils (91%), sugar (90%), beef (57%) and poultry (63%).
- Food prices remained high or "sticky" in many countries for an extended period after world primary commodity prices fell following the price surge of 2007/08. In 2009, the contribution of food price increases to inflation fell considerably from 2008, particularly in OECD countries, but still remains significant in some developing and emerging countries.
- Since the price spike of 2006-08, short term price volatility has increased considerably. However, the evidence is inconclusive as to whether and how price volatility has changed over the long term for the major food crops examined in this Outlook. What is clear is that the extent to which world prices are transmitted to domestic markets varies markedly by country and depends on the level of market integration. The transmission of international prices to domestic markets can be impeded by border measures, domestic price supports and infrastructure weaknesses.
- The Outlook's relatively stable price projections result from the assumption of "normal" conditions. Uncertainties around weather, macroeconomic factors, policy interventions, and especially energy prices suggest that commodity prices will remain unpredictable. Many governments are concerned about price volatility even in the very short term, because it threatens both farm viability (low prices) and food security (high prices). High uncertainty also affects investment decisions. There are a number of policy options to consider at both the domestic and international levels.
- Governments can underpin farmers' risk management strategies by focusing on those unpredictable and unavoidable risks that may be rare, but have large consequences, and which farmers cannot manage themselves. Governments can also empower farmers to manage their own business risk and can provide good risk governance, including by creating effective markets and by not creating incentives for rent seeking in the form of *ad hoc* support and assistance.
- National and local emergency stockholding of key food security commodities, for food emergencies, particularly for low-income food importing countries, may increase confidence in the access to food in times of crisis and help stabilise local markets. Increased research, capacity building, and sharing of best practices to improve the functioning of emergency stock schemes are required. Whatever actions governments consider taking, it is always important to keep in mind the full set of policy measures, risks and possible responses for the targeted population.

- Market price support for agricultural commodities is a policy option that has clearly shown in many countries and over several decades to be inefficient and its use has declined. Price support masks market signals to producers, destabilises world markets and acts as a regressive tax on the poor by raising prices to consumers. Price support also skews support towards large producers and, encourages intensification with potentially adverse effects on the environment, and much of the benefit is either capitalised into fixed asset values (such as land or quotas), thereby raising production costs over time, or is transferred outside the farm. Such measures should be assessed against other less distorting alternatives, such as targeted direct income supports, investments in productivity enhancements, etc.
- At the international level, the unco-ordinated policy actions of governments during the 2006-08 price spikes exacerbated volatility and impeded access to markets. There is a need for greater assurance of unimpeded access to global supplies and improve confidence in market functioning. While experience with international efforts to manage stocks has not been positive, options to reduce the unpredictability of food import bills should be explored.
- Organised commodity exchanges are useful and time-tested price discovery and hedging institutions, if they are regulated properly and attract sufficient volume to avoid monopolistic practices. They have facilitated commodity marketing in many developed countries and their expansion in developing countries is a welcome institutional development and a sign of market deepening.

OECD-FAO Agricultural Outlook 2010-2019

Chapter 1

Overview

Introduction

The *Agricultural Outlook* is a collaborative effort of the Organisation for Economic Co-operation and Development (OECD) and the Food and Agriculture Organization (FAO) of the United Nations in Rome, bringing together commodity, policy and country expertise of both Organisations, and national government agencies to produce, with the aid of the Aglink-Cosimo model, a consensus view on a longer term assessment of global commodity markets. This assessment is not a forecast about the future, but rather a conditional scenario of what can be expected to happen under certain key assumptions concerning the macroeconomic environment over the coming ten years, and a continuation of current agricultural policy and trade settings around the world and specific external factors. The projections of production, consumption, stocks, trade and prices for the different agricultural products described and analysed in this report cover the years 2010 to 2019.

The setting

The last two years have witnessed considerable volatility in international commodity markets and global economic conditions. During this period many agricultural commodity prices rose to historically high levels that effectively shook the world out of its longstanding complacency about food availability and re-ignited concerns with food security. This was followed in rapid succession by an unprecedented global financial crisis, the start of a slowdown in global activity and then the rapid descent of the world economy into the deepest recession since the 1930s.

A pathway out of this economic abyss has now emerged with global economic recovery starting in late 2009. However, while the start of an economic turnaround is undeniable, growth is still not strong enough to conclude that a durable expansion is underway. The world is currently undergoing a two speed recovery. A rapid rebound in activity is taking place in the large developing countries, which is gradually extending to the rest of the developing world, while a more hesitant and fragile turnaround is underway in much of the OECD area. With the recovery primarily being driven by huge monetary easing and extensive fiscal stimulus packages, its sustainability will depend on how quickly confidence and private demand is restored, in a context of continuing high unemployment and rising energy prices. In any event, the transition to sustainable growth within the OECD area will be slow and likely protracted as robust recovery is still far from assured in a number of countries.

Agriculture has been more resilient to the economic downturn than many other sectors of the economy, but has not gone completely unscathed. Demand and trade in most commodities fell with the fall in GDP and this impacted more heavily on income sensitive products and consumption, as well as investments with tight credit, in the developing countries than in the OECD area. By the same token, with recovery advancing faster in the developing countries than in the developed world, this has led to a more rapid turnaround in agricultural demand and world trade. However, the aftermath of the recent turmoil in commodity, economic and financial markets will continue to be felt over coming years and this situation increases the uncertainty in assessing market prospects around the world over the coming decade. Agricultural markets, in the near term, will thus reflect ongoing adjustments to the period of peak prices, the lingering effects of the deep recession and the start of economic recovery. All these adjustments effectively cloud the picture for the Outlook in the short term.

Despite this uncertainty, there remains in place a dynamic and highly predictable element at the heart of world agricultural markets. This is the inexorable shift underway at the core of agriculture towards an increasing role, and rising importance, of the developing and emerging economies in world agricultural production, consumption and trade. By and large, these countries are rebounding strongly from recession and with population growth rates that remain more than double those of the OECD area, will represent the major growth markets that will drive world agriculture forward over the next and coming decades. A return to higher global economic growth over the projection period together with continuing population gains, are expected to increase demand and trade and underpin prices for all agricultural products over the medium term Growth and activity remains particularly dynamic in much of Asia and Latin America, with domestic demand, production and trade expansion in China, India and Brazil driving growth in their regions not only in the near term, but throughout the period covered by this Outlook.

The underlying economic conditions for agriculture are now more favourable than they were at this time last year, following the start of economic recovery. However, considerable uncertainties remain in the short term concerning the strength and pace of recovery in returning to a period of sustainable growth. The key macroeconomic and other assumptions underlying the *Agricultural Outlook* are summarised in Box 1.1.

Box 1.1. **The main underlying assumptions**

Macroeconomic

- From the start of the Outlook, economic growth in the world and OECD area are in a recovery phase from the financial crisis and economic recession. The rebound in growth is expected to be more rapid in the developing countries and initially tepid and fragile in the OECD area and involving a longer transition period. Once the transition to sustainable gains is reached, the EU and US economies are expected to grow by 2% and 2.5% *per annum*, respectively to 2019; and with faster growth in some other OECD countries such as Korea, Turkey and Australia. Among the developing countries, the leading Asian economies have fared better than most and lead the world economic recovery. China and India's GDP are projected to grow by nearly 8%and 6.6% *per annum*, respectively. GDP growth in Brazil and Argentina averages 4.5% and 3% *per annum* to 2019. Agricultural trade is estimated to have declined sharply in 2009, along with general merchandise trade, and is expected to bounce back sharply from the beginning of the Outlook in 2010 and continue to grow in following years.
- The timing and implementation of government exit strategies to remove excessive liquidity from national economies and to reduce excessive budget deficits and restore fiscal balance are expected to occur in a period of strengthening private demand so that GDP growth is unimpaired.
- World population growth is expected to average 1.1% *per annum* to 2019, compared with 1.2% in the previous decade. Only slow population growth of 0.4% *per annum* is expected in the OECD area. Higher growth is expected in the developing countries, with the population of Africa as a whole growing at over 2% *per annum*. Continuing urbanisation trends and rising per capita incomes, emerging large middle classes and underlying population demographics collectively reinforce higher food demand in these countries.

Box 1.1. **The main underlying assumptions** *(cont.)*

Figure 1.1. **Macroeconomic trends**

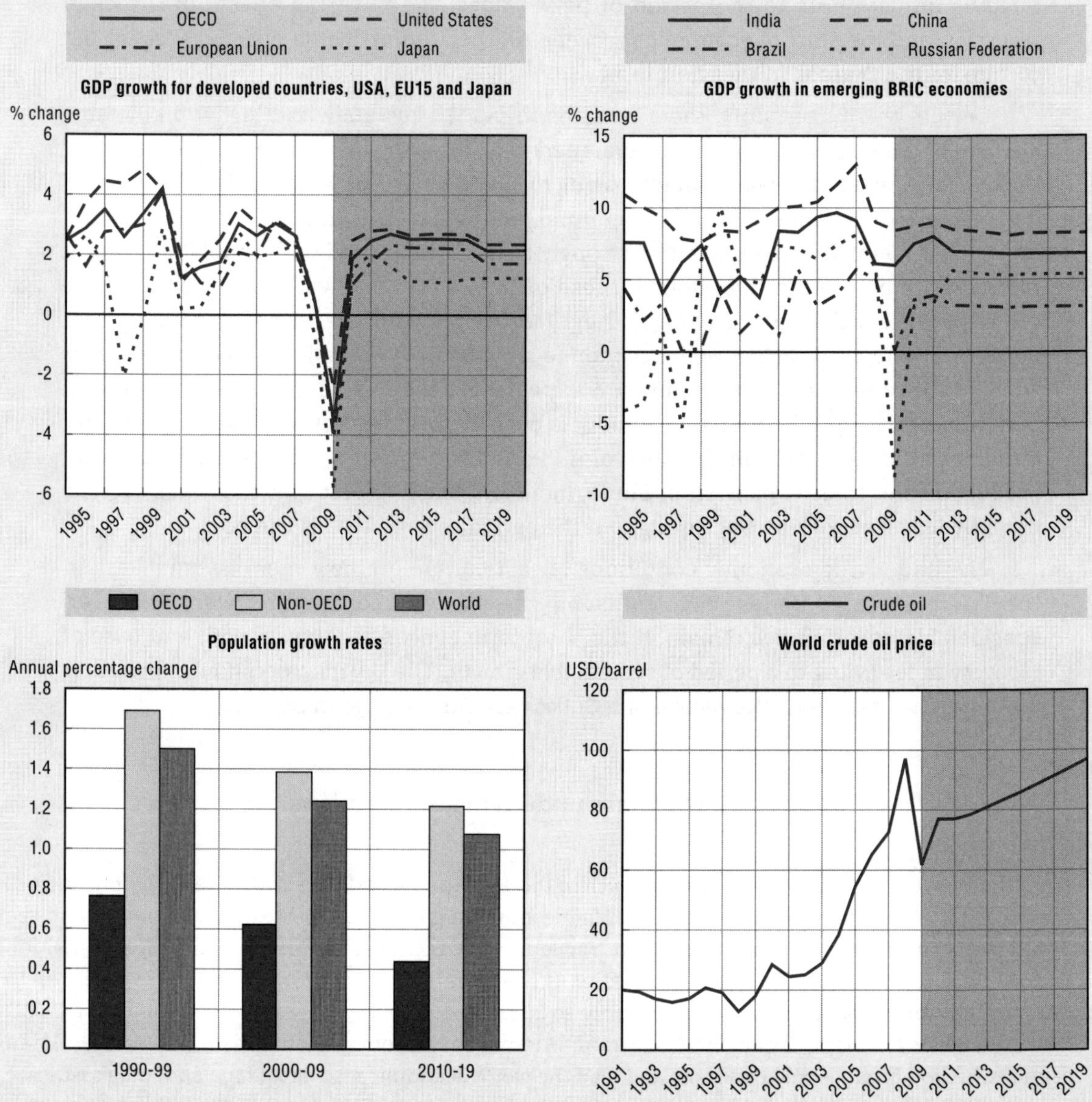

Source: OECD and FAO Secretariats.

StatLink http://dx.doi.org/10.1787/888932285981

- The global economic downturn has dampened inflationary pressures. As the OECD and world economy moves back into growth, Inflation is expected to remain subdued. Inflation is projected at levels close to 2% *per annum* throughout most of the OECD area to 2019. Higher inflation is expected in a number of emerging and developing economies such as the Russian Federation, Argentina, India and South Africa.
- Under the assumption of constant real exchange rates, the US dollar strengthens against most currencies. The currencies of high inflation countries will depreciate most relative to the US dollar. Crude oil and energy prices are assumed to increase over the coming decade as global economic activity is restored. Crude oil prices are expected to reach over USD 96 per barrel in 2019 and to remain above the average level of the decade prior to the oil price spike.

Box 1.1. **The main underlying assumptions** *(cont.)*

Policy considerations

- Agricultural and trade policies play an important role in both domestic and international markets for agricultural commodities and food products. OECD and emerging economies have gradually reformed their agriculture policies over the past two decades. At the same time, non-agricultural policies, such as energy, environmental and rural development measures, have a growing impact on the agri-food sector.
- Provisions of current legislation concerning agricultural and trade policy are assumed to remain in effect over the Outlook period. These include the provisions of the Food, Conservation and Energy Act of 2008 in the United States. For the European Union the outcome of the CAP Health Check of 2008 will continue in force in the European Union. Other provisions include mandates for renewable fuels such as in the EU and US based on agricultural feed stocks, blending provision for renewable fuels as mandated in Brazil for ethanol. In the US the Energy Independence and Security Act of 2007 is complemented by the Renewable Fuel Standard Program (RFS2) Final Rule of the EPA (Environmental Protection Agency). For the EU, the Renewable Energy Directive (RED) specifies the share of renewable energy sources (including non-liquids) should increase to 10% of total transport fuel use by 2020. Countries are also assumed to comply with all existing bilateral and multilateral agreements, such as NAFTA and WTO agreements in effect in late 2009. Other assumptions included a continuation of longer term trends in productivity growth and average or normal weather conditions *i.e.* absence of weather related supply shocks.

World markets at a glance

Prices to remain on a higher plateau

As was correctly anticipated in last year's *Agricultural Outlook*, international market prices for most agricultural products have retreated considerably in 2009 in response to a strong production response and lower demand due to the recent high prices and with the onset of the global recession. In a context where energy prices remain generally high by historical standards and expected to rise further with global economic recovery that is underway, the Outlook projects that most crop prices will remain at or above 2010 levels in the longer term. These will continue to exceed, in nominal and real terms (once adjusted for inflation), the average price levels in the decade preceding the price hikes of 2007/08. Apart from pig meat, this is also true for livestock prices which remain above the average levels for the last decade, in real terms (Figures 1.2 and 1.3).

Overall, the Outlook foresees that nominal price of all commodities covered in the Outlook will be on a higher plateau over the projection period, 2010 to 2019. However they will tend to remain below the recent peak levels of 2007/08. In the case of wheat, rice, protein meals, cheese and skim milk powder, average prices over the projection period will be around 50% or more below the peak levels reached in 2007/08. As some commodities, such as sugar, beef and pig meat did not undergo the same rapid run up in their prices in 2007/08, average nominal prices for the decade ahead will be about the same or exceed the levels of 2007/08 by 10-20%. For the entirety of agricultural products covered by this Outlook, however, average nominal prices over the projection period will exceed those of the previous decade prior to the period of peak prices. These price gains are expected to be highest for vegetable oils and butter at over 85% above those in achieved in 1996-2006. Of the products at the lower end of the nominal price increase scale, pig meat prices will show the smallest increase, rising by just over 21% above the 1997-2006 average level, on average, over the period to 2019.

Figure 1.2. **Nominal commodity prices to remain above average levels of the previous decade but lower than 2007/08**

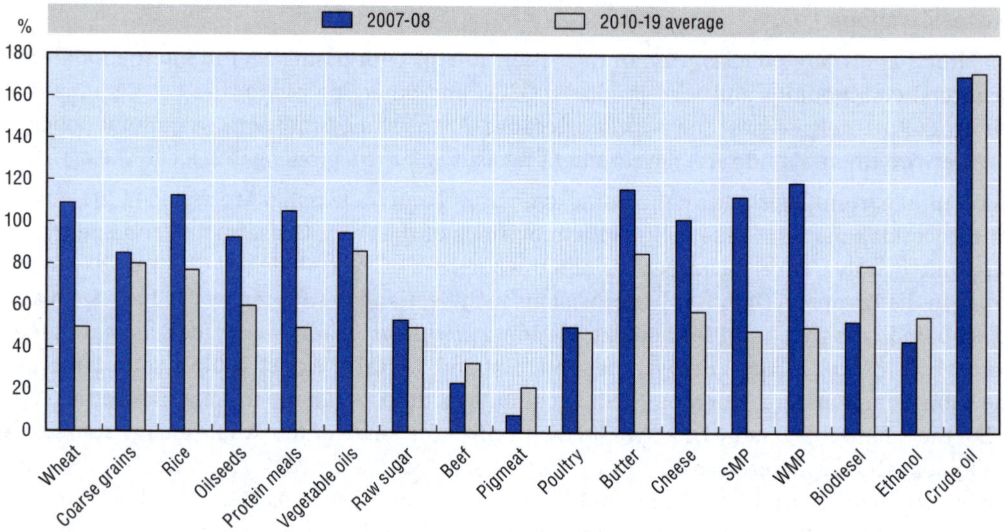

Note: For biodiesel and ethanol the base period is 2001-06.The crude oil price projection is an Economic Department exogenous assumption.

Source: OECD and FAO Secretariats.

StatLink http://dx.doi.org/10.1787/888932285525

Figure 1.3. **Most commodity prices in real terms to remain above the last decade's levels**

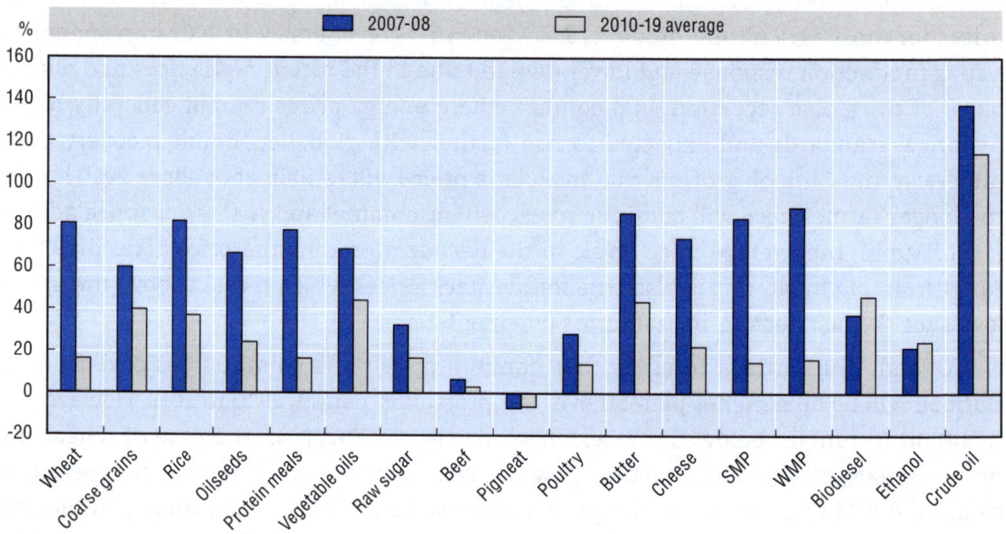

Note: For biodiesel and ethanol the base period is 2001-06. The crude oil projection is an Economic Department exogenous assumption.

Source: OECD and FAO Secretariats.

StatLink http://dx.doi.org/10.1787/888932285544

After allowing for inflation, prices in real terms are also expected, on average, to be much below their 2007/08 peak levels (Figure 1.3). The agricultural products that show the largest fall in real prices, when compared to their 2007/08 level, are: wheat, rice, oilseeds, protein meals, butter, cheese and skim milk powder. However, over the Outlook period, real prices of all products other than pig meat are expected to be above their average 1997-2006 level. Pig meat prices are not anticipated to increase much beyond 2014 due to a fast increase in supply with high productivity gains in Brazil and China. The price increases, in real terms, range for crops from around 16% to over 40% above their average for the last decade. In the case of livestock products the increase over the average of the last decade are largest for dairy products. The average level of the crude oil price, in real terms, that is exogenously projected for the coming decade is also substantially below its 2007/08 peak, but still remains relatively high at 114% above the 1997-2006 average level.

Agricultural commodity markets increasingly driven by developing countries

Increased market integration, globalisation and rapid income growth over a number of years prior to the recent economic crisis, have enhanced the role and importance of key developing and emerging economies of the non-OECD region on world agricultural markets. This increasing influence associated with rising affluence and feeding expanding populations is being manifested in different ways on international markets with continued economic development and the resulting transformation of their economies. Initially the momentum arising from strong income growth boosts food demand and imports for a range of agricultural products and processed foods to feed large concentrations of people migrating from rural to mega urban centres. Subsequently it provides the impetus for the development of domestic production capacity, financed from either domestic savings or from growing foreign direct investment flows to these developing and emerging economies. Investment in manufacturing, processing and domestic production capacity is expected to be particularly strong in the "expanded" BRIIC countries of Brazil, The Russian Federation, India, Indonesia and China. It is also becoming a generally shared priority of other high growth emerging countries. One of the motivations behind such investments is to capture a growing share of the higher value added component of domestically consumed agricultural products.

At the same time OECD area agriculture is undergoing reforms that reflect changing circumstances and priorities and which are gradually modifying production incentives towards increased market orientation and any underlying comparative advantages. These evolving developments and trends are raising the profile of lower cost agricultural industries and sectors in the developing and emerging economies relative to their counterparts in the mature economies in the OECD area. In fact, for the OECD area as a whole, both production and consumption growth prospects for all the 15 agricultural commodities covered in this Outlook and listed in Table 1.1, are expected to be increasingly eclipsed by the group of developing, least developed and emerging countries comprising the non-OECD region. As revealed in this table, the largest growth differentials are consistent across both production and consumption, reflecting the strong expansion in the high value added livestock meat products of beef, poultry and pig meat, and of dairy products covering butter, cheese and milk powders. Apart from these products, high growth differentials also favour the non-OECD region for coarse grains, protein meals and sugar. In fact, in almost all cases, the majority of production growth for these products will be coming from outside the OECD area, where growth rates are 2-3 times larger than for OECD countries. The same situation applies for much of the consumption growth as well. As a result, the OECD area's share in global output and use for these products continues to decline over the projection period from 2010 and 2019.

Figure 1.4 shows the percentage increase in production of crops and livestock products over the projection period. Figure 1.4 indicates that that global production of crop products will increase by over 13% by 2019, when compared to the base period 2007-09, suggesting

Table 1.1. **Production and consumption annual growth rates (least squares), 2010-19**

	Production (%)			Consumption (%)		
	Total	OECD	Non-OECD	Total	OECD	Non-OECD
Wheat	1.1	0.8	1.3	1.2	1.0	1.3
Rice	1.0	0.3	1.1	1.1	0.6	1.1
Coarse grains	1.6	1.0	2.1	1.5	0.9	2.1
Oilseeds	1.9	1.3	2.2	1.9	1.4	2.2
Protein meal	2.2	1.5	2.5	2.2	1.0	3.2
Beef	1.5	0.5	2.2	1.5	0.6	2.1
Pig meat	1.7	0.7	2.3	1.8	0.7	2.3
Poultry meat	2.4	1.3	3.0	2.4	1.6	2.8
Milk	2.2	0.8	3.1	..	..	..
Butter	2.2	0.7	3.0	2.1	0.4	2.9
Cheese	1.8	1.3	3.1	1.8	1.3	2.9
Skim milk powder	1.0	0.3	3.0	1.0	0.2	1.9
Whole milk powder	2.5	0.7	3.8	2.5	1.0	2.9
Vegetable oils	2.9	1.7	3.2	2.8	2.3	3.1
Sugar	1.4	0.0	1.8	1.8	0.5	2.2

Source: OECD and FAO Secretariats.

StatLink http://dx.doi.org/10.1787/888932287843

more ample supplies in the period ahead. The increases in projected production range from 14% for wheat to nearly 39% for vegetable oils. In terms of the OECD and non-OECD regions, production of oilseeds and vegetable oils increase the most for the former group of developed countries, and this is also true for the non-OECD countries, but with sugar also included amongst the crops showing the largest production expansion. For livestock products, the largest increase in global production is of whole milk powder which is projected to increase by 31% and for poultry and butter which grow by 29% and 28%, respectively, over the projection period, relative to the base period. Within the OECD area, cheese (14%), whole milk powder (12%) and poultry (11%) show the largest increases relative to the base period. For the non-OECD group of countries, the leading growth products are comprised of whole milk powder (48%), poultry and skim milk powder (43%), butter (42%), and with cheese (38%) also showing rapid production increases to 2019.

It is interesting to see what the projections imply for agricultural production when disaggregated by countries and regions. This is shown in the next section.

Agricultural production by country and region

The Outlook provides extensive detail on balances for various commodities on a global, regional and country basis. Aggregation across commodities provides a summary measure of how the agricultural sectors of countries and regions or economic groups are performing.[1] In terms of the commodities covered in this Outlook, agricultural commodity sectors are performing quite differently across these groups, as noted in Figures 1.5 to 1.8.

Based on commodities of this Outlook, Brazil is the fastest growing agricultural sector by far, growing by over 40% to 2019, when compared to the 2007-09 base period. The Russian Federation and Ukraine are projected to grow 26% and 29%, provided plans and support measure by the respective governments proceed as indicated and bear fruit, marking a significant recovery in production levels. China and India may also grow significantly by 26% and 21%, respectively. While Australia is projected to grow some 17%,

Figure 1.4. **Change in production of crop and livestock products (per cent change 2019 compared to 2007-09)**

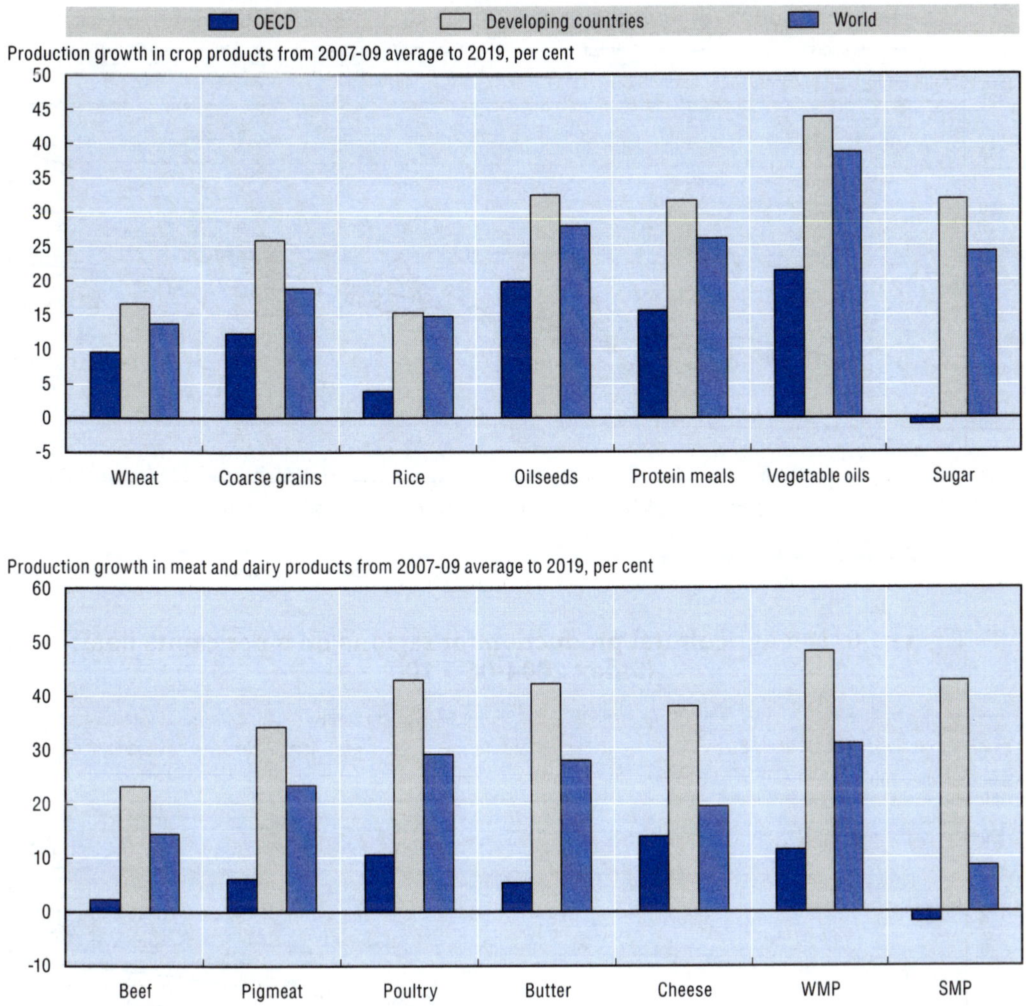

Source: OECD and FAO Secretariats.

StatLink http://dx.doi.org/10.1787/888932285563

this growth reflects an assumed return to more normal yields; over a longer period of comparison, Australia's production by 2019 is only some 7% higher than in 2000. Production growth in the US and Canada is projected in the 10-15% range over the same period. In contrast, over the same period, net agricultural output in the EU27 will have grown less than 4%. These diverse trends reflect important developments in these countries which may be generating or inhibiting growth.

By region, production measures, on a per capita basis, provide an interesting viewpoint on longer term developments in global agriculture and their potential implications for food supplies. As noted in Figure 1.6, per capita output has fallen in North Africa and the Middle East, largely due to limited water availability and policies in some countries such as Saudi Arabia to reduce highly subsidised wheat production. Production in the sub-Sahara region of Africa is expected to be stagnant in per capita terms, as production barely keeps pace with population growth still averaging around 2.2% per year. In Western Europe, production is also stagnant. Growth in consumption on a per capita

Figure 1.5. **Net agricultural production for selected countries (index 2004-06 = 100)**

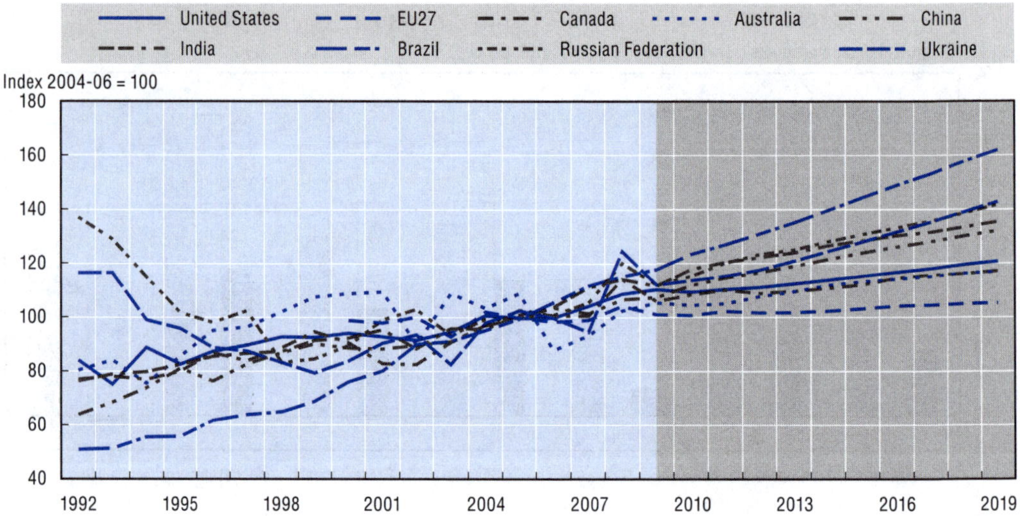

Source: OECD and FAO Secretariats.

StatLink http://dx.doi.org/10.1787/888932285582

Figure 1.6. **Net agricultural production for regions on a per capita basis (index 2004-06 = 100)**

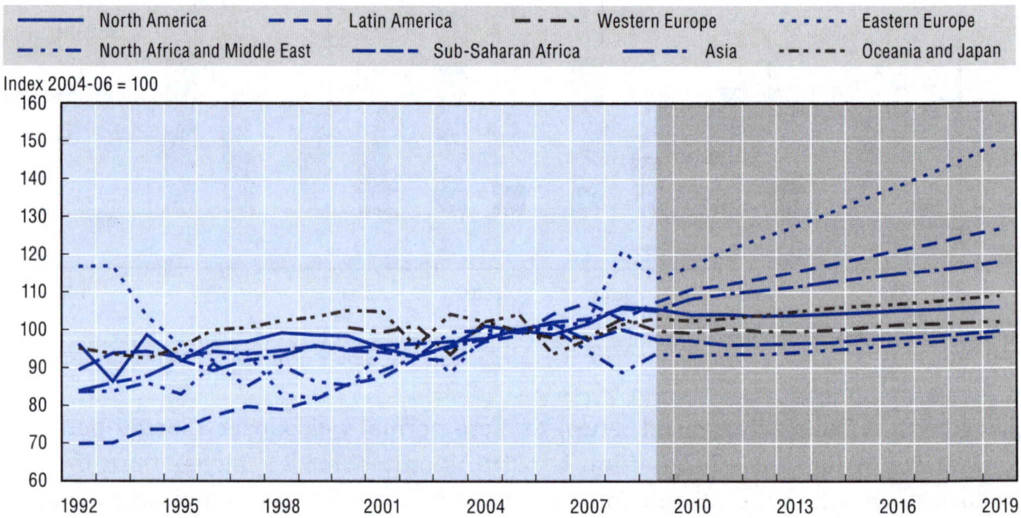

Source: OECD and FAO Secretariats.

StatLink http://dx.doi.org/10.1787/888932285601

basis in this region will need to be met by imports. Latin America is the fastest growing production region, but in per capita terms Eastern Europe, is the fastest growing, because projections assume that in this region's population numbers will actually decline by over 3% over the Outlook period.

Agricultural production is growing much faster outside the OECD area (Figure 1.7). While world net production of commodities covered in this Outlook will have grown 22% over the period to 2019, production in the OECD area is projected to grow only 10%. Some country groupings grow about three times as fast with Brazil, The Russian Federation,

Figure 1.7. **Net agricultural production for world and economic groups (index 2004-06 = 100)**

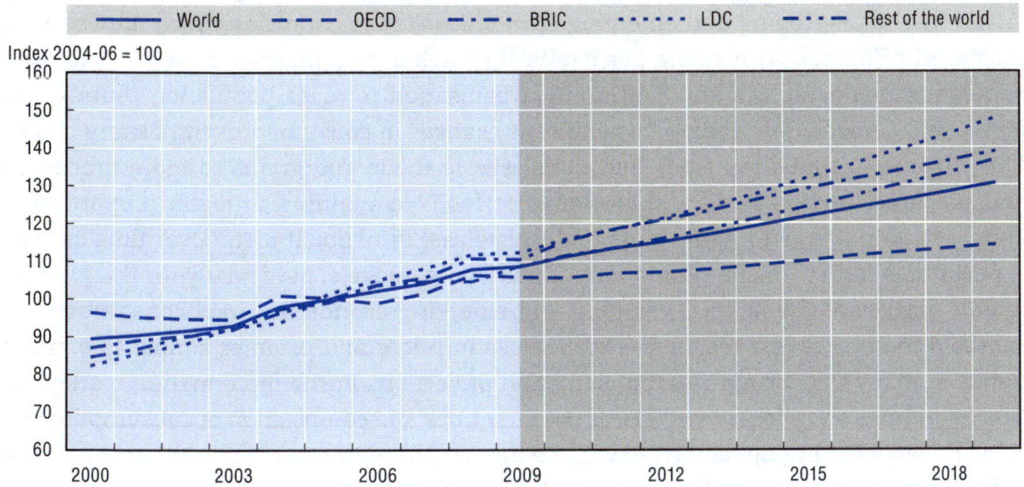

Source: OECD and FAO Secretariats.

StatLink http://dx.doi.org/10.1787/888932285620

Figure 1.8. **Net agricultural production for economic groups (percentage change)**

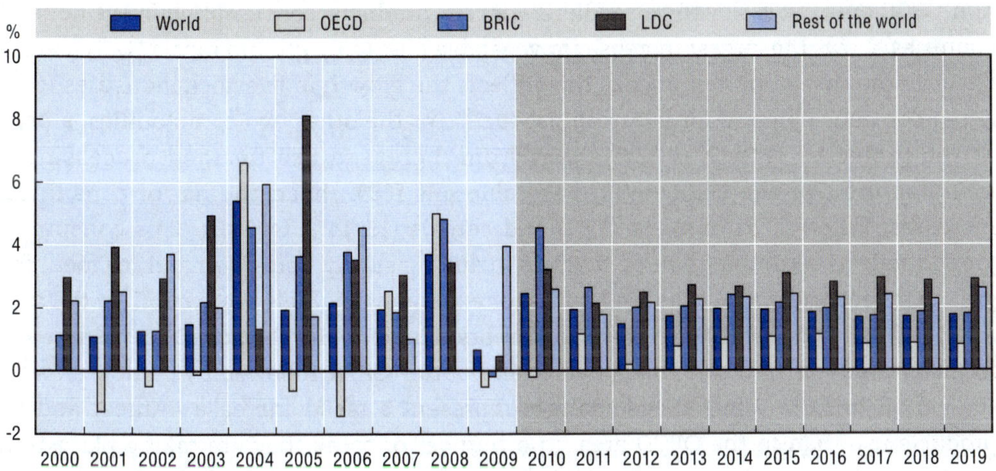

Source: OECD and FAO Secretariats.

StatLink http://dx.doi.org/10.1787/888932285639

India, China (BRIC) group growing by 27%, LDCs by 33% and other developing countries by 29% to 2019. If measured in per capita terms, OECD agricultural production growth is minimal as is production growth by the LDC group.

If assessed in terms of annual changes in net production, some interesting patterns emerge, as seen in Figure 1.8. First, the historical annual variation of production in the OECD area exceeds the variation by other country groups. Second, it is notable that in response to the high prices of 2007/08, OECD production response was the largest, followed closely by BRIC countries, and LDCs. Production by the remaining countries of the rest of the world as a whole declined in 2008/09, largely due to weather elated production problems in larger countries such as Argentina. Third, it is also notable that agricultural production growth was also largely stagnant in most countries during the global recession, but declines were by and large less in agriculture than in the rest of the economy.

Consumption trends

As in the case for production, world consumption of agricultural products is also being driven by the developing and emerging economies. These countries which are enjoying increased affluence with rising per capita incomes over a number of years, and with population demographics and continuing urbanisation to mega population centres also reinforcing demand, are leading to significant changes in consumption and dietary habits. This involves a transition from traditional staple foods and grains to more processed and prepared food products and convenience foods, containing a greater proportion of animal protein and with more fruits and vegetables, in national diets. Over time as food expenditures form a smaller proportion of household budgets, particularly for the swelling middle classes of the large developing and emerging economies, food consumption is expected to become less responsive to changes in prices and incomes similar to existing trends in many OECD countries. Consumption projections in the more mature markets of OECD countries show less growth. Here the quantities and composition of consumption are being driven more by population growth and its changing demographics with ageing as well as by concerns over diets and general health issues than by price or income considerations.

Figure 1.9 shows the percentage increase in consumption of crop and livestock products over the projection period to 2019, when compared to 2007-09. It clearly illustrates that the consumption of agricultural products continues to grow rapidly in the non-OECD area but is slowing elsewhere. For crop products, consumption in the non-OECD countries shows the largest increase for vegetable oils (44%), protein meals (42%) and sugar (30%). In the case of protein meals, this reflects the growth in livestock industries to meet rising domestic demand for livestock products. For the other two commodities, a portion of the projected growth in use arises from increased use for food processing and manufacture. For the OECD area, vegetable oils (28%) head the list of consumption increases, followed by oilseeds (16%) and cereals (13-14%) by 2019. This consumption growth reflects a combination of changes including slowly rising demand for food, faster growth in feed use and also as feedstocks for expanding biofuels production. In the case of meat and dairy products, the fastest consumption increase for the non-OECD region occurs for dairy products of whole milk powder and butter (38%), followed by poultry (37%) over the period to 2019. While these changes represent a faster increase in meat and dairy products use than in the OECD area, in a number of cases they are taking place from a smaller consumption base.

Trade developments

Trade flows and trade patterns in agricultural products also continue to evolve with increased south-south trade in addition to the traditional north-south trade. On the import side, the developing countries are becoming increasingly integrated into world agricultural trade and regional markets. While other countries such as China with a well established presence on international markets seek to diversify their sources of supply resulting in new economic interactions and trade between developing and other countries that are effectively reshaping world-wide flows of agricultural trade. In addition, developing country groupings which have become well integrated into the world trade infrastructure, are assuming greater prominence in international trade negotiations and their agendas. Apart from dominating import growth for most of the commodities in the Outlook, non-OECD countries also exhibit, with few exceptions, the strongest growth in exports; albeit from a low base for some commodities. Although exports may be growing rapidly for

Figure 1.9. **Change in consumption of crop and livestock products (per cent change 2019 compared to 2007-09)**

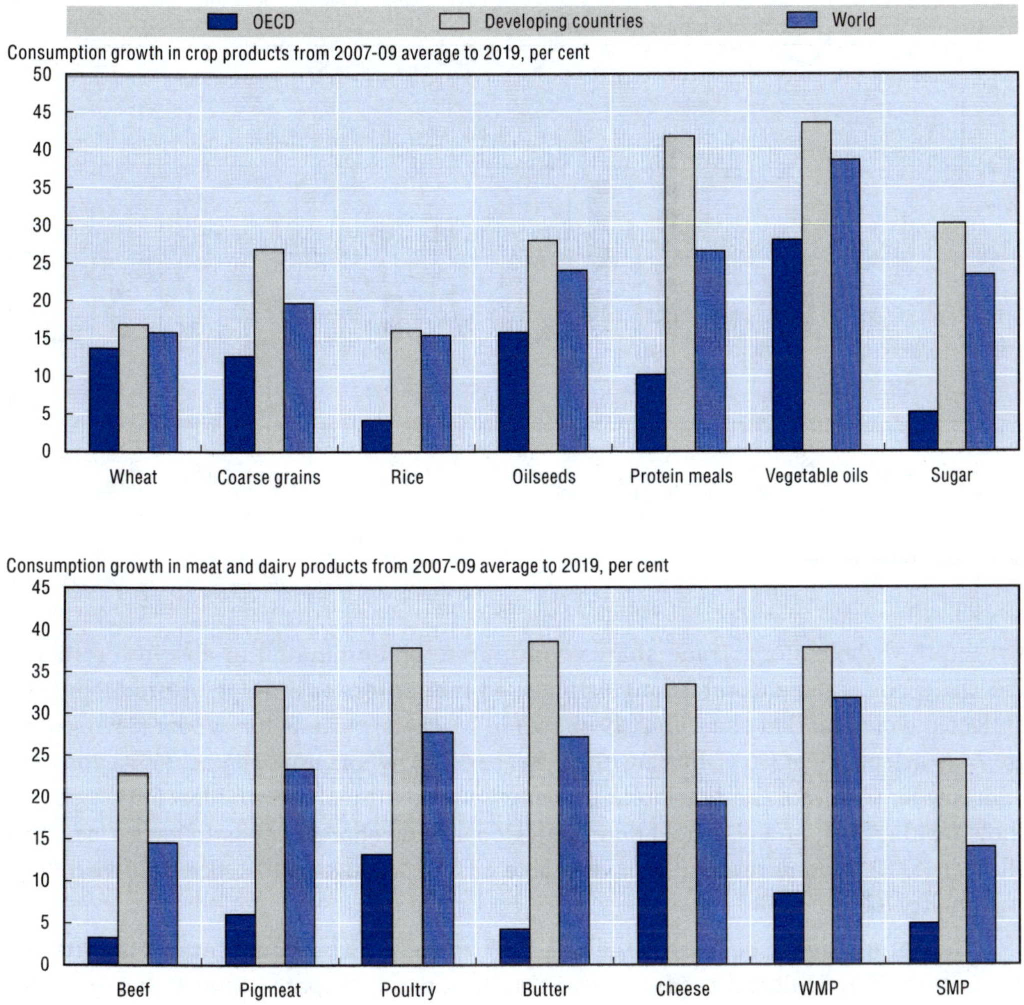

Source: OECD and FAO Secretariats.

StatLink http://dx.doi.org/10.1787/888932285658

commodities such as dairy products, the non-OECD countries still remain very large net importers over the Outlook period. Increasing export shares in almost all commodity markets is being achieved by a combination of strategies. Not only are they displacing traditional exporters and competitors with lower cost products or by growing faster, they are also becoming dominant in regional markets which can provide a springboard for further international market expansion. From Figure 1.10, non-OECD countries are projected to show the strongest percentage increase in exports by 2019, relative to the 2007-09 base, for oilseeds (59%), poultry (54%), wheat (50%), skim milk powder (43%), and cheese and vegetable oils (39%). On the other hand, OECD countries as a whole will tend to lose export shares in many commodities to non-OECD countries over the Outlook period. For the OECD area the largest gain in exports by 2019 compared to 2007-09, are for protein meals (49%), vegetable oils (38%) and rice (16%). While the gains illustrated in Figure 1.10 underscore the dynamic growth underway in the trade of developing and emerging economies as a whole, relative to the OECD area's general contraction, they do

Figure 1.10. **Exports of OECD and non-OECD countries to 2019 (per cent change)**

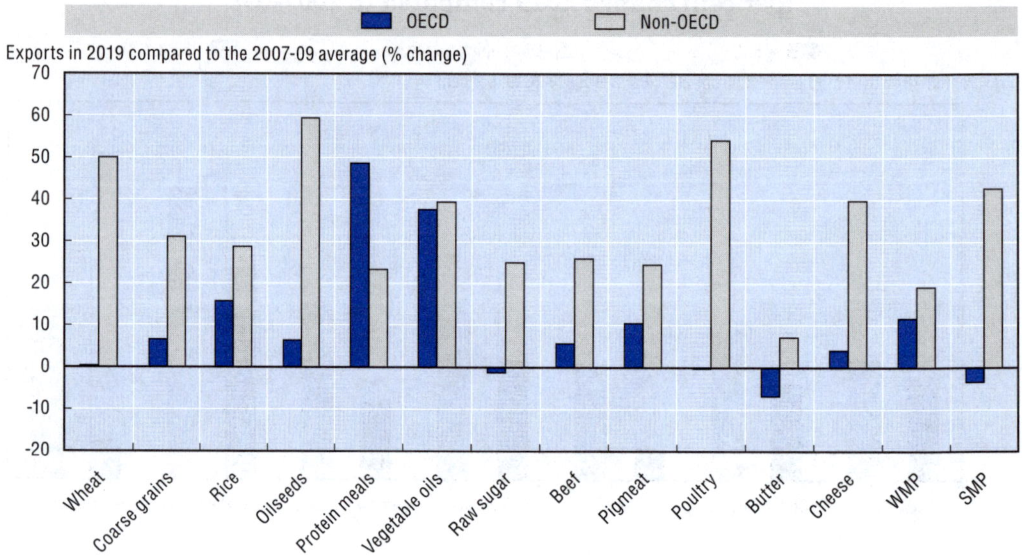

Source: OECD and FAO Secretariats.

StatLink http://dx.doi.org/10.1787/888932285677

not tell the whole story. Trade shares continue to be dominated in absolute terms by the OECD countries and their long established industries for a range of products (with projected global market share in 2019 shown in brackets) such as for: wheat (54%), coarse grains (60%), pig meat (80%), butter (79%), cheese (65%), whole milk powder (66%) and skim milk powder (73%). For the developing and emerging countries, the products for which they show rapid growth and also hold dominate global trade shares, are as follows: rice (89%), oilseeds (57%), protein meals (81%), vegetable oils (92%), sugar (89%), beef and veal (56%) and poultry (66%).

The foregoing discussion suggests that there will be considerable additional production of agricultural products available to meet anticipated higher domestic and import demand for food and feed purposes over the coming decade to 2019. OECD agriculture will continue to supply a large share of the additional world food and feed supplies. However, in many instances, their lower growth prospects, higher cost and more limited resource bases relative to the new players on the block from the non-OECD region suggests that the contribution of the OECD area to global food balances will continue to decline and with an increasing orientation towards the higher value-added components of the food chain.

Main trends in individual commodity markets

World commodity prices to remain high

With the exception mainly of sugar, agricultural commodity markets covered in this Outlook have calmed considerably with a return to more normal conditions following the turbulence of the last two years. Most commodity prices have fallen from price peaks at the start of the Outlook as a result of larger supplies becoming available and with continuing weaker demand in the aftermath of the economic crisis. With the start of economic recovery, the economic environment has now turned more positive than this time last year. Provided growth is durable and increases to levels offering sustainable gains over coming

Figure 1.11. **Imports of OECD and non-OECD countries to 2019 (per cent change)**

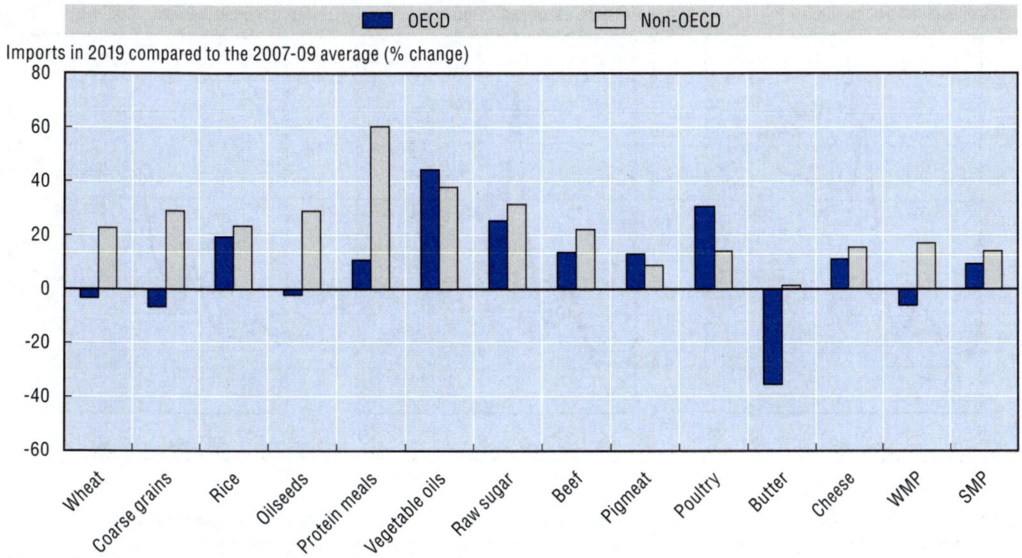

Source: OECD and FAO Secretariats.

StatLink http://dx.doi.org/10.1787/888932285696

years this should be supportive to a general strengthening of demand, trade and commodity prices over the Outlook. Stronger demand, with an anticipated return to higher growth following economic recovery and from increasing populations, should outpace production growth, on average, over the projection period to maintain all commodity prices on a higher plateau relative to the average of the last decade prior to the 2007/08 price surge. The projection of crop prices in nominal and real terms (once adjusted for inflation) is illustrated in Figure 1.12 and those for livestock products in Figure 1.13.

A longstanding feature of international commodity prices, including for agricultural products, is their high volatility in comparison to industrial goods and manufactures. This arises from the characteristics of agricultural products and comes primarily from the production side. Supplies of agricultural products are characterised by low responsiveness in the short term with realised production fluctuating due to weather and changes in relative prices, while demand grows steadily and remains fairly inflexible. Low elasticities mean that small shocks to production can have a large impact on price. The extent to which this price volatility is passed through to domestic markets depends on the degree of price transmission. The co-movement of world and domestic prices can be obstructed by border measures, domestic price supports and infrastructure weaknesses that affect the degree to which domestic markets are integrated with world markets. The issue of price transmission and possible policy responses to international price risk and volatility are discussed more fully in Chapter 2 of this report.

The surge in crop prices to near record highs in 2007/08 was due to the contemporaneous occurrence of a panoply of contributing factors, which are not likely to be repeated in the near term. However, if history is any guide, further episodes of strong price fluctuations in agricultural product prices cannot be ruled out nor can future short-lived crises. This is particularly clear when considering the heightened linkages between crop and energy prices. The increased variability in crude oil prices should impact on crop prices through both demand and supply, even though the demand link is weakened in this Outlook by the increasing

Figure 1.12. **The Outlook for world crop prices to 2019**

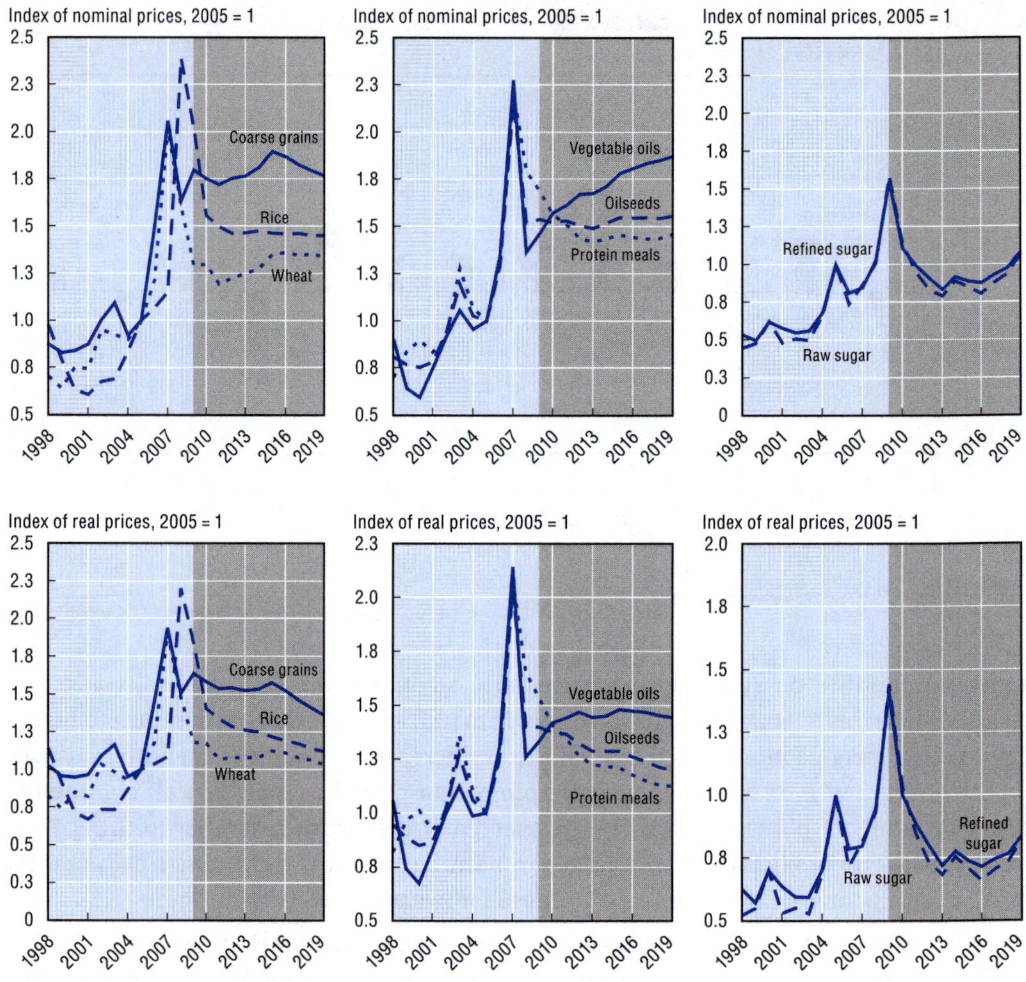

Source: OECD and FAO Secretariats.

StatLink http://dx.doi.org/10.1787/888932285715

importance of quantitative biofuel mandates. In addition, with trends underway towards greater macroeconomic integration and increased globalisation, world financial and economic shocks will be increasingly transmitted through exchange rates onto domestic markets. These changes when sustained can profoundly affect the competitive position of nations wishing to trade on international markets, or to build domestic production capacity and thus also increase variability on world commodity markets.

Large supplies to keep cereal prices under pressure

The world wheat and coarse grain markets at the start of the Outlook are marked by a return to normality after two exceptionally turbulent seasons. A combination of a sharp recovery in supply, with bumper crops replenishing stocks and a slowdown in demand, driven by the high prices, reduced policy supports for biofuels in some countries, troubled financial markets and recession, forced international prices to lower levels at a fast pace. Stocks of wheat and coarse grains are expected to increase over the Outlook period, although much of these will be located outside the traditional exporters and the stocks-to-use ratio should remain relatively low when compared to the previous decade.

Figure 1.13. **The Outlook for world livestock prices to 2019**

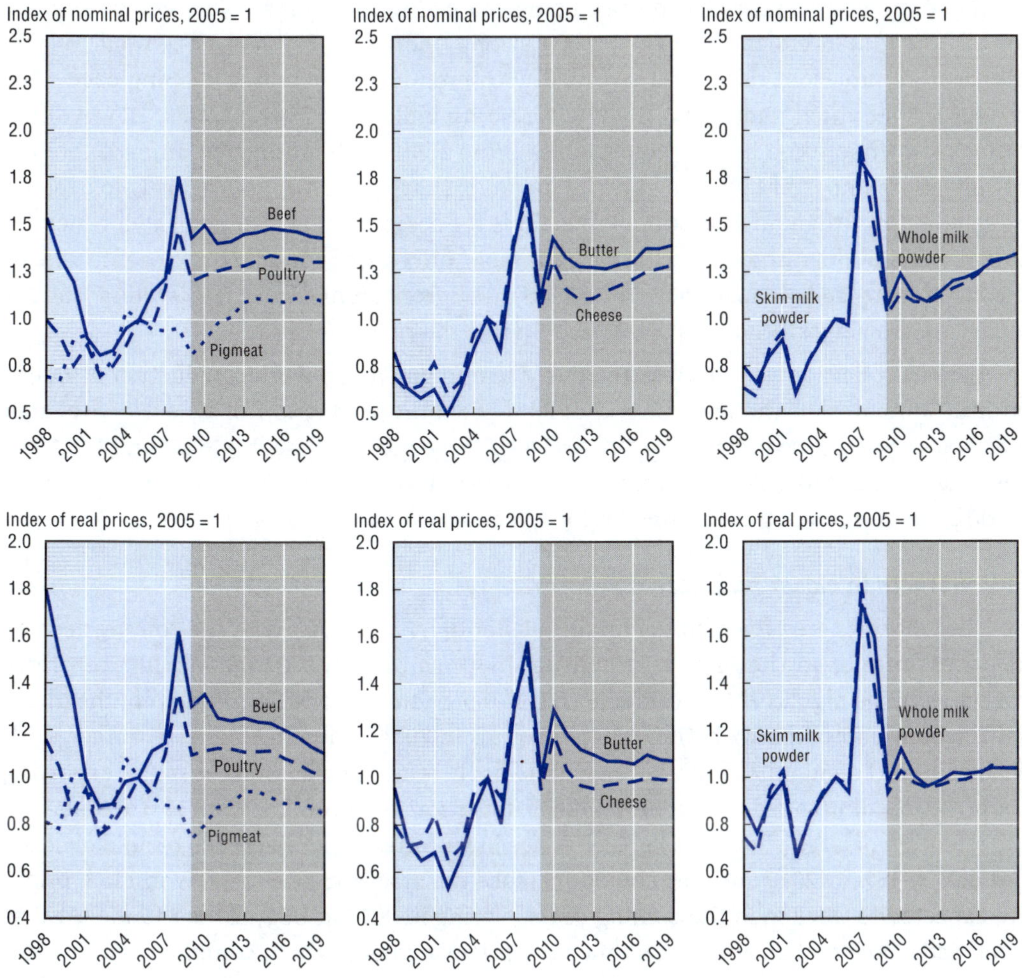

Source: OECD and FAO Secretariats.

StatLink http://dx.doi.org/10.1787/888932285734

This development is expected to underpin wheat prices to some extent but may also make them more unstable. The price of the benchmark US wheat (No. 2 Hard Red Winter, f.o.b. Gulf) is projected to increase to USD 225 per tonne by 2019, up 3% from the average in 2007-09. In real terms, however, wheat prices are likely to continue their long-term decline, albeit falling less rapidly and from higher levels. In the case of coarse grains, current projections for stocks and utilisation point to a somewhat tighter supply and demand balance during the early years of the projection period resulting in prices rising the fastest until 2016. The situation is likely to improve thereafter because of slower growth of use for ethanol production. By 2019, the price of the benchmark US maize (No. 2 Yellow, Gulf) is projected to reach USD 187 per tonne, almost unchanged from the average in 2009. A noteworthy feature is the drop in wheat to maize price ratio to a low ratio of 1.1-1.2, compared to 1.3-1.6 in the past, indicating a stronger upside potential for maize prices than for wheat.

World producers of cereals may take comfort in the fact that prices are likely to remain relatively strong compared to the previous decade, and consumers will find that prices are unlikely to reach the highs that provoked so much of the recent turmoil in food markets.

However, one feature which will not be of any benefit to producers or consumers is a continuation of price volatility in the coming years. Continuing instability will be a factor for all cereal markets as the linkages are strong enough to influence them all. Wheat markets are projected to be well supplied with production increases keeping pace with consumption which should even allow for some build-up of inventories. However, as regions known for their erratic yields, which bring about high production unpredictability, become more important players in world markets, sharp price swings are likely to become more of a norm than an exception. Projections for coarse grains also point to a generally balanced situation. Although, in the case of maize, prices are seen to move closer to wheat, the faster increase in maize prices reflects demand from biofuels and feed sectors, both of which are growing, albeit at a slower pace than in the previous decade.

The next decade will witness relatively strong growth in world production of major grains. Compared to the base period of 2007-9, world production of wheat and coarse grains are projected to increase by 14% and 19% respectively, to reach 746 Mt and 1 311 Mt, over the next decade. These projected production levels are likely to match or outstrip world consumption of these grains for food, feed and industrial use.

Rice markets in closer balance

Rice markets are projected to remain substantially in balance over the coming decade, at prices inferior to the relatively high levels prevailing in 2007/08. World rice prices continued to be high in 2010, averaging USD 570 per tonne in the first three months of the year, which compares with USD 335 per tonne in 2007, prior to the price surge. With weaker import demand, the market strength largely reflects policies in the major exporting countries, tending to reduce export supplies, for instance, through export restrictions and the retention of large public stocks. Global rice inventories are projected to rebuild over the Outlook by 18% by 2019 and this should increase the stocks-to-use ratio. World rice prices are expected to weaken in the coming years, ending in 2019 at USD 422 per tonne. At that level, rice would still be almost 1.9 times the price of wheat, which compares with a ratio of 2.7 in 2009.

A continuation of policies in support of production together with renewed interest by private investors in large scale production in land and water-rich countries are expected to be important drivers of the rice sector in the coming decade. For instance, the large production gains anticipated in Myanmar, Cambodia and Laos are likely to allow these countries to emerge as important players in the export market, which would widen the choice of origins for rice trade and reduce the dependence on traditional suppliers such as Thailand and Viet Nam. Because important Asian countries are expected to engage in the commercial production of genetically modified rice by 2015, the questions of product segregation and labelling may also acquire increased prominence in the commercialisation of rice both in domestic and export markets. Compared to the base period 2007-09, world production of rice is projected to increase by nearly 15%, or 67 Mt, to reach 522 Mt by 2019. World consumption is projected to grow by 1.1% *per annum* to reach 521 Mt by 2019. Particularly fast growth in consumption is foreseen in Africa amid relatively strong expansion of population and growing preference for rice in diets. In China, the largest consuming country, consumption should continue to decline as consumers become more affluent and shift to higher protein-based diets.

High vegetable oil demand and prices drive the world oilseed economy

Following the sharp fall in prices towards the end of 2008, values for oilseeds and derived products have since moved at levels above those prevailing prior to the recent price spike. With sustained food vegetable oil demand in developing countries, robust mandates for biodiesel consumption and strong use of protein meal by the expanding livestock sector, oilseeds and oilseed products markets are expected to continue to undergo further expansion over the projection period. Compared to the levels prior to the food crisis, oilseeds and protein meal prices are projected to remain firm over the projection period as global stock-to-use ratios are expected to stay at low levels. With sustained crush demand and increasing production, oilseeds prices are expected to increase in nominal terms to USD 419/t in 2019 but to decrease in real terms (when adjusted for inflation) over the entire projection period. In line with other feed commodities, protein meal prices are expected to decrease in the early years of the Outlook period before marginally increasing over the rest of the projection period to reach USD 288/t in 2019, about on par prices at the start of the Outlook. In a context of rising food and biodiesel use, demand for vegetable oils rises faster than for oilseeds and protein meals. The renewed firmness in crude oil prices over the projection period also contribute to a gradual strengthening in oils and fats values. Vegetable oil prices are projected in nominal terms to reach USD 1043/t in 2019; well above the base period 2007-09, and levels in nominal and real terms for the decade prior to the price surge.

While still high relative to other crops, the rate of growth in oilseed production over the next 10 years will not match that observed during the previous decade. Much of the foreseen expansion will be concentrated in Brazil, the EU and Argentina, supported by land reallocation from other uses and new land entering production. The US should remain the major oilseed producer over the projection period. When compared to the 2007-09 average, world oilseed oil production should increase by 30% over the projection period. However, higher marginal costs of area expansion and growing environmental concerns in many key producing regions means that global production growth rates will be lower than over the previous decade.

At the world level, vegetable oil production should increase by almost 40% over the Outlook period. Global production of palm oil remains very concentrated, with Malaysia and Indonesia producing over 85%. Combined they are expected to reach 70 Mt in 2019. Because of environmental concerns and area restrictions the growth rates are projected below recent trends, especially in Indonesia. The share of vegetable oil consumption used for biodiesel production is estimated to increase from 9% during the base to 15% in 2019, driven by binding mandates in many countries. During the Outlook period, annual growth in protein meal consumption is projected at 1% in OECD economies, compared with 3.1% in non-OECD economies.

Despite increased production, steady demand growth underpins sugar prices

World prices of sugar, historically one of the more volatile of agricultural commodities, have been following a different rhythm of late than those of the major temperate zone crops. World sugar prices were relatively low in 2007/08 when other commodity prices surged and then reached 29 year highs in February 2010, while the others declined. The price surge was partly due to bad weather in key producing countries, such as Brazil and India. These production shortfalls lead to sharply higher imports by India and reduce export availabilities for Brazil. After spiking in early 2010, world raw and white sugar prices have since fallen back to pre-peak levels on expectations of increasing supplies particularly in Brazil. A broader supply response is expected by 2010/11 as many countries boost their production in response to the recent high prices. Once production has time to adjust, the world sugar market is likely to switch from a deficit to an overall surplus situation, increasing export availabilities and leading to sharply lower prices as current price pressures are released.

Beyond this period, steady growth in global sugar consumption along with expanding demand over time for alternative products of sugar crops such as biofuels and particularly bioethanol, along with rising production costs in major supplying countries, are expected to lift raw sugar prices to USD 372 per tonne and white sugar to USD 439 per tonne in 2019 and to maintain them at elevated levels in nominal terms relative to prices at the turn of this century. World sugar prices by 2019, however, remain below the peaks reached at the start of the Outlook. When adjusted for inflation, sugar prices are expected to fall less rapidly than in the past. Both raw and white sugar prices are expected to continue to be variable over the projection period, particularly in response to the Indian production cycle, and this situation will also be reflected in the white sugar premium.

World sugar production is projected to increase to just over 200 Mt in 2019-20, some 39 Mt or 24% above the average level for the 2007-09 period. The longstanding production cycle in some countries of Asia, an important dynamic in the world market, will influence the growth in, and pattern of, global production in particular years. With low production costs and the potential to bring substantial additional land into production, Brazilian sugar production is expected to grow by some 11 Mt or 31% to reach 47 Mt in 2019, and this will propel exports to new heights. The growth underway in Brazil implies further concentration in sugar production and trade that is not without risks to sugar users and a potential source of additional price instability. World sugar consumption has slowed in the beginning of the Outlook in response to high prices and the lingering effects of the recession, but is expected to return to growth of an average of 1.8% *per annum* in following years to reach nearly 198 Mt in 2019-20. Global sugar stocks which are rundown at the start of the Outlook, rebuild in the near term as production outpaces consumption in response to the high prices and then gradually fall by 2019 with further strengthening of demand. The stronger demand and lower stocks-to-use, help to lift prices by the end of the projection period.

Biofuel prices set to rise as policy mandates drive demand

Weaker energy prices and lower profit margins and reduced investment following the economic crisis, slowed the expansion of the biofuel sector in 2009. As a result, when compared to their peak 2008 levels, ethanol and biodiesel prices decreased, respectively, by 6% and 26% in 2009. For the projection period, biofuel markets are projected to be highly influenced by mandates and other incentives in countries all over the world, with the US, Brazil and the EU playing major roles, respectively, on ethanol and biodiesel markets. Based on sustained political support for biofuels, the Outlook projects increasing world biofuel prices and these will also be underpinned by rising crude oil and energy prices. The world ethanol price[2] should follow an increasing trend to reach USD 54.4 per hl in 2019 supported by demand conditions in the US market where the Conventional Renewable Fuels mandate is assumed to be binding over the entire projection period. The world biodiesel price[3] is projected to increase up to 2015 and then to remain at a plateau of almost USD 144 per hl as second generation biofuel will increasingly become available in the EU in the latter years of the Outlook period and thus will diminish the pressure on supply globally.

Mandates calling for growing ethanol use and higher crude oil prices mean that global biofuel production should increase and is projected to reach 200 bnl in 2019, and comprising 159 bnl of ethanol and almost 41 bnl of biodiesel. These quantities are far above the average 2007-09 base levels. In the context of the new Renewable Fuels Standard (RFS2) in the United States, ethanol use for fuel is projected to increase continuously over the projection period to reach 77 bnl by 2019, but to remain below the 2019 mandate of 102 bnl. Cellulosic ethanol production is indeed only projected to expand significantly in the latter years of the projection period to reach 9 bnl in 2019 and to remain far from meeting the RFS2 32.2 bnl mandate. Ethanol use for fuel should represent an average share of 8.8% in gasoline types for transport fuel by 2019.[4]

In the case of the European Union ethanol production is mainly from wheat, coarse grains and sugar beets. It should increase to 18 bnl by 2019. Under the Renewable Energy Directive (RED) fuel ethanol use is projected to increase to reach 21.2 bnl in 2019 representing an average share of almost 8.5% in gasoline types for transport fuels by 2019. With rising domestic demand for domestic use, by a growing fleet of flexi-fuel vehicles, and for exports, ethanol production in Brazil is projected to grow by almost 7.5% *per annum*, on average, to reach 55 bnl in 2019, while ethanol exports expand to reach 13.3 bnl by the close of the projection period. As biomass based 2nd generation ethanol and biodiesel are only expected to take-off in the latter years of the Outlook, reaching respectively 7% and 6% of global production, most of biofuels are expected produced from agricultural commodities. On the trade side, Brazil will be the major international ethanol supplier. Trade of biodiesel should remain marginal; Argentina is expected to be the main supplier on international markets.

On the biodiesel market, the major player should be the EU where total biodiesel use is expected to reach almost 24.4 bnl by 2019 given mandates and tax reductions by member states and the RED. The share of biodiesel in diesel type fuels is projected to grow to 8% (almost 10% in volume terms) on average[5] by 2019. In the US, the mandate defined in the RFS2 calls for 3.8 bnl of biodiesel to be used by 2012, driving the initial growth in US biodiesel use. The Outlook assumes biodiesel use to be held constant over the remaining years although no explicit mandate for biodiesel is legislated thereafter, the subsequent mandates referring only to advanced biofuel. Trade of biodiesel should remain low; Argentina is expected to be the main supplier on international markets. While many developing countries, especially in sub-Saharan Africa and in South East Asia, have initiated ambitious renewable energy programmes, many have been put on hold during the economic crisis, credit constraints and with the more uncertain market prospects. The *OECD-FAO Agricultural Outlook*, therefore, presents a conservative view on biofuel prospects in many of the developing countries over the projection period.

Meat markets and prices trends are driven by developments in non-OECD countries

Although there was no run-up in meat prices similar to the experience of many crop products, these prices were adversely affected by increasing meat supplies with herd liquidation due to the rapid rise in feed costs and lower demand with the onset of recession. With renewed economic growth now underway, all meat markets are set to recover quickly in the near term of the projection period. Nominal prices for beef and pork increase by 21% and 17%, respectively, to reach USD 3562/t d.w. and USD 1681/t d.w., respectively, by 2019, relative to the base period 2007-09. Poultry prices are expected to be on average 32% higher reaching USD 1 638/t p.w. by 2019. When expressed in real terms (*i.e.* adjusted for inflation) prices are expected to trend higher than those observed during the decade of the 1990s, as high feed costs will somewhat constrain the expansion of output. Beef prices are anticipated to be firm for the first half of the projection period, mainly due to a tight meat supply with the start of a herd rebuilding phase. However, expansion of meat output in following years, coupled with a reduction of imports by the Russian Federation, are anticipated to exert downward pressure on prices. Pig meat prices in both the Atlantic and Pacific markets which can be substitutes in certain markets are not anticipated to be sustained beyond 2015 due to an increase in supply from Brazil and China, both of which are experiencing high productivity gains. Sheep meat prices are anticipated to be weak during the early years, but with an anticipated reduction of sheep flocks in New Zealand, the tighter supplies exerts upward pressure on world sheep meat/ lamb prices in later years of the projection period to reach USD 2 830/t d.w. by 2019. Poultry prices expressed in nominal terms are to remain relatively firm throughout the Outlook as demand continues to favour white meats.

The economic downturn triggered by the financial crisis in mid 2008 impacted severely the meat sector in 2009. Falling consumer demand and difficult access to credit affected both demand and supply. All meats were affected, although beef suffered the most compared to others, as consumers preferred cheap beef cuts and cheaper alternative sources of animal proteins. A renewed expansion of the meat sector is expected by and large for non-OECD countries from the start of the Outlook, and these will be responsible for much of the growth in the sector. Improved producer returns are anticipated to boost global meat output, with the shorter cycle of the pig and poultry sectors likely to respond rapidly to renewed demand. However, reduced cattle inventories may constrain beef production in the short term. World meat production growth is projected to increase by 1.8% *per annum* during the Outlook period, somewhat slower than in the past decade as the sector is increasingly constrained by the availability of natural resources. Meat production in the OECD area is anticipated to expand just short of 1% *per annum*, as most farmers already benefit from technological advances, and face increasingly stringent animal welfare and food safety regulations.

World meat consumption continues to experience one of the highest rates of growth among the major agricultural commodities. Much of the increase in demand is accounted for by the large non-OECD countries with their growing wealth and affluence. Poultry meat consumption in this area is projected to grow by 38%, pig meat by 33%, beef by 23% and sheep meat by 31% by 2019, when compared to the 2007-09 base period. Measured on per capita basis, meat consumption in the OECD area rises by 4% in the same period, the equivalent of one-fourth of the non-OECD countries rate of growth. World meat exports, driven mainly by increased shipments of poultry and beef, are projected to expand by 22% by 2019 relative to the base period. OECD country exports are projected to increase by slightly more than 7%, while those of non-OECD countries increase by 29% to 2019. The bulk of growth in meat traded is expected to originate largely from outside the OECD area, in particular from Brazil which will single-handedly account for nearly 60% of all the meat exported from non-OECD countries in 2019.

Dairy prices driven by stronger demand and rising supply costs

International dairy markets have experienced a dramatic rise and fall cycle in recent years. The dairy Outlook, like that for meats, is more optimistic than last year. In the course of 2009, international prices started to strengthen, rebounding rapidly at the end of the year. The strong recovery in prices was triggered by increased demand mainly from China and oil exporting countries but also by lower supplies that contracted in some regions in response to low profitability in the previous year and weather impacts on production in the Southern hemisphere producing countries. The European Union also has been restrained in the export of dairy products from higher intervention stocks. In 2010, fundamentals indicate a recovery in demand with improved economic prospects and market confidence.

The dairy sector is expected to remain one of the fastest growing sectors covered in the Outlook with strong potential as the popularity of dairy products rise mainly among developing country consumers and as demand expands with increasing affluence. The return to global economic growth and increasing population are expected to underpin international dairy markets and prices over the Outlook.

In the near term, dairy product prices are expected to be dampened by stock reduction mainly in the US and the EU. Nominal dairy prices are expected to rise steadily by 2-3% *per annum* on average, from 2012, driven by rising demand but also increasing production costs. In real terms, the longer term downward trend in prices is expected to abate, with world prices remaining relatively flat over the projection period. On average, world market prices in real terms are expected to stay 15-40% higher when compared to the decade preceding the 2007/08 peak. Butter prices in real terms are expected to register the highest gains. These are linked to continuing high energy and vegetable oil prices and to the fact that considerably less butter will be exported from countries such as the European Union or the United States. The new emerging exporters of dairy products are expected to concentrate their efforts on milk powder rather than butter which entail less sophisticated logistic requirements than butter exports.

World milk production is expected to increase by 2.1% *per annum* to reach 170 Mt by 2019 relative to the 2007-09 base period. The vast majority of the additional milk is projected to be produced outside the OECD area. World production of WMP, butter, cheese and SMP is expected to grow from the base period by 31%, 28%, 20% and 9% respectively, by 2019. The OECD area continues to dominate global cheese consumption, accounting for nearly three-quarters of the total consumption which is expected to increase by 20% over the Outlook period. In non-OECD countries demand growth is expected for all dairy products with WMP and butter consumption growing the strongest (both by 38%), followed by cheese (33%) to 2019. World exports of dairy products are anticipated to recover and grow mainly for cheese and WMP (both at 14%).

Structural change in the dairy sector is expected to intensify over the Outlook. Milk producers and the dairy industry will increasingly need to take a more proactive role to adapt to changing input markets, demand conditions, price fluctuations and increased pressure to assure quality, safety and traceability of their products. Environmental constraints and climate change related issues, and policy, will pose further challenges for the sector in the future.

Main developments in food prices

Growth in consumer food prices slows

Rising commodity prices, particularly for cereals in 2007/08 contributed to sharply increasing food prices, especially in the developing countries for less processed foods that make up a larger part of diets. While world prices have since fallen, agricultural prices and food prices inside many countries did not reflect this change with the same cadence or rhythm having remained "sticky" at high levels for an extended period of time. While the Outlook does not project food prices, there remains considerable interest in how fluctuations in commodity prices translate into changes in the cost of food. For this reason the following section discusses how food prices around the world have changed between the period of peak prices and 2009 as commodity prices have declined.

Food price inflation

Food price increases as measured by the food component of the consumer price index (CPI) slowed markedly over 2009 in most countries, though they continued in general to outpace overall inflation. In OECD countries, food prices increased by 1.6%, a significant reduction from the 6% plus increase registered in 2008. There were, however, significant differences in food price inflation among OECD as well as non-OECD countries. For most countries, food price increases slowed significantly in 2009 compared to 2008 yet for others double digit increases continued.

The food price movements discussed here refer to the food component of the CPI which measures the cost of a fixed basket of foods at the retail level. The basket reflects actual consumption patterns, thus it provides a good indication of overall change in the cost facing consumers when making food purchases. This means that food prices and commodity prices will differ substantially because retail food prices include additional costs such as processing, transportation and distribution. The share of commodities in the cost of the food basket varies across countries, for instance in the United States the account for only 20-25% of the total, with the remainder attributed to labour, energy and distribution costs.[6] In low income countries the commodity share in food prices is likely to be larger since the share attributed to energy, distribution and processing is often smaller.

In most OECD countries food prices increased by less than 5%, while in eight countries they decreased. This is in stark contrast to 2008 when two-thirds of the OECD countries experienced increases of between 5% and 10%. See Figure 1.14 for an overview of food price increases over the 2006-09 period for selected OECD countries. In the non-OECD countries food price increases were larger than in the OECD with many between 5% and 10%, but less than in 2008. For instance, in Brazil food prices increased by 5.8% in 2009 compared to 13.1% in 2008 and in Indonesia they rose 7% compared to 17% in 2008. India also had double digit food price inflation in 2009. An equally remarkable slowing of food price increases was experienced by China, where they rose by less than 1% compared to 14.4% in 2008 and 12% in 2007. A number of countries, such as Pakistan, Bangladesh and the Russian Federation as well as a Rwanda and Ghana, however did continue to experience double digit increases. Other countries such as Sri Lanka, Guatemala and China had relatively small price increases, that is less than 3%, while countries such as Senegal and Estonia, experienced net declines after increases of 9% and 14% respectively in 2008. See Figures 1.15 and 1.16 for changes in food price indices for the years 2006-09 in selected non-OECD and African countries, respectively.

Figure 1.14. **Percentage change in the food prices: Selected OECD countries, 2006-09**

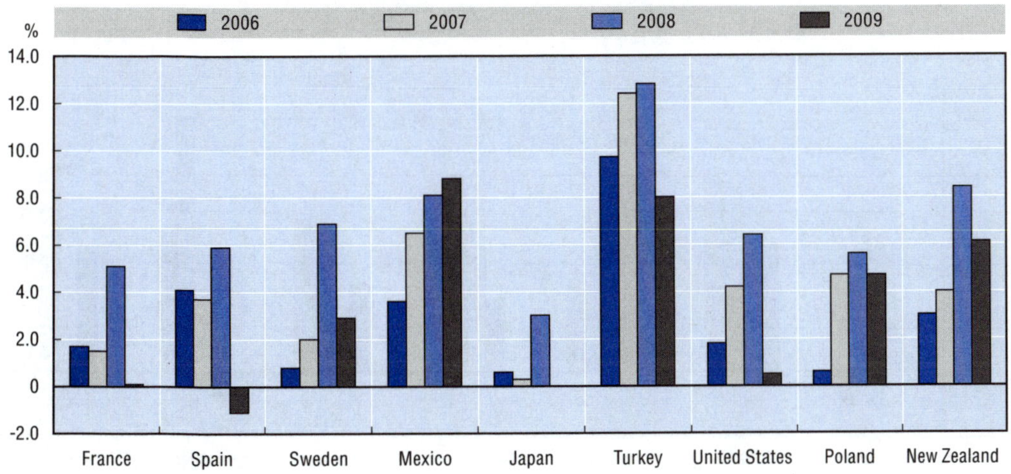

Source: *Main Economic Indicators*, OECD.

StatLink http://dx.doi.org/10.1787/888932285753

Figure 1.15. **Percentage change in the food prices: Selected non-OECD countries, 2006-09**

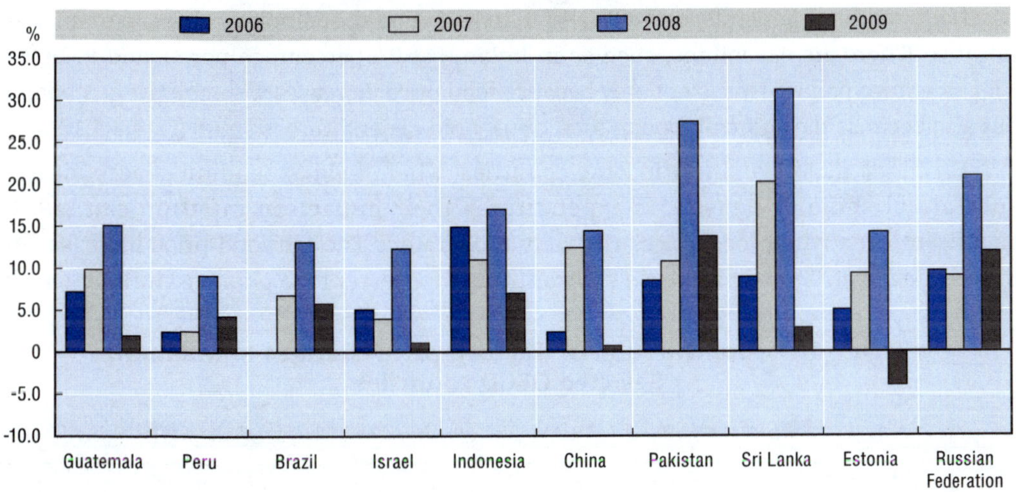

Source: National Statistical Institutes, see Annex B for data sources.

StatLink http://dx.doi.org/10.1787/888932285772

What has been the contribution of food prices to inflation?

The weight of the food component in the CPI varies widely across countries reflecting the structure of household expenditures. In high income countries, the share of food in the CPI ranges from less than 10-20% but in the middle and low income countries it is substantially higher, generally in the 30-60% range. For example, the food component accounted for 47% of the CPI in Sri Lanka, 58% in Malawi, 55% in Tanzania, 38% in Peru and 28% in Brazil. In contrast, in the United States it is but 8.2%, 10.4% in Switzerland and Germany and 11.8% in the United Kingdom.

Figure 1.16. **Percentage change in the food prices: Selected African countries, 2006-09**

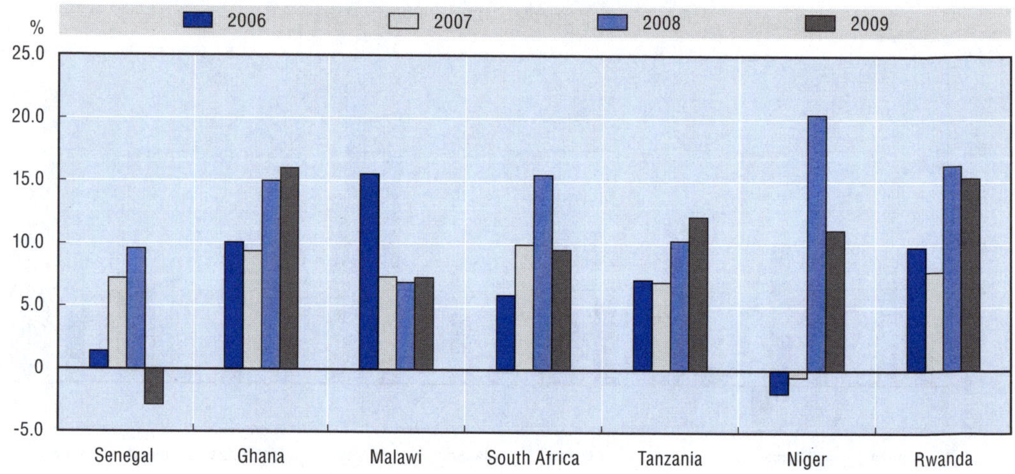

Source: National Statistics Institutes, see Annex B for data sources.

StatLink http://dx.doi.org/10.1787/888932285791

For OECD countries the contribution of food price increases to inflation, measured by the percentage change in the CPI, has been very small this past year (2009), with contributions being generally between less than half of one percentage point and many were negative. There are exceptions, such as in Ireland, at 2.5 percentage points and Poland at 1.6 percentage points. This is not only because food price increases were relatively moderate but also because the share of food in total household expenditure is small (Figure 1.17).

For many middle and low income countries, where food expenditures account for a substantial share of household expenditures their impact on inflation can still be significant even when food prices rise only moderately. Though food price increases did indeed slow in 2009, their increase still contributed 4 percentage points to inflation in the

Figure 1.17. **Contribution of the food price changes to inflation: Selected OECD countries**

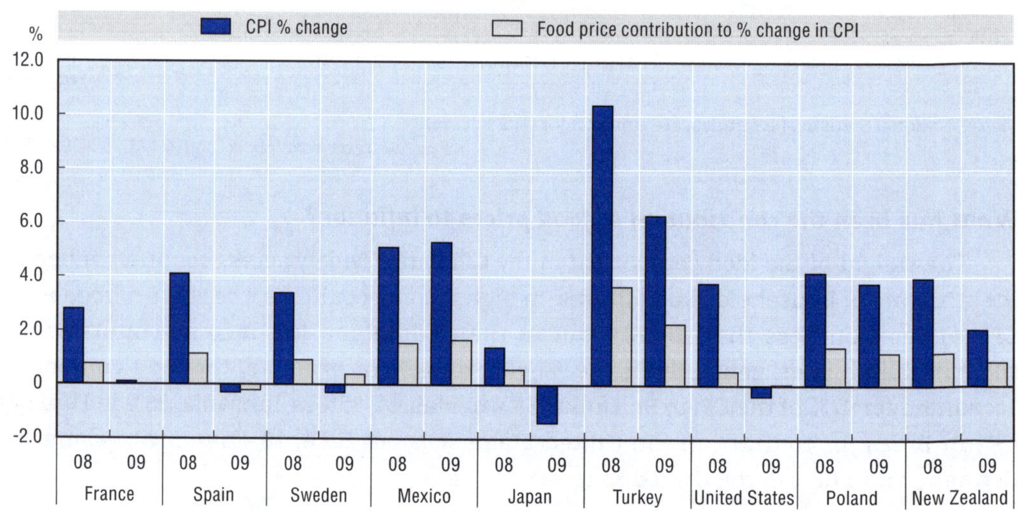

Source: MEI and OECD Secretariat.

StatLink http://dx.doi.org/10.1787/888932285810

Figure 1.18. **Contribution of the food price changes to inflation: Selected non-OECD countries**

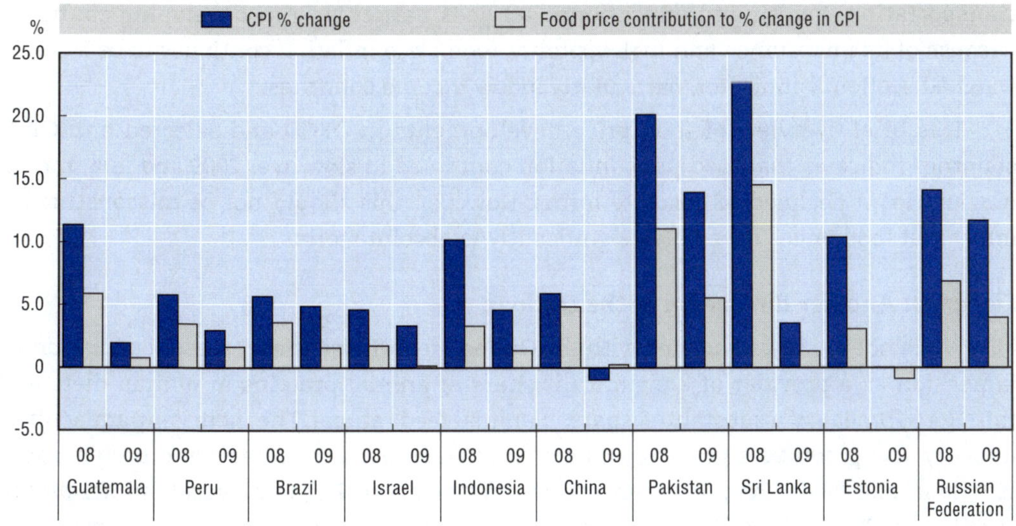

Source: OECD Secretariat.

StatLink http://dx.doi.org/10.1787/888932285829

Russian Federation and 5.5 points in Pakistan. However, it only contributed less than 2 percentage points in Brazil, Peru, Guatemala, Indonesia, China, Sri Lanka and Israel. The contribution of food price increases to inflation has come down significantly from 2008 as can be seen in Figure 1.18.

For the set of sub-Saharan African countries examined, the contribution of food price increases to overall inflation remains significant as seen in Figure 1.19 even if it has come down from 2008 levels. For example, in 2009 it contributed over 9 percentage points in Tanzania and 7 percentage points in Ghana and over 4 percentage points in Rwanda, Malawi and Niger, although it was negative in Senegal.

Figure 1.19. **Contribution of the food price changes to inflation: Selected African countries**

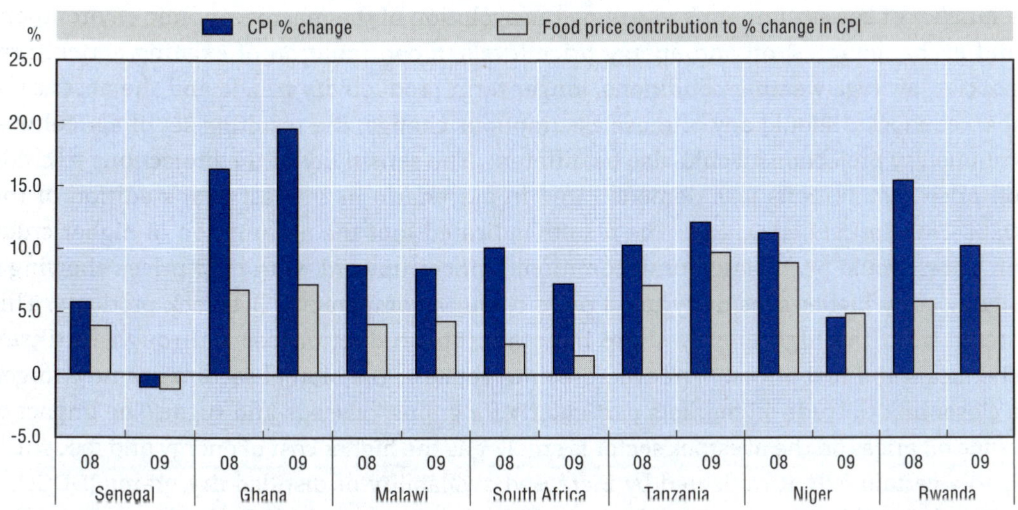

Source: OECD Secretariat based on National Statistics.

StatLink http://dx.doi.org/10.1787/888932285848

In countries where a large share of household expenditures are devoted to food, rising food prices mean that there is less money available for non-food items, such as housing, transportation, health and educational services particularly in developing countries. Because of its importance and high visibility, food price inflation continues to be a closely watched economic indicator, particularly in low income countries.

This brief overview of food price developments in OECD and selected non-OECD countries indicates that food price inflation continued to slow over 2009 and in a number of countries it declined in absolute terms. However, this should not be misconstrued to imply that food prices have fallen significantly in absolute terms.

Fisheries: Another dimension of the Outlook

While not usually associated with the Outlook report, fisheries represent an important dimension as a provider of a significant share of animal proteins in human diets and, through fishmeal, of a substantial share in animal feed rations. The increasing aquaculture industry also produces a growing and competitive demand for cereals and protein meals for use in fish feed. Fish proteins accounted for about 16% of total world animal protein supplies in 2008. Global fish production has increased about eight times in volume since 1950 to reach some 142 Mt in 2008. Capture fisheries production has stabilised at 88-94 Mt over the past decade while aquaculture production has increased significantly and now contributes 37% of the total fish production and 46% of the total fish destined to human consumption. FAO and other organisations have projected total fish production to increase by 10-15% over the next ten years (Box 1.2).

The future potential of the industry is linked to the ability of policy makers to provide a conducive policy landscape for sustainable and profitable operations. In recent years, national and international policy debates have focused on sustainable and responsible fisheries and stock rebuilding, recognising that major fish stocks are either overexploited or at very high levels of exploitation.

Issues and uncertainties

The agricultural market projections through to 2019 discussed in this chapter are a representative scenario based on a consensus view of what may happen in the future given a number of key assumptions regarding the evolution of the macroeconomic environment and exchange rates, oil and energy price levels, a continuation of existing agricultural policies, average weather conditions, longer term productivity trends and the absence of market shocks. Should any of these assumptions change, the resulting set of agricultural commodity projections would also be different. The sensitivity of the projections to crude oil price assumptions was demonstrated in a scenario in the last year's edition of the *OECD-FAO Agricultural Outlook*. The results indicated that the assumption of higher crude oil prices would push agricultural commodity prices upward, with crop prices showing a significantly higher sensitivity to oil price changes compared to livestock products. This stems from the high energy share in total crop production costs through fertiliser, chemicals and fuel prices. Moreover, the emergence of the biofuel sectors has now forged a closer link to crude oil markets particularly for grains, oilseeds and sugar. The impact of crude oil prices on the livestock sector is smaller as the higher cost of energy and feedstuffs is to a certain extent mitigated by increased availability of distilled dry grains (DDGs), a by-product of bioethanol production, which can be used in animal feed.

Box 1.2. **The importance of the fisheries sector and its links with agriculture**[a]

The fisheries sector plays a significant role in global food security providing a valuable dietary source of proteins, minerals, micronutrients and essential fatty acids. In addition, the sector contributes to economic activity, employment and in generating foreign exchange. World per capita fish consumption is estimated at about 17.1 kg, with fish providing about 3.0 billion people with 15% of their average per capita intake of animal protein.

Fish is widely traded, with about 38% of production entering international trade as various food and feed products. Trade of fish and fishery products has significantly increased in the last decades, reaching a record USD 102 billion in 2008. In 2009, following the global economic recession, there was a contraction in demand, with a slight decline of fishery trade in both value and volume terms. However, trade is again expanding and the Outlook for 2010 is generally positive as is the longer term trend for fishery trade. Developed countries absorb about 80% of world fishery imports in value. Developing countries play a crucial role in fishery exports with a share of about 50% by value and 60% by quantity (live weight equivalent) of the total. The fishery net exports of developing countries (*i.e.* the total value of their exports less the total value of their imports) has shown a continuing rising trend in the last decades, growing from USD 9 billion in 1986 to USD 27 billion in 2008. These figures were significantly higher than those for agricultural commodities such as rice, sugar, coffee and tea.

At present, about 80% of total fishery production is used for direct human consumption. The remaining 20%, entirely from capture fisheries, is destined for non-food products, mainly for production of fishmeal and fish oil, as well as direct feed in aquaculture and livestock. In 2008, total world fish production (capture and aquaculture), excluding aquatic plants, reached 142 Mt. It should be mentioned that this figure might underestimate the effective amount due to the incomplete recording of subsistence fisheries as well as of illegal, unreported and unregulated catches. Estimates for 2009 show a slight increase from the previous year. Compared with production figures a decade ago, the current amount represents a growth of more than 28 Mt. This additional supply is mainly due to increases in aquaculture production. Capture fisheries production, which reached its peak in 1996 with 93.8 Mt, has stabilised within a range of 85 and 94 Mt, with variations mainly caused by the *El Niño* climate pattern phenomenon. In the last three years, capture fisheries production remained close to 90 Mt. Forecasts indicate that any major increase in future supply will come from aquaculture. Aquaculture production is already playing a substantial role in supplying fish for human consumption, growing from a share of 17% of per capita fish consumption in late 1980s to an estimated 46% currently. It is estimated to reach 50% within the next decade.

With an annual average rate growth of about 9% in the last two decades, aquaculture is currently growing faster than all other food-producing sectors. Aquaculture is expanding in all continents, in new areas and species, intensifying and diversifying the product range. Although the rate of increase in aquaculture production has slowed to about 6% per year in the 2000s, growth prospects for the sector are still good. Several more species and new product forms will be commercially produced in the near future and further technological development will contribute significantly to improve productivity, yield, quality and consumer acceptability. However, it is also evident that, in many countries, significant challenges remain in order for the aquaculture sector to reach its full potential and become economically, environmentally and socially sustainable. Important factors which will have a future impact on the aquaculture industry include climatic changes, environmental issues, access to sites and water, raw material supply for feed, pandemics and fish health management, integration and ownership structures, governance, food safety and traceability. The OECD Workshop on Advancing the Aquaculture Agenda, held in April 2010, underscored the importance of ensuring a solid governance system for the sector with a view to ensuring future growth.

Fisheries, and in particular aquaculture, interact in several ways with agriculture. One evident example is in integrated farming,[b] but more important is their impact on ecosystems, markets, products and prices, as well as on innovations and technology. Competition between the fishery and agriculture sector may arise for water and land resources, especially for irrigated agriculture, as well as on the availability and relative efficiency on the use of feeds between terrestrial animals and farmed fish. Fishmeal and fish oil obtained from capture fisheries and from fisheries by-products are used as feed in aquaculture as well as in the feed rations of pigs, poultry, ruminants and pets. With the expansion of aquaculture, supplies of fishmeal have been largely directed to this sector, as major constituents of aquatic carnivorous/omnivorous species feeds, diverting away from livestock, which now mainly use fish meal in starter and breeder diets for poultry and pigs. It is estimated that in 2007 aquaculture consumed about 68% of world fishmeal and 81% of world fish oil supplies. The growth of the aquaculture industry and the increasing competition with

Box 1.2. The importance of the fisheries sector and its links with agriculture[a] (cont.)

the livestock sector generated upward pressures on prices of fishmeal and fish oil and has also led to an increasing demand for additional or substitutive sources of feed. Livestock and agriculture by-products, used traditionally to feed terrestrial animals, are now increasingly employed by the aquaculture sector, especially for the farming of non-carnivorous aquatic species. Continued growth in demand for livestock and fish products, as well as for biofuels obtained from agriculture by-products, has raised concerns over the competition for feed supplies, in particular for the finite fishmeal and fish oil resources, and the impacts of such growth on the environment and on the aquaculture and livestock developments.

Figure 1.20. World fish utilisation and supply

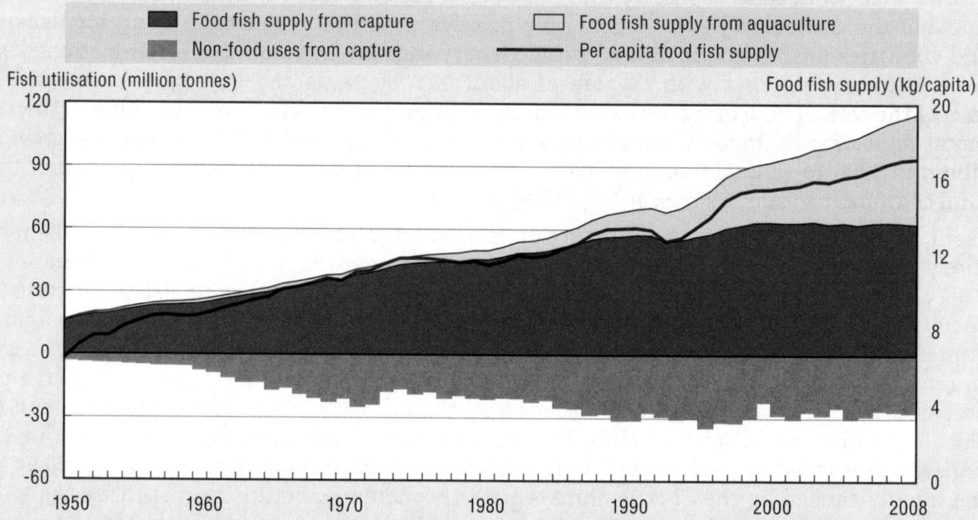

Source: OECD and FAO Secretariats.

StatLink http://dx.doi.org/10.1787/888932286000

Figure 1.21. World production (quantity)

Aquaculture versus selected agricultural products – Average annual percentage growth rate

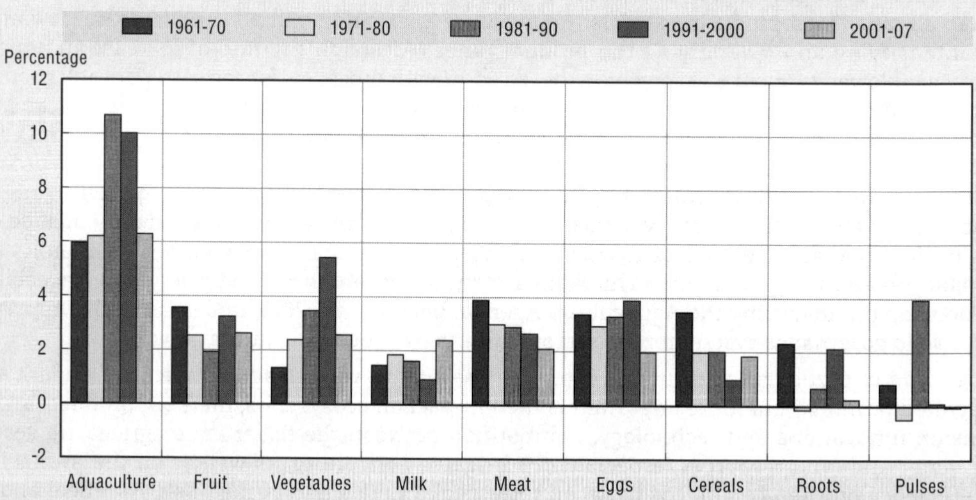

Source: OECD and FAO Secretariats.

StatLink http://dx.doi.org/10.1787/888932286000

a) All statistics quoted in this box are from FAO, Fisheries and Aquaculture Department.

b) The term "integrated farming" refers to integrated resource management of different activities, such as crop, livestock and fish subsystems. When involving aquaculture, it is the concurrent or sequential linkage between two or more activities, of which at least one is aquaculture.

A number of major uncertainties remain. At the beginning of the Outlook, the lingering effects of the global financial market turmoil and the deep economic recession are intermingled with the onset of a strong recovery in the large developing countries and a more fragile turnaround in much of the OECD area. This has been made more fragile by the unfolding Greek crisis and fears of contagion to other countries with large budget deficits in Europe that now threaten to slow the global recovery. Many OECD countries are starting to address excessive budget deficits to restore fiscal balance and to soak up excess liquidity in their economies in order to prevent asset bubbles and inflation. The issue for governments for 2010 and beyond is to devise exit strategies as to when and how to reduce the liquidity and fiscal stimulus without reversing the nascent recovery. With globalisation and closer economic and financial integration between nation states, many external factors such as variable oil and energy prices, freight rates, inflation, interest rates, credit availability and exchange rate changes can have profound impacts on the competitive positions of national agricultural industries and their trade performance. Among other developments are the continuing migration of production to areas of the world which suffer higher yield variability together with more frequent weather disturbances associated with climate change may render global yields much more variable, leading to greater instability in production and trade flows. Finally, future changes in agricultural and trade policies and the eventual outcome of the current Doha Round of international trade negotiations and bilateral agreements that may be under consideration can be expected to have an important influence on agricultural markets. All of these factors play an ever increasing role in commodity price formation suggesting a continuation of volatile and uncertain agricultural commodity markets.

The past decade has been one of constant change, altering the environment in which the sector operates. The most recent years have been characterised by significant price volatility and a sharp rise in the numbers of malnourished people. Coming years will also be characterised by continuing economic, demographic, market and environmental pressures that will bring both opportunities and challenges to farmers, food businesses, consumers and governments. Some of the issues that were identified as major challenges:

Food security: Hundreds of millions of people remain food insecure. Although the world now produces enough to feed its population, the number of undernourished has increased since the mid 1990s, reaching more than one billion persons in 2009, in part as a result of recent price spikes and the global economic recession. Paradoxically, many of the world's food insecure people are themselves farmers. The population and income dynamics in emerging economies will continue to increasing demand for food in the decades to come, while the growing biofuels market is a new source of demand impacting on food markets through related land use changes. Production and productivity will need to be increased while a well functioning, rules-based multilateral trading system will be crucial in ensuring food can move from where it can be abundantly produced to where it cannot. These issues are further discussed in Box 1.3.

Climate change: There is a broad scientific consensus that less-resilient agricultural production areas will suffer the most, as temperatures rise further, for example in semi-tropical and tropical latitudes, and as already dry regions face even drier conditions. Climate change may also increase food safety risks that might result from heat-related and water borne diseases with temperatures rising and more flooding. Production variability and uncertainty of supplies are expected to rise as a result of likely increases in the frequency of

Box 1.3. **Food security remains high on the international policy agenda**

The *World Food Summit on Food Security*, held in Rome on 16-18 November 2009, called for action to reduce hunger and malnutrition. Objectives, cited in the Summit's Declaration included better co-ordination at the global, regional and national levels, a reversal of the decline in domestic and international funding for agriculture and a proactive approach to the challenges of climate change. As background to the Summit, the FAO released a number of background studies and held a high-level expert forum in October 2009 on *How to Feed the World in 2050*.

The challenge is to feed a growing, more urban and, on average, richer population while adopting more efficient and sustainable production methods and adapting to climate change. World population is expected to grow by 2.3 billion people between 2009 and 2050 with nearly all this growth from developing countries. The population in sub-Saharan Africa is expected to grow the fastest, by some 114%. It is estimated that feeding a population of 9 billion would require a 70% increase in global food production between 2005-07 and 2050. Production in the developing countries would need to almost double. Of course, the need to increase output would decline significantly if progress could be made in reducing production losses (*e.g.* from disease, pests, storage) and food waste (*e.g.* during processing, transportation and consumption). The projections of this Outlook indicate global production growth which is on track with estimated longer term food requirements.

Demand for cereals for food and feed is projected to rise by one-third to 3 Bt by 2050, and possibly higher due to a growing liquid biofuel market. Net cereal imports into the developing countries would increase almost three-fold to nearly 300 Mt by 2050, some 14% of their total cereal consumption. Demand for more income-responsive vegetable oils, meats and dairy products are expected to rise even faster. Livestock is one of the fastest growing sub-sectors in agriculture with over 80% of the projected growth in the next decade taking place in developing countries, particularly in Asia and the Pacific (especially China) and Latin America, outpacing growth in the OECD area by a factor of 2:1 over the next decade.

To support the necessary expansion in output in developing countries, FAO estimates the required average annual investment in primary agriculture and necessary downstream services (*e.g.* storage, processing) at USD 209 billion in 2009 prices (or USD 83 billion net of depreciation), much of which would come from private sources. Still, this amount represents a 50% increase from current levels and does not include the public investments required in such areas as roads, irrigation, electricity and education. In general since the 1970s, those countries with higher net investment per agricultural worker have been more successful at reducing hunger.

As suggested in the 2009 Outlook report and supported by the FAO Food Summit background studies, the technical ability to produce enough food can be achieved, given sufficient investment and sustainable resource management. However, this alone will not alleviate hunger which is primarily a question of poverty. It is more a problem of food *accessibility* than of food *availability*. A comprehensive approach to food security needs to include investments to promote income generating activities for the poor and thus improve their ability to purchase food. Importantly, the payoff from hunger reduction in terms of economic growth can be substantial. Throughout the 1990s, the value-added per worker, in countries where 2.5% of the population was undernourished, was 20 times higher than in countries where more than 35% of the population was undernourished.

Sources: *www.fao.org/wsfs*; *www.fao.org/wsfs/forum2050/wsfs-forum* and *www.oecd.org/agriculture*.

extreme events such as droughts and floods. In more extreme cases, production zones might shift. It was also recognised that agriculture will be required to make an important contribution to reducing greenhouse gas emissions (see Box 1.4).

Pressure on resources: Agriculture must increasingly compete with other parts of the economy for the available land, water, mineral and energy supplies. Water scarcity is a particular concern, with agriculture currently accounting for 70% of world freshwater withdrawals (45% in OECD countries). Some predictions suggest that 36 countries with a combined population of 1.4 billion will be either cropland or freshwater scarce by 2025. Depletion of non-renewable resources (including oil and minerals) and degradation of arable land add to the resource pressures. Increased efforts to better manage water resources including through water prices that reflect scarcity of supplies is urgently required. Some countries are now making foreign investments to secure food supplies from other countries where resource constraints are lower, but guidelines for such investments are critical to respect local economies.

Global food chains: The food industry is becoming more vertically integrated, globalised and concentrated. These developments have enabled the industry to respond well to a wide range of changing consumer preferences, while maintaining relatively low prices. But there are also concerns about their growing market power, about price transmission, transparency and what is a "fair" distribution of profits across the food chain. Both public food safety standards and private quality standards have been raised in response to consumer demands, but these imply higher compliance costs and some farmers, particularly in developing countries, may have difficulty in meeting the more stringent standards.

Innovation: The capacity of the global food and agriculture system to continue to provide adequate supplies for food, feed, and non-food uses depends in large part on technology and innovation. In some cases there remains considerable scope for improving productivity through more widespread adoption of available technologies. Progress can also be made to make better use of what is produced – as much as one-third of food "disappearance" has been estimated as "waste". Waste occurs at the farm level, in the storage and distribution system, in food service, and at home. Governments, in partnership with the private sector, need to increase investment in research and development, while keeping markets open to allow the free flow of innovation and technology. A major challenge remains in the development and approval of genetically modified crops, and the extent to which asynchronous action by Governments results in trade diversions among countries and regions.

The OECD Committee of Agriculture met at Ministerial level on the 25-26 February 2010 and discussed many of these same issues. What Ministers' had to say on these different issues is covered in the Ministerial *Communiqué* from the meeting, an extract of which is shown in Box 1.5.

Box 1.4. **To what extent is climate change included in the *Agriculture Outlook*?**

Climate change refers to a change in weather patterns over relatively long periods of time, usually at least a decade. While recent climate change concerns are mainly related to increasing global mean temperature (global warming), in the broader sense it encompasses changes and variability in temperature, precipitation, atmospheric pressure, humidity and wind. Climate change also affects the frequency of storms, floods, droughts and other extreme weather events.

The global warming hypothesis is supported by a statistically significant increase in average global temperatures over several decades, as measured by a variety of sensors based on land, sea and satellites.[a] According to the Intergovernmental Panel on Climate Change (IPCC) in its fourth assessment report, "*Warming of the climate system is unequivocal*".[b] Climate change is expected to impact significantly on agriculture as weather evolves in the future. While science is not yet capable of providing precise predictions, there is wide agreement in the scientific community on some climate change trends.

Average temperature increases will be unevenly distributed around the globe. The IPCC projects that warming will be greater at the equator and the poles than at the mid-latitudes.[b] In very general terms, this may translate into more favourable agricultural conditions for temperate regions (*e.g.* warmer, longer growing season), but less favourable conditions (*e.g.* heat stress) in the already hot tropics.

Rising temperatures will accelerate the hydrological cycle,[c] changing the temporal and spatial distribution of fresh water. Global water availability is likely to remain constant. While in the short-term the accelerated melting of glaciers may result in increased risk of floods, over the longer term it implies reduced flows to areas which rely on such water supplies. It is also conceivable that even if annual precipitation remains unchanged, rainfall may occur more frequently when it is not needed (*e.g.* during the summer harvest rather than in winter and spring).

Higher CO_2 concentrations are expected to have a positive "fertiliser effect" on plant yields, especially rice, wheat and soybeans. However, this hypothesis is largely based on laboratory experiments and may be less in reality. Moreover, a number of important crops like maize and sugarcane belong to a plant family where this fertilisation effect is smaller, even in the laboratory.

All these factors will have both positive and negative consequences for crop and livestock production, depending on many environmental conditions prevailing in a particular geographical location, production management systems, and the impact of mitigation and adaptation policies and practices. However, most analysis of climate change focuses on 25-50 year time horizons so it is difficult to incorporate such analysis into a medium-term *Agricultural Outlook*. There are no climate change variables or equations in the Aglink-Cosimo model used to generate projections for the OECD-FAO medium term *Agricultural Outlook*. To date, there are no explicit links between the physical models used to access climate change and the economic models used for market analysis, although this is an area likely to be addressed in the near future.

While the OECD-FAO medium-term projections may implicitly incorporate certain elements of climate change for some regions, this impact cannot be separated out or quantified. One of the assumptions underlying the projections is "average" weather or growing conditions in the different countries and regions. Climate change impacts, through changed average weather conditions, would be expected to appear through yield performance/variation and average pasture conditions. The initial phase of the Outlook process involves commodity experts adjusting projected crop yields and output, based on a number of factors such as input use, innovation, technology. National experts from all major producing regions respond to a detailed, commodity-specific questionnaire so that global projections are built up from country level analysis.

While it may not be possible at this stage to provide quantitative estimates of the medium-term impact of climate change on agriculture, this will be a topic of increased analysis over the next few years. Agriculture Ministers, at the February 2010 meeting in Paris noted that climate change presents challenges and opportunities for the agricultural sector in reducing green house gas emissions, in carbon sequestration, and the need for adaptation. They asked for further analysis of the likely impact of climate change on agriculture and on agro-forestry, the role of the sector in mitigation and adaptation, and the appropriate policy responses.[d]

a) NASA Goddard Institute for Space Studies, *http://data.giss.nasa.vog/gistemp/*.
b) IPCC Fourth Assessment Report: Working Group I: The Physical Science Basis.
c) World Meteorological Organization, *www.wmo.int*.
d) The Ministerial *Communiqué* is available at: *www.oecd.org/agriculture/ministerial*.

Box 1.5. **2010 OECD Agriculture Ministerial Meeting**

Ministers' discussions were wide-ranging and forward looking. A focus of discussion was the question of food security. Will the food and agriculture system be able to respond as population growth causes demand for food to increase, in a world where pressure on land, water and other natural resource is already evident and where climate change will bring additional challenges? The task for governments is to make sure that the right policies and institutions are in place.

Ministers agreed to build on and complement the policy principles agreed in 1998 acknowledging that the main priority is the need to provide an adequate supply of safe and nutritious food, on a sustainable basis, for the world's growing population. Specifically, ***Ministers recognised***:

a) That an ***integrated approach to food security*** is needed involving a mix of domestic production, international trade, stocks, safety nets for the poor, and other measures reflecting levels of development and resource endowment, while, poverty alleviation and economic development are essential to achieve a sustainable solution to global food insecurity and hunger in the longer term.

b) That "***green growth***" ***offers opportunities*** to contribute to sustainable economic, social and environmental development, that agriculture has an important role to play in the process, as do open markets that facilitate the sharing of technologies and innovations supportive of green growth, and that, in this context, care needs to be taken to avoid all forms of protectionism.

c) That ***climate change presents challenges and opportunities*** for the agricultural sector in reducing green house gas emissions, in carbon sequestration, and the need for adaptation.

and Governments should ensure that:

d) Farmers and food suppliers, in developed and developing countries, are able to respond effectively to changing consumer and societal demand, and that the transmission of price signals along the food chain is improved locally, regionally and internationally.

e) The necessary institutional, regulatory and policy frameworks are in place to enable markets for food and agricultural products to function efficiently, effectively transparently and fairly.

f) Appropriate policies are developed to facilitate the management of risk at the farm and farm household levels and throughout the agro-food sector, including, where appropriate, in response to the impacts of extreme price volatility on farmers, while maintaining an efficient distribution of responsibilities between private and public actors.

g) Policies for the food and agriculture sector are coherent with general macroeconomic, trade, industrial, environmental, energy, consumer and social policies (including health and nutrition), and that there is coherence between country policies and efforts to assist developing countries.

h) Trade play a role in matching global supply and demand, as a reliable source of supply for countries dependent on imports and a reliable outlet for competitive suppliers, through an efficient well-functioning rules-based multilateral trading system, to which an ambitious, balanced and comprehensive conclusion of the Doha Development Agenda would be an important contribution;

i) Policies are supportive of the efforts of farmers and other participants in the supply chain to effectively manage natural resources to supply sustainably produced commodities.

j) Incentives and disincentives can be effectively and transparently designed to reflect the total costs and benefits to society, with a view to improving environmental performance, in consistency with multilateral trade rules and commitments; facilitating adaptation to and mitigation of climate change; allowing the food and agriculture system to respond to resource pressures particularly those affecting land and water; reducing losses and waste in the food supply chain; ensuring the provision of public goods and services such as rural amenities, biodiversity, maintenance of landscape and land eco-system functions and contributing to the development of rural areas.

Box 1.5. **2010 OECD Agriculture Ministerial Meeting** *(cont.)*

k) There is a supportive investment climate in particular with respect to foreign direct investment in emerging and developing countries, in line with internationally agreed guidelines.

l) Innovation, including transfer of technologies, is fostered in order to increase productivity, enhance efficiency, improve sustainable resource use, respond to climate change and reduce waste including through balanced protection of intellectual property rights, and a regulatory environment conducive to innovation and new technology, and to public-private partnerships.

m) Consumer protection is enhanced through further development and implementation of efficient, science-based food and feed safety standards, consistent with international agreements.

n) Policies are explicitly connected to specific objectives or intended beneficiaries, while also limiting the administrative burden on the sector so that total costs to the public are minimised, and that policies are monitored and evaluated regularly for continued relevance, cost-effectiveness and efficiency.

Note: The text is extracted from the *Communiqué* from the Ministers whose complete text can be consulted at *www.oecd.org/agriculture/ministerial*.

Notes

1. One aggregate measure is the net production index, which is a summary measure of the growth in gross value of production of all commodities included in the Outlook, net of seed and feed costs which are internal to the sector, all measured at constant international reference prices of 2004-06.
2. Brazil, Sao Paolo (ex-distillery).
3. Producer price Germany net of biodiesel tariff.
4. All biofuels use are expressed on the basis of energy contained unless otherwise specified.
5. All biofuel use shares are expressed on the basis of energy contained unless otherwise specified.
6. With the decline in commodity and oil prices, food price increases across the globe subsided in 2009 compared to 2008.

OECD-FAO Agricultural Outlook 2010-2019

Chapter 2

Price Volatility and Price Transmission

Increased interest in price volatility has been aroused by the rapid rise in food commodity prices in 2007/08, and their subsequent fall. As in former periods of rapid price change, there are different implications for various agents across the food chain. Producers (sellers) clearly benefit and consumers (buyers – food or feed) lose from a rise in prices; the benefits are reversed for a fall in prices. These variations also have different implications for producers and consumers in developed *versus* developing countries.

When prices spike, consumers in developing countries who spend a high share of their incomes on food are most seriously affected; they may also have less options in their food choices. When prices fall, producers in developed countries often have access to various support schemes and credit markets, while producers in developing countries may face large income fluctuations, often for commodities for which they are highly dependent for their incomes. Moreover, high volatility – an unexpected large rise and fall in prices – imposes large costs throughout the food chain, as uncertainty hinders investments and sectoral development. Governments, who are concerned that price volatility may be increasing, or if not increasing, remaining at unacceptably high levels, are looking for the means to prevent or minimise high price volatility and its harmful effects on their populations.

This chapter examines two dimensions of the question as to whether agricultural commodity prices are becoming more volatile. The first is volatility at the global level. Are the fluctuations in world commodity prices greater now than in the past? The second is market integration and the transmission of international prices to domestic markets. How are shocks in international markets absorbed and by whom? These complex issues are not evident in a projection of the nature provided in this *Outlook*, which assumes an inherently stable trajectory for key driving forces such as crop yields, input prices, energy prices and policy environment. The price spike of 2007/08 is revisited and measures of price volatility are presented. The focus then shifts to a discussion of market integration and price transmission to domestic markets, again looking at the implications for price volatility. Finally, some policy options and a research agenda are discussed.

The global price spike of 2007/08

The agricultural commodity price spike of 2007/08 has been widely examined.[1] In this period international food commodity prices rose to unprecedented levels in nominal terms, as witnessed by the FAO food commodity price index which reached a peak in June 2008, before retreating back to 2006 levels by early 2009. As shown in Figure 2.1, this price surge in primary food commodity prices followed what has been described as the longest and largest surge in global commodity prices in over a century.[2] The factors underlying this broad surge appear largely global and macroeconomic in nature, including the rapid economic growth of developing countries during the period, particularly in Asia, but also monetary factors including money supply growth, financial laxity and exchange rate movements (particularly depreciation of the US dollar). Given a substantial

Figure 2.1. **Co-movements of commodity prices 2000-10**

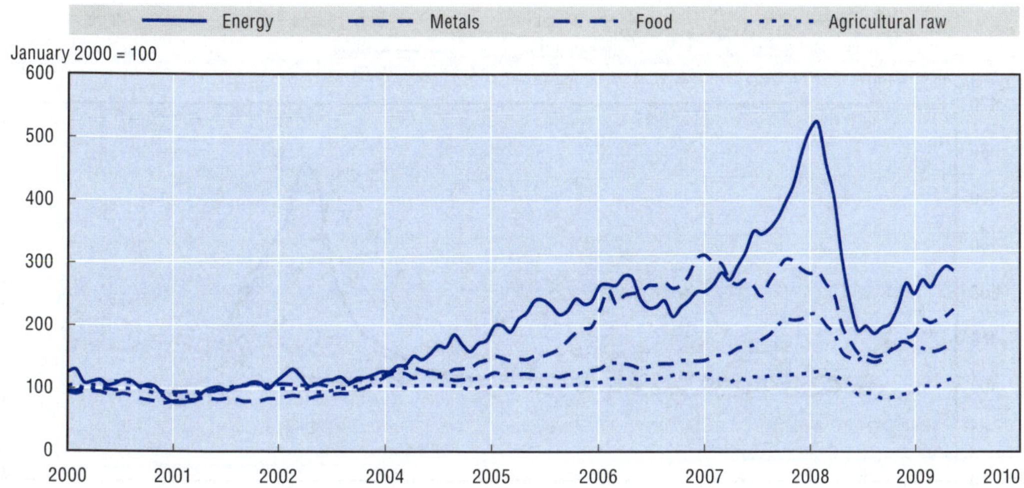

Source: IMF International Financial Statistics (2010).

StatLink http://dx.doi.org/10.1787/888932285867

co-movement among primary commodity prices during the period, food commodity prices, despite their huge implication for food security, were relatively more restrained than many other commodity prices.

In the context of the broader commodity price surge, the food price hike was affected by a series of drought-induced crop shortfalls at a time of low stocks. It was also influenced by the increasing integration of agricultural markets to energy markets, and the important impact, both intended and unintended, of government policies. Importantly, energy prices, which experienced the largest price spike, underpinned production costs of agricultural products relying on energy and fertilisers. Coupled to this impact was the emerging demand for feedstocks to support production of biofuels. This impact was largely crop-specific and included maize in the United States, vegetable oils in the EU, and to a lesser extent, sugar in Brazil. Mandated consumption targets for biofuels, and other support policies further re-enforced the links between energy and feedstock prices. Additionally, increased production of feedstocks was to the detriment of other crops whose cultivated areas decreased (*e.g.* wheat and soybeans). Fears about food price inflation incited further policy reaction by food commodity (including rice) exporters and importers alike who were keen to assure food supplies, and in combination put additional upward pressures on prices (Figure 2.2).

While the energy factor explains an important and controversial part of the increase in agricultural commodity prices, other factors were at play too. Agricultural supply initially exhibited sluggish responsiveness to the increase in demand, not only due weather related production shortfalls and its inherent production lags, but also after having undergone a long period of low investment given the low real prices in the previous decade. Commodity stock levels fell to critically low levels in 2006 and 2007. Macroeconomic factors such as the depreciation of the US dollar and monetary expansion also influenced the crisis, including agriculture. The depreciation of the US dollar improved the purchasing power of many importing countries, causing an increase in prices of commodities which are denominated in dollar terms.

Figure 2.2. **Co-movements of agricultural food crop prices**

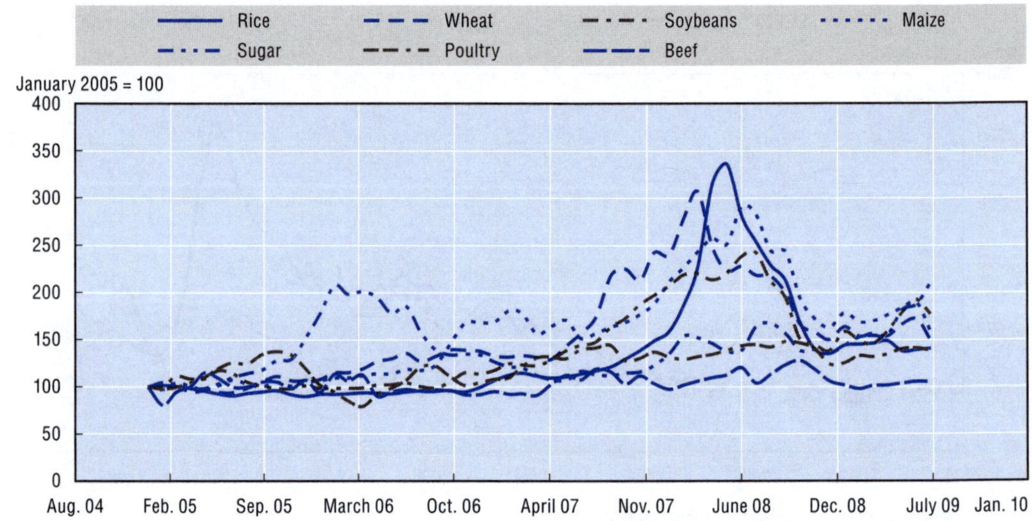

Source: www.fao.org/worldfoodsituation/wfs.

StatLink http://dx.doi.org/10.1787/888932285886

The role of speculation in financial markets encounters vigorous debate. Some analysts argue that low interest rates and low returns in other markets attracted noncommercial investors into agricultural and other commodity markets, fueling higher prices. Of course the causality is debatable – higher prices more likely attracted speculators, rather than the other way round. Anecdotal evidence suggests the number of traders in futures markets increased as prices increased. For example, institutional investment funds, which trade on large, long-term commodity-indices rather than specific markets, may have had a role in rising futures prices. Various studies, such as by Irwin and Saunders (2010) and Gilbert (2009) provide differing conclusions as to whether index funds have caused the 2006-08 bubble in commodity prices.

Volatility and uncertainty in agricultural markets

The nature and causes of commodity price volatility is complex, and many contributing factors can be identified. Price volatility refers to unpredictable price movement. The nature of volatility, in terms of periodicity of movements, either daily, monthly, yearly, or season of a year may have differing implications for producers or consumers depending on the commodity.[3] Agricultural prices have followed a long run stagnant trend, punctuated by high spikes; at least one study has identified three major spikes over the period 1970 to 2008.[4] In terms of causes, volatility fundamentally follows from the dependence of agricultural production on weather conditions, which can have direct impacts on the variability of yields. It is also compounded by low short run supply responses to price changes (*i.e.* low supply elasticity), when production decisions have already been taken. Demand responses for food are often also typically low given that food itself is an essential product. Under these basic circumstances, prices often react strongly under seemingly small shocks to demand and supply. The low elasticity of supply can be compensated partly by stock holding, which enables continued supply to meet demand. Hence, stocks play a critical role, unless they are reduced to low levels which limit their cushioning effect. Low levels are associated with periods of higher price volatility [see Deaton and Laroque (1992), Balcombe (2008), and Chavas and Kim (2006)].

Shocks can be transmitted from other markets. They can originate from the demand side such as economic expansions or recessions; or from the supply side through reduced supply of inputs, availability of substitutable/complementary products or, typically, energy markets. Finally, shocks can be transmitted through macroeconomic variables such as exchange rates and terms of trade. Balcombe (2008) finds that the volatility of most individual food commodities is positively related to the general volatility of other agricultural commodities.

One common measure used to gauge price volatility is the coefficient of variation (CoV) of a given price series, which expresses an estimation of the variability of the series as a ratio to its average value. This permits comparison across commodities with different average prices. A traditional measure of variability used in this calculation is the standard deviation of observed prices. This measure refers to *ex post* observations of actual prices. But it implicitly considers all price variability to be unexpected. Clearly, some variability can be predicted (*e.g.* seasonal variation, business cycles, or other trending behaviour) such that results from using the simple standard deviation may overstate the degree of volatility or uncertainty (for more discussion see Moledina *et al.*, 2004). Therefore, in order to have a better measure of the unpredictability or uncertainty faced by the market, it is common to take into account only movements of the series that cannot be predicted on the basis of its previous values. Table 2.1 presents the CoV of prices for a selection of commodities, after the predictable component has been removed from the observed values.[5] Values close to 0 indicate small volatility, and higher values denote greater volatility.

Table 2.1. **Coefficients of variation of commodity prices in selected countries: 2006-10**

	Maize	Rice	Wheat
International	0.110	0.117	0.080
Argentina	0.084		
Australia			0.080
Bangladesh		0.055	0.056
Brazil	0.103	0.085	0.118
Burkina Faso		0.102	
Cambodia		0.107	
China		0.012	
Ecuador	0.046		
Egypt			0.081
European Union	0.061		0.096
India		0.034	0.043
Israel			0.061
Japan		0.027	
Kenya	0.093		
Mexico	0.042		
Peru			0.026
Philippines		0.046	
South Africa	0.078		
Sudan			0.128
Thailand		0.117	
Uganda	0.157		
United States	0.110	0.110	0.080

Source: FAO (2010). Prices are wholesale monthly prices per tonne, converted in USD, for the period January 2006-January 2010. International reference prices are US for maize, wheat; Thailand for rice.

StatLink http://dx.doi.org/10.1787/888932287862

The results indicate a wide diversity of experience in price volatility. International price volatility is indeed large. For the selected countries and commodities displayed below, domestic price volatility is generally lower than for international markets. Important exceptions include some developing countries such as maize in Uganda, or wheat in Sudan, for which prices have been more volatile. For markets more open to trade, volatility in prices converges to values close to those in international markets. But for countries which use price stabilising policies extensively, such as India, China and Japan, price volatility is substantially less than international markets. It should be noted, however, that comparison across countries is complicated by many factors, including product quality or attribute differences which may affect the diversity of movement in market prices.

Due to recent price fluctuations, the question of whether price volatility has increased is being debated. Figure 2.3 plots the nominal annualised historic volatility for the same three crops over 1957-2009. While one can distinguish spikes in the volatility due to the 1971-73 and the 2007/08 price hikes, there is no apparent trend in the series, suggesting that volatility of observed prices may increase for certain periods, but then return to normal levels. Table 2.2 documents the estimated CoVs for these crops over the period 1967-2009 using monthly data. Similarly, no clear pattern emerges when making comparisons across decades. In the past twenty years, historic price volatility of wheat and maize has increased, but over a longer time span there is little discernable evidence of such a trend. In fact, prices appear to have been less volatile in the most recent decade than throughout the 1970-80s. Note also that for rice, Table 2.2 shows declining price volatility in decade averages over the

Figure 2.3. **Nominal annualised historic volatility: Cereal commodities**

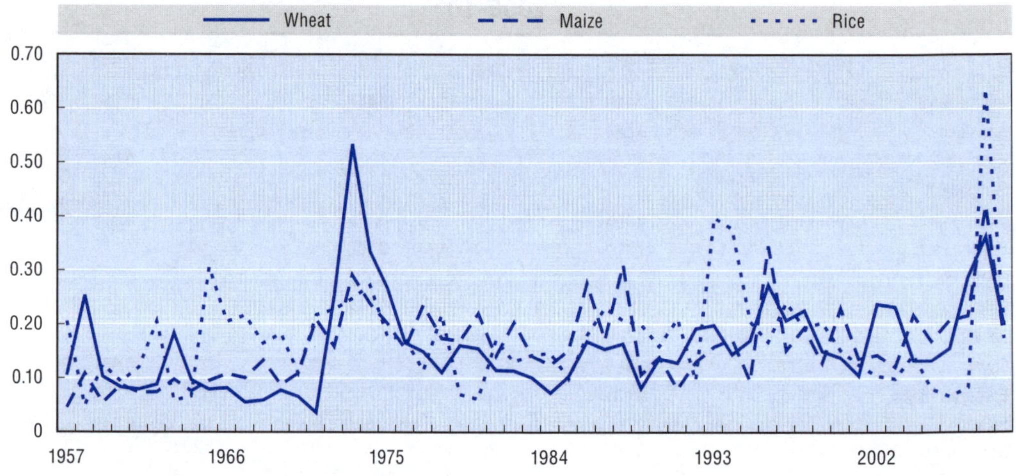

Source: FAO Secretariat.

StatLink http://dx.doi.org/10.1787/888932285905

Table 2.2. **Historic annualised volatility of international grain prices**

	Maize	Rice	Wheat
Average 1967-69	0.088	0.154	0.104
Average 1970-79	0.194	0.183	0.208
Average 1980-89	0.185	0.157	0.123
Average 1990-99	0.089	0.121	0.096
Average 2000-09	0.135	0.116	0.112

Source: FAO (2009).

StatLink http://dx.doi.org/10.1787/888932287881

past thirty years. Using more formal time series analysis methods Balcombe (2008) also investigates the dynamics of international commodity prices over a long period. Using monthly and yearly observations, the author does not find evidence of a trending behavior in price volatility. A recent study by Gilbert and Morgan (2010) also confirms this general result. However it finds an important exception in the case of rice prices, for which price volatility may be increasing. Other evidence, however, based on analysis of implied volatility inferred from options market prices, which offer an indicator of market uncertainty about the subsequent futures and cash prices, for maize and wheat, suggests that market uncertainty and unpredictability for these commodities has risen over the past twenty years.[6] Given this mixture of results, based on differing methods and time spans, the evidence is inconclusive as to whether and how price volatility has changed. More research is required.

Even if the assessments are mixed, high price volatility is an on-going characteristic of agricultural markets. The current debate centers on whether there are now new factors which may cause higher price volatility in the future. First is the increased linkage between energy and commodity markets. As agricultural production increasingly relies on energy inputs on the supply side, and is increasingly used as feedstock for energy production, commodity prices will tend to be increasingly linked with oil prices and the volatility of energy prices will be transmitted to them. Uncertain energy futures in the context of a recovering and expanding global economy, and apparently fixed supplies of conventional fuel sources, raises fears of "food *versus* fuel" in times of shortages. It is expected that if energy prices rise further, the link with food prices will be stronger.

Another new element is the presence of large institutional investors in futures markets, and concerns over the potential for speculation to influence food commodity prices in the short term. In terms of policy, according to Chavas and Kim (2006) the removal of stabilisation policies and the reduced public stocks of food grains have created conditions favorable to greater price volatility. This last point is particularly important with respect to China, which has greatly reduced its stocks over the last 10 years. The common view is that low world stock/use ratios translate into an inability of world markets to help stabilise prices when there are adverse shocks to short-term supplies. Finally, there is concern that extreme weather events, associated with climate change, are increasing in frequency and impact. High production variation, particularly by major suppliers/ exporters, would induce higher price volatility in the future.

The transmission of international prices to domestic markets

Movements of prices in global markets are arguably less important than what happens inside domestic markets, as trade is a low percentage of global transactions.[7] The issue is how global price movements affect domestic prices and markets for agricultural products inside countries (and, potentially, *vice versa*). This is the concept of "spatial price transmission", which is an economic process by which prices in spatially separate jurisdictions may be influenced by arbitrage/trade between them. In terms of global to domestic markets, it is a measure of the extent to which domestic markets are integrated with global markets; a high degree of price transmission is indicated by co-movement in prices, and a low degree of price transmission is indicated by prices which move independently of each other. Price transmission may also be assessed within domestic markets, measuring the impacts of price movements vertically in the food chain, and this also impacts how international prices may be transmitted differentially to producers and consumers. Assessment of vertical price transmission involves a host of important domestic issues which are beyond the scope of the current discussion and is omitted from the analysis presented.

Price transmission is affected by the movement, or the potential movement, of imports or exports to take advantage of price differentials in spatially separated markets. Hence price transmission must take account of transaction, transport and informational costs between markets, as well as differences in product attributes. Importantly, domestic policies often play the critical role in price transmission, as import or export restrictions, variable tariffs or export taxes, and import risks due to sanitary or phyto-sanitary inspection affect the movement of goods. International pricing schemes, such as minimum support prices or intervention policies affect price transmission among markets. Market structures also may play an important role if importing or exporting agencies exercise market power in conducting their operations.

Figures 2.4, 2.5 and 2.6 provide evidence of price movements experienced for the major traded food grains in selected developing and developed countries relative to those of international reference prices during the price hikes of 2007/08, and the consequent recession of 2009.[8] The evidence of these figures confirms the results of a large number of studies which indicate that the impact of higher global prices varied substantially across countries.

For rice, markets of key exporters, such as Thailand (the international reference point for rice) and the United States, face, by the fact of their high export exposure the full brunt of international price fluctuations. The level of trade can affect price transmission. Market prices for certain importers, such as Bangladesh, reacted less than international prices, showing similar movements in terms of timing but with considerably less amplitude. Prices in some large markets such as Japan, India and China barely reacted at all to international price movements. In fact, prices in these markets appear largely independent of international markets, and often evolved in opposite directions over the period. The presence of support price systems, and managed trade environments where varying

Figure 2.4. **International and wholesale prices of rice**

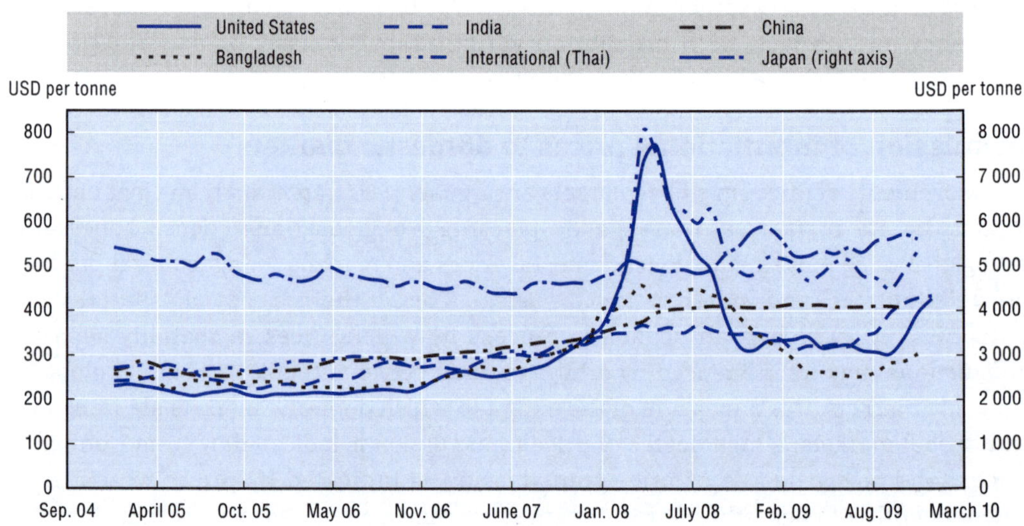

Source: FAO, see *www.fao.org/giews/pricetool/*; and Ministry of Agriculture Fisheries and Forestry, see *www.maff.go.jp/e/tokei/kikaku/monthly_e/index.html*.

StatLink http://dx.doi.org/10.1787/888932285924

Figure 2.5. **International and wholesale prices of wheat**

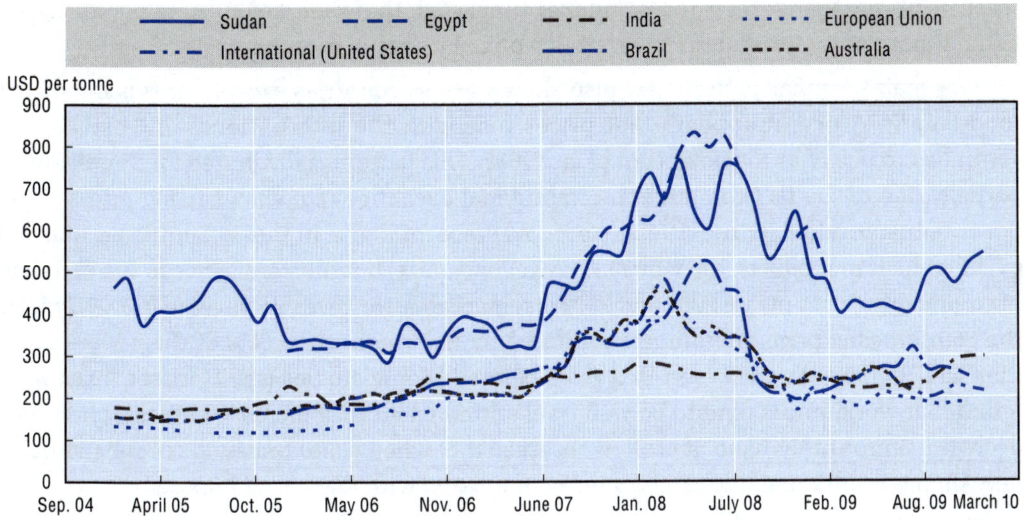

Source: *www.fao.org/giews/pricetool/.*

StatLink http://dx.doi.org/10.1787/888932285943

Figure 2.6. **International and wholesale prices of maize**

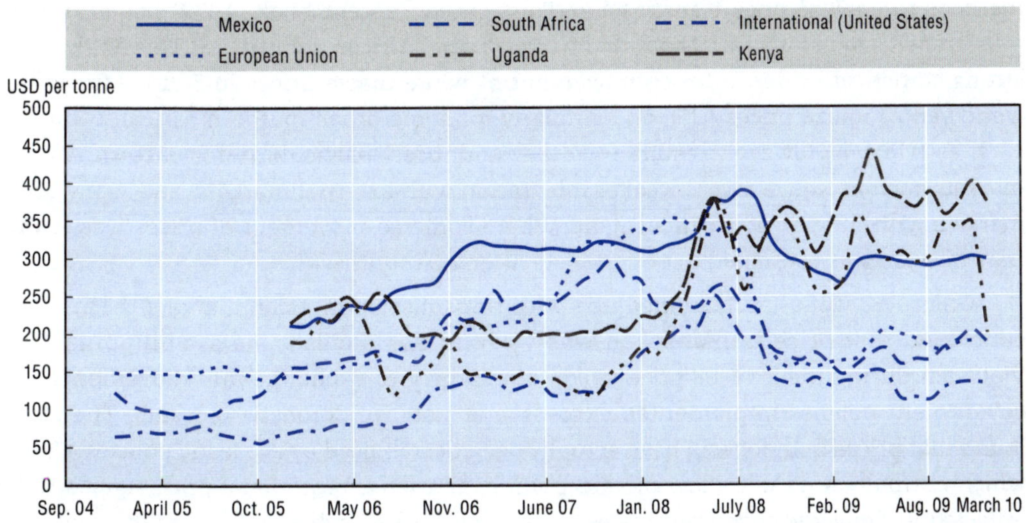

Source: *www.fao.org/giews/pricetool/.*

StatLink http://dx.doi.org/10.1787/888932285962

import tariffs, tariff quotas, export bans, or export taxes were imposed played a key role in reducing the transmission of high international prices for rice to domestic markets.[9] At the same time, analysis suggests that imposition of export bans or tariffs by key exporters also played an important role in exerting upward pressure on international prices by restricting supplies. In fact, most major rice producing and consuming nations have various forms of price intervention systems in place.

In wheat markets which are larger and less protected, data for selected countries appear to show stronger co-movements with international prices than for rice, although similar patterns remain for certain countries such as India which appear largely independent of other prices. While greater coherence in prices may be noticeable, some markets illustrate that their reaction takes time, such that their key turning points lagged behind those of the international reference price by several months.

For maize, similar patterns are also shown across countries except, for this important food crop in Africa, it appears that prices rose with the international price spike, but continued to rise after the downturn of late 2008. This pattern is illustrated for Kenya, and is partially due to the fact that an appreciating real exchange rate shows rising prices in US dollar terms. In addition, the diversity of movement for maize in these countries is affected by differences in products (*e.g.* white *versus* yellow maize) but their resilience to the deflation of commodity prices in late 2008 and 2009 is remarkable. However, in November 2009, among the 860 domestic price quotations monitored by FAO, more than 60% of them were more than 25% higher than their level two years before.[10] Some studies (see Morisset, 1998) have found that world prices tend to be well transmitted when they increase but less when they decrease. On the other hand, it is often observed that when world prices do fall substantially, developing country markets can be quickly inundated with lower cost imports.[11]

It is difficult to draw firm conclusions from visual inspection of data presented in figures for selected countries and commodities. Preliminary tests (not reported) on these series suggest low levels of co-movement between domestic and world prices.[12] More rigorous research is becoming available which uses sophisticated time series analysis methods to analyze price transmission. For example, recent work at FAO has examined price transmission for several commodities in several African countries.[13] This work found strong statistical evidence for co-movement of white maize prices in South Africa with world yellow maize prices (US), and similarly for some other southern African countries such as Malawi. However, results for other countries, particularly for eastern African markets such as Kenya reveal weaker integration with external markets, and many other markets with little or no co-movement. Where some co-movement of prices was found, transmission was estimated to be complete after 4-9 months.

Other FAO work has examined rice price transmission in Asian markets.[14] This work concluded that price transmission was weak in Bangladesh, India, Philippines and Viet Nam during the 2006-08 price spike, due largely to exchange rate movements and government policies implemented expressly to stabilise domestic markets. The study found that prices in Indonesia have actually been destabilised by domestic policies. China, which normally does not allow private trading, allowed international price signals to be reflected in domestic consumer prices. The study further indicated that as international rice prices soared in 2008, domestic prices in most Asian countries also increased despite stabilisation programmes.

There is a substantial literature on price transmission in agricultural markets, most of which has been undertaken before the price spike of 2007-08. For example, Rapsomanikis (2009), Rapsomanikis, Hallam and Conforti (2003, 2004), and Conforti (2004) shed light on the degree of integration for developing countries. The findings emerging from this work also assess the various conditioning factors in price transmission, in particular the fact that non-tariff trade barriers, domestic policy and lack of domestic infrastructure can significantly obstruct the transmission of international price fluctuations.

Other studies trying to test more formally the link between policy and price transmission identify key policy shifts undertaken during structural adjustment reforms (Conforti, Baffes and Gardner). By testing for structural breaks in the data, it is possible to evaluate whether price transmission improved as a result of policy reforms. One often cited example is that of maize in Egypt, a country that underwent significant agricultural liberalisation measures in the late 1980s.[15] There, the transmission of world to domestic prices went from being non-existent to near 100% in years following structural adjustment reforms (Baffes and Gardner). But such clear cut examples are the exception. In that same study, only 11 instances of structural breaks due to reform were identified. Of those, the transmission of world prices to domestic markets increased significantly in six cases (Ghana's rice, Madagascar's wheat and rice, Egypt's maize, Colombia's maize, and Argentina's wheat). The majority of the country-commodity cases did not detect a significant effect of reform on either short-run transmission or longer-term adjustment of domestic to world prices.

The relationship between market integration and volatility is a complicated one, and largely dependent on the causes and sources of a given shock. On one hand, the more fluid the flow of trade, the greater capacity markets have to dissipate a shock. For a drought in a country, for example, imports can limit domestic price increases, and for a bumper crop, exports can limit price declines. If such events occur in a very large country, stronger integration can transmit domestic shocks to international markets, affecting volatility in all countries. However, by the same token, domestic stabilisation policies may destabilise world markets, as was seen when export bans in major rice exporting countries were put into effect in 2007/08.

If the study of price transmission points to the role of domestic policies in affecting the integration of world and domestic markets, many of those policies have been put in place because global prices are considered to be too volatile. Volatility is an important concern both at the macro level for the government and at the micro level for both consumers and for producers. Indeed, the delay between production decisions and actual production creates risks for the producers, who often must establish decisions based only on an estimation of future prices. On the demand side, high prices affect consumers whose food expenditures may constitute a high percentage (in some countries, as much as 70%) of household income. As a result of these concerns government policies attempt to stabilise international price fluctuations by border policies, stock holding policies and price intervention schemes. In some cases consumer prices are mandated in the effort to contain price volatility.

Private measures, such as futures markets, exist in most developed countries offering the possibility to mitigate risk, but such institutional structures are not easily implemented in developing countries. For producers, a normal strategy to circumvent uncertainty is to diversify production to encompass different commodities whose prices are not correlated. Of course, while diversification can help reduce risk, it can also reduce advantages related to increasing returns, and might lead to investment in less risky but less profitable crops. But in general, risk averse producers are likely to produce less than under more certain situations [Holt and Aradhyula (1990)]. This may be particularly true if the costs of investments required for production are high. The use of production contracts, as a means to spread risk within the food chain has increased markedly over time in certain markets, particularly but not exclusively in developed countries. For consumers, strategies to avoid large variations in food costs are difficult. Diversification of diets may be possible, but with low incomes, and the lack of means to undertake precautionary savings, low income consumers face difficult circumstances under a price surge, and often must reduce consumption.

On the macro scale, price volatility plays an important role especially for developing countries that have a less diversified production base, and where import or export of a commodity represents a large share of the trade balance. Indeed, in these situations, large changes can induce considerable disequilibria on trade balances, including export revenues or, importantly the food import bill, both of which may threaten food security.

Price volatility: What are the policy options?

Governments and policy-makers remain concerned about future price fluctuations, and in particular, about the impact of such shocks on their domestic economies. Price fluctuations have important implications for resource allocation in agriculture and the food security of the world's poor who typically spend an important share of their income on food. The degree of market integration has important implications for the transmission of these shocks to domestic markets, but also for the magnitude of the fluctuations themselves. Terms-of-trade effects of higher food commodity prices have also important implications especially for resource-poor countries; and even in cases where price fluctuations do not affect a country's balance of payments, adverse effects on the poor and food insecure have been observed following the 2007/08 price increases. High food prices clearly place a burden on poor net food importers; however, they also present an opportunity to encourage food production and enhance the contribution of agriculture to poverty reduction and economic growth.

Recent policy responses to higher food prices

A large number of emerging and developing countries, including those covered in the OECD policy monitoring report on emerging economies, made various policy interventions in response to the higher food prices in 2007/08.[16] The most common policy response was to reduce or suspend taxes and import tariffs on food products (albeit in some cases at a fiscal cost), followed by the imposition of export barriers. The wide range of interventions also included the release of government held stocks, measures to stimulate domestic production, retail price controls, consumer subsidies for staples specifically targeted at the poor, changes to biofuel policies and direct income transfers.

Most short-term policy responses in key emerging economies to international agricultural price surges in 2007/08 were dismantled in the second half of 2008 and in 2009. But food security considerations frequently associated with objectives of high levels of self-sufficiency in production of selected commodities, further enhanced by the price surge, remained an important driver for agricultural policy measures applied by emerging economies' governments in 2009. For example, while China eliminated export taxes on grains in mid-2009, VAT rebates on grain and soybean exports had not been reintroduced by the end of 2009 and the objective of 95% self-sufficiency in grain production led the government to increase minimum prices for grains and to add to already high input subsidies in 2009.

A large part of India's policy responses remained in place. While some export restrictions on selected grains and pulses were lifted in 2009, the export ban on rice, reduced tariffs on imports of selected grains, pulses and oils, limitations on private stockholding, de-listings from futures trading on rice continued to be implemented in 2009. Moreover, to stimulate production, minimum support prices for over 20 commodities listed were increased significantly in the season 2008/09, but then left mostly unchanged for the 2009/10 season. While fertiliser subsidies reached record high levels in 2008/09, they were budgeted to decline by one-third in 2009/10 due to the decline in the prices of imported fertilisers.

Some other Asian countries, such as Indonesia and Viet Nam, lifted all short-term export restrictions, but continued to apply reduced tariffs on imports of a wide range of food products (Viet Nam) and continued to provide yield-enhancing input subsidies in particular for rice producers (Indonesia). The Russian Federation and Ukraine lifted short term policy responses limiting grain exports, but The Russian Federation's drive to increase self-sufficiency in livestock production led the government to impose further limitations on meat imports and to stronger budgetary support for livestock producers in 2009 and early 2010. Argentina lifted some export restrictions such as maximum export prices for dairy products and lowered its export taxes on cereals and soybeans. Chile and South Africa are among very few emerging economies that focused on one-off direct support to consumers and did not apply any policy responses directly affecting the price or increasing the supply of agricultural commodities on domestic markets. In Brazil, all trade-related measures were lifted in 2008 while cash transfers to the poor population and concessional credits for agricultural producers gained importance in 2009. Analysis is underway at both the OECD and FAO into the efficiency and effectiveness of these policy responses.

In most OECD countries, poverty rates are below those in the emerging economies and the impact of price fluctuations on producers, rather than consumers, dominate governments' concerns. Many OECD countries maintain policies to protect farmers from low prices: the marketing loan assistance and countercyclical programmes in the United States, the intervention price mechanisms in the European Union, the rice diversion programme in Japan and the supply management and agri-stability programmes in Canada. Those are also part of the policy response to price fluctuations and of the price transmission between global and domestic markets.

Policy alternatives toward a holistic risk management

Governments are concerned about the impacts on consumers and food security (particularly poor consumers when prices are high) and impacts on producers (particularly when price are low). Fluctuations in prices and production are common in agriculture and traditionally part of a farmer's risk management strategy. The broader policy question is how can policy underpin farmers' and consumers' risk management strategies?

Governments have a role to play in facilitating access to market and non-market strategies, while empowering farmers to take responsibility for managing their own business risk[17] (as well as assisting poor consumers with food expenditures). Good risk management practices require a diversified government strategy to facilitate the management of the impacts of different agricultural risks on targeted populations. The distinction needs to be made between normal risks that are frequent but generate limited damage and catastrophic risks that are rare but have large consequences for individuals or regions. These latter risks should be the main focus of policy actions, keeping in mind the pre-existing policy environment and the whole set of risks affecting the targeted population.

From the point of view of farming risk management, most OECD countries offer market price support and technical and investment support, such as water management and inspection services. *Ex ante* measures for risk mitigation, in particular income tax smoothing systems for agriculture are also used. Some countries go further by providing payments that are countercyclical with respect to prices or revenue, and provide subsidies for insurance policies or futures contracts. Support for income diversification strategies is rare, but in some countries rural development and social policies may provide alternative sources of incomes. *Ex post* risk-related measures, such as disaster relief, social policy, and other *ad hoc* assistance like debt relief and labour replacement are also available in most countries. Typically countries with lower levels of price support have larger shares of risk-related payments.

A great diversity of sectoral and non sectoral policies, sometimes addressing part of the risk, affects agricultural risk management. This may have unintended effects due to important correlations between different sources of risk, policy instruments and risk management strategies. Countercyclical payments may discourage farmers from taking advantage of natural hedging due to negative production/price correlations; make market instruments less attractive; and contribute to the incompleteness of markets. Insurance subsidies may discourage farmers' diversification strategies. Generous disaster assistance may displace other risk management strategies. Good risk management policies for the agricultural sector need good risk governance through: creation of markets by addressing market failures such as missing asymmetric information; avoidance of rent seeking incentives in support and disaster assistance; accounting for trade-offs between different government objectives that most reduce risk may not have the largest positive impact on farmers' welfare.

Tools for increased market information should be enhanced. At national levels, governments should promote mechanisms to encourage price discovery and tools for hedging of market risks by local agents. Organised commodity exchanges are useful and time tested price discovery and hedging institutions, if they are regulated properly and attract sufficient contract volume to avoid monopolistic practises. They have facilitated commodity marketing in many developed countries and their expansion in developing countries is a welcome institutional development and a sign of market deepening.

The case of price support

Recent events have brought the discussion of the desirability of price support for agricultural commodities back to the forefront of debate on agricultural policy. Policy in OECD countries was largely dominated by price support mechanisms in the past – and price support still accounts for a significant share of total support to the sector. There is therefore a wealth of information and analysis concerning their effects.[18] One of the most notable effects is of course to mask price signals to producers. Usually governments have set prices higher than market clearing levels leading to the kinds of market and trade disruption that characterised the 1970s and 1980s when surplus production was disposed of through export subsidies and dumping. Price support has also had the effect of raising prices to consumers. This is equivalent to a regressive tax to the extent that poorer consumers tend to spend a higher share of their household budget on food.

It is also well established that price support can have perverse distributional effects on the producer side, raising prices and incomes in direct proportion to production, so that the distribution of support is heavily skewed towards the biggest producers. Clearly, this problem is most serious in countries where the distribution of production (or land) is itself heavily skewed. Price support has also encouraged intensification with sometimes adverse effects on environmental outcomes as farmers used increasing quantities of fertiliser and pesticides in pursuit of higher output and fragile lands were brought into production, also with adverse effects on environment, soil quality and biodiversity. Finally, price support is found to have an extremely poor transfer efficiency – that is a relatively small proportion of the price increase it generates is actually captured by farmers. Instead, it goes to input suppliers, is captured by processors and distributors or ends up capitalised in land values, benefitting land owners who are often not themselves farmers.

International policy considerations

While price volatility may not be increasing, it is certain that price surges will take place periodically, and given that the causes and consequences tend to be global in nature, a co-ordinated international policy response is appropriate. A major aspect of any commodity price spike is a fast and sudden erosion of confidence in the workings of the market, both national and international, with the result that unco-ordinated operations, by private and public agents alike, for individual protection leaves all worse off. In this context, there is a need for enhanced rules and disciplines, particularly those which concern export bans and taxes, which enhance confidence in market access in times of crisis.[19] The most efficient way to reduce the probability of future surges in global food markets is to promote market information, transparency and competition and at the same time to create or enhance institutions to ensure confidence in the markets.

At the international level, policy options to stabilise prices are limited. The complex mechanisms by which world market price surges arise and the individual country reactions which follow render international interventions difficult. International stock management schemes, such as those characterizing the International Commodity Agreements, require continuing commitment and are vulnerable to changing market conditions. Indeed, the experience of international food reserves has not been promising. As one example, the ASEAN *Food Security Reserve*, established in 1980 with an initial stock of 50 000 tonnes of rice, has been used infrequently, if at all. Moreover, the quantities in the Reserve are very small and would only be sufficient to deal with localised shocks. Establishment of a larger scheme, by extending to more countries or holding higher levels of stocks would likely encounter even larger collective action problems.[20]

The experience with public buffer stocks suggests that, often, such interventions have been disruptive, rather than stabilising. Given the current state of knowledge about markets and previous experiences with collective action problems, it is not clear how such initiatives could present practical solutions on a multilateral basis. The same concerns arise for what has been termed "virtual stocks" which are designed to alter the fundamentals of the futures rather than the cash markets.[21] Any attempt to publicly influence the prices in futures markets might become extremely expensive and could lead to a withdrawal of the agents who use the futures markets for hedging purposes, thus rendering futures market purely speculative.

Stockholding programmes which specify rules for public stock levels and release for during emergency situations may underpin confidence necessary to prevent panic and hoarding. However, past experience suggests that a study of best practices in stockholding for emergency situations may provide useful information for capacity building in those countries most concerned about food security. One of the major international responses to commodity market volatility in the past has been compensatory financing, such as what was provided through the European Union's *Système de Stabilisation des Recettes d'Exportation* (STABEX) to ACP countries and the *Compensatory Financing Facility* of the International Monetary Fund (IMF). Both programmes aimed to provide compensatory finance to help countries avoid a negative impact on growth from sharp commodity price changes.[22] During the recent price surge, a number of countries which experienced significant increases in their food and fertiliser import bills, resorted to the *Exogenous Shock Facility* (ESF) of the IMF. ESF provides for liquidity to mitigate the negative impact of exogenous shocks on developing countries' balance of payments, international reserves position and inflation.[23]

Although compensatory financing mechanisms can be used to stabilise the economies of developing countries during price surges, they may not be appropriate for addressing short-term food financing difficulties. The need for such food financing facility to assist low income net food importing developing countries was foreseen by the Marrakesh Decision and the World Trade Organization (WTO) Ministerial Conference at Doha.[24] On the basis of analysis by FAO, it was suggested that a *Food Financing Import Facility* (FIFF) be created to enable a country to finance food imports when there was a need, rather than to compensate balance of payment losses after the fact.[25] Very little has been pursued on FIFF or similar alternatives, perhaps due to the low food price period that ensued. However, an international food import financing programme could have provided some relief to the affected countries during the recent period of soaring food prices. The rationale for this suggestion remains valid.[26] The DDA draft modalities text (WTO, 2008) has refined rules on how food aid would be governed, as between emergency and non-emergency situations, to facilitate more effective and market responsible assistance.

An international research agenda

A number of institutions or arrangements could ensure more confidence in global markets, and assure smoother flows of food supplies. The FAO has identified a number of areas for further analysis and discussion.[27]

- An enhanced system of global market information, in particular more accurate and timely information on national stocks of commodities.
- A system of timely advance notice of agricultural trade policy measures affecting the supply of agricultural exports and the demand for imports, and possible disciplines on such measures.[28]
- Multilateral or regional agreements among major exporters and major importers to assure normal flow of supplies during crises.
- A reliable system of assurance of supplies for the most vulnerable countries which could also be enhanced by guarantees of trade finance.
- A market based insurance system for imports of the most vulnerable countries which could provide fast disbursing funds in such cases.
- A linking of organised commodity exchanges across different countries to ensure that all transactions are executed and avoid the problems of counterparty risk experienced by some food importing countries during the recent crisis.

Ministers for Agriculture from the OECD countries, and from non-OECD countries that are major players in food and agricultural markets met in Paris on the 25-26 February 2010. In their *Communiqué*, Ministers[29] also expressed concern about the significant price volatility in recent years and recognised that, in this context, "*an integrated approach to food security is needed*" while "*governments should ensure that appropriate policies are in place to facilitate the management of risk*". Among other areas identified for further analysis, "*they requested that OECD:*

i) Distinguish areas where farmers and the agro-food sector can address challenges and exploit opportunities on their own, from areas where government policy responses might be required.

ii) Harness OECD's broad expertise and capacity for horizontal work to explore policy options that would contribute to growth and development, to poverty alleviation and to building global food security on a long term and sustainable basis, building also on experiences and expertise in the developing countries themselves.

vi) Analyse the functioning of markets and the extent to which the changing physical and market environment is generating new or increased risk and volatility affecting the agriculture and food system, and define appropriate individual, market or public responses to manage risk; and ensure transparency and efficient functioning of markets.

viii) Explore how trade policy, on both the import and export side, can contribute to building global food security and sustainable resource use, paying particular attention to policies that might be needed to facilitate adjustment and to ensure outcomes that are equitable as well as efficient."

The policy issues related to price volatility identified above provide a useful research agenda for international organisations, national governments and academic institutions.

Notes

1. See FAO (2009), OECD (2008, 2009), Abbott and Borot de Battisti (2009), Sarris (2009c).
2. See World Bank (2008).
3. For illustrative purposes, the analysis of volatility in this chapter refers to average monthly data.
4. See FAO (2009a).
5. We use here such a measure in Tables 2.1 and 2.2, where the decomposition between predictable and unpredictable part is done with an AR(p) process which best fits the data in each case. The volatility then refers to the standard deviation of the unpredictable part (the residuals of the AR) only, and may remove a substantial amount of the variation in the raw data. For example, this method reduces the standard coefficient of variation from 0.32 to 0.11, from 0.39 to 0.08, and from 0.41 to 0.12 for the international prices of maize, wheat and rice, respectively. Consequently, the way volatility is measured affects the results obtained and care must be used in interpretation.
6. See FAO (2009), "Implied Volatilities", *Food Outlook*, December, p. 98.
7. For example, global trade (exports) as a ratio to production in 2007/08 was about 7%, 10% and 19% for rice, coarse grains and wheat respectively.
8. Comparing in different currencies is difficult. The figures use US dollars, but it may be more appropriate to adjust for relative price inflation in the non-food sectors of each country to normalise for other factors influencing commodity prices. Data for Japan and the EU come from their respective websites. Prices were converted from local currencies to US dollars using monthly average exchange rates reported by the IMF (2010), *International Financial Statistics Database*.
9. See Gilbert and Morgan (2010).
10. See FAO (2009b).
11. See FAO Briefs on Import Surges at *www.fao.org/economic/est/publications/import-surges/en/*.
12. Only Australian and EU wheat prices showed evidence of co-integration with world reference prices, over 2005-10.
13. See Rapsomanikis (2009).
14. See Dawe (2008).
15. In 1986, the government of Egypt underwent structural adjustment reforms which included the liberalisation of domestic prices, exchange rates and interest rates, and removal of import and export restrictions. For additional information see Raspsomanikis *et al.* (2006).
16. For a detailed description of the policy responses in Brazil, Chile, China, India, the Russian Federation, South Africa and Ukraine, see OECD (2009), *Agricultural Policies in Emerging Economies: Monitoring and Evaluation 2009*. For a survey of policy responses in developing countries see FAO (2009), *The State of Commodity Markets*, Part 3.
17. For a detailed discussion, see OECD (2009), *Managing Risk in Agriculture; A Holistic Approach*.
18. For a discussion of policy design and reform, see OECD (2002), *Agricultural Policies in OECD Countries: A Positive Reform Agenda* and OECD (2008a), *Agricultural Policy Design: A Synthesis*.

19. Various proposals have been made in the context of current WTO negotiations to correct the current imbalance of import and export rules to imports, including the binding and elimination of export taxes, and prohibition of export restrictions.
20. The purpose of the ASEAN Food Security Reserve, as stated in the original agreement, is to provide for a supply of rice in emergency situations when a member country, having suffered a natural or man-induced calamity, is unable to cope with such state or condition through either its national reserve stocks or normal international trade. See also Dawe (2005).
21. Proposals for intervention in the futures markets can be found in von Braun and Torero (2009).
22. The IMF Compensatory Financing Facility has not been used since 2000 due to very tight conditionalities. See IMF (2004).
23. Countries which made use of the Exogenous Shock Facility to mitigate the impact of the food and oil price surge include Ethiopia, Malawi, the Kyrgyz Republic, Senegal, Mozambique and Kenya. Other countries resorted to the facility due to the impact of the global economic downturn.
24. Report of the Inter-Agency Panel on Short-Term Difficulties in Financing Normal Levels of Commercial Imports of Basic Foodstuffs, Document G/AG/13, WTO Committee on Agriculture, 28 June 2002.
25. For a recent analysis of this proposal see Sarris (2009b).
26. For more on trade rules that maybe required in the medium term in light of envisioned developments, see Sarris (2009a).
27. FAO (2010), *Management of Wide International Commodity Price Movements – National and International Experiences and Policy Responses*, presented to the 68th session of the FAO Committee on Commodity problems.
28. The current Agreement on Agriculture in the WTO does not prevent governments from reducing or banning exports.
29. The full text of the *Communiqué* can be found at *www.oecd.org/agriculture/ministerial*.

References

Abbott, P.C. and A. Borot de Battisti (2009), "Confronting Food Price Inflation: Implications for Agricultural Trade and Policies", *Recent Global Food Price Shocks: Causes, Consequences and Lessons for African Governments and Donor*.

Ahmed, S. (2008), "Global Food Price Inflation: Implications for South Asia, Policy Reactions, and Future Challenges", *Technical Report*, No. 4796, the World Bank.

Baffes, J. and B. Gardner (2003), "The Transmission of World Commodity Prices to Domestic Markets under Policy Reforms in Developing Countries", *Journal of Policy Reform*, Vol. 6(3), pp. 159-180.

Balcombe, K. (2009), "The Nature and Determinants of Volatility in Agricultural Prices", *Technical Report*, Report to the FAO.

Chavas, J.P. and M.T. Holt (1996), "Economic Behavior under Uncertainty: A Joint Analysis of Risk Preferences and Technology", *The Review of Economics and Statistics*, Vol. 78(2), pp. 329-35.

Chavas, J.P. and K. Kim (2006), "An Econometric Analysis of the Effects of Market Liberalization on Price Dynamics and Price Volatility", *Empirical Economics*, Vol. 31(1), pp. 65-82.

Conforti (2004), "Price Transmission in selected Agricultural Markets", *FAO Commodity and Trade Policy Research Working Paper*, No. 7, FAO, Roma.

Dawe, D. (2005), *The Role of Food Reserves in Achieving Food Security*, presented at the East Asia Emergency Rice Reserve (EAERR) Pilot Project Workshop on Rice Reserve System, March.

Dawe, D. (2008), "Lost in Transmission", *Rice Today*, July-September.

Deaton, A. and G. Laroque (1992), "On the Behaviour of Commodity Prices", *Review of Economic Studies*, Vol. 59(1), pp. 1-23.

FAO (2009a), *State of Commodity Markets*, 2009.

FAO (2009b), *Food Crops and Shortages*, November 2009.

FAO (2009c), "Crop Prospects and Food Situation", in *Global Information FAO-GIEES, Early Warning System on Food and Agriculture*.

FAO (2009d), "Trade and Markets Division", *Technical Brief*, No. 2, March.

Gulatti, Dutta (2010), "Rice Policies in India in the Context of the Global Rice Price Spike", in David Dawe (ed.), "The Rice Crisis: Markets, Policies and Food Security", FAO and Earthscan.

Gilbert, C.L. (2008), "How to Understand High Food Prices", *Technical Report*, No. 0823, Department of Economics, University of Trento, Italy.

Gilbert, C.L. (2009), "Speculative Influences on Commodity Futures Prices, 2006-2008", *Working Paper*, Department of Economics, University of Trento, Italy.

Gilbert, C.L. and C.W. Morgan (2010), "Has Food Price Volatility Risen?", *UK Foresight Working Paper*, April.

Holt, M.T. and S.V. Aradhyula (1990), "Price Risk in Supply Equations: An Application of GARCH Time-Series Models to the US Broiler Market", *Southern Economic Journal*, Vol. 57(1), pp. 230-242.

IMF (2004). "Review of the Compensatory Financing Facility", IMF Policy Development and Review Department, February.

Irwin, S.H. and D.R. Saunders (2010), "Commodity Price Volatility: The Role of Index Traders and Swap Dealers", *Working Paper Series*, Trade and Agriculture, April.

Moledina, A.M., T.L. Roe and M. Shane (2004), "Measuring Commodity Price Volatility and the Welfare Consequences of Eliminating Volatility", presented at the AAEA annual meeting on 1-4 August 2004, Denver, Colorado.

Morisset (1998), "Unfair Trade? The Increasing Gap between World and Domestic Prices in Commodity Markets during the Past 25 Years", *World Bank Economic Review*, Vol. 12(3), pp. 503-526.

OECD (2002), *Agricultural Policies in OECD Countries: A Positive Reform Agenda.*

OECD (2008), *OECD-FAO Agricultural Outlook 2008-2017.*

OECD (2008a), *Agricultural Policy Design: A Synthesis.*

OECD (2009), *OECD-FAO Agricultural Outlook 2009-2018.*

OECD (2009a), *Managing Risk in Agriculture: A Holostic Approach.*

PPRC (2009), "Food Price Inflation: Impact and Response, Lessons from Recent Experiences", *Technical Report*, PPRC Policy Paper.

Rapsomanikis, G. (2009), "The 2007-2998 food price Swing", *Technical Report*, No. 12, FAO Commodities and Trade Technical Paper.

Rapsomanikis, G. and A. Sarris (2008), "Market Integration and Uncertainty: The Impact of Domestic and International Commodity Price Variability on Rural Household Income and Welfare in Ghana and Peru", *The Journal of Development Studies*, Vol. 44(9), pp. 1354-1381.

Rapsomanikis, Hallam and Conforti (2006), "Market Integration and Price Transmission in Selected Food and Cash Crop Markets of Developing Countries: Review and Applications", in Sarris and Hallam (2006), *Agricultural Commodity Markets and Trade*, Edwar Elgar Publishing, United Kingdom.

Saifullah (2010), "Indonesia's Rice Policy and Price Stabilization Program: Managing Domestic Prices During The 2008 Crisis", in David Dawe (ed.), *The Rice Crisis: Markets, Policies and Food Security*, FAO and Earthscan.

Sarris (2009a), "Evolving Structure of World Agricultural Market Instability and Requirements for New World Trade Rules", in A. Sarris and J. Morrison (eds.), *The Evolving Structure of World Agricultural Trade*, Rome, FAO.

Sarris (2009b), "Hedging Cereal Import Risks and Institutions to Assure Import Supplies", *FAO Policy Research Working Paper*, No. 30, Rome, December.

Sarris (2009c), "Factors Affecting Recent and Future Price Volatility of Food Commodities", in *Risiken in der Agrar und Ernahrungswirtschaft und ihre Bewaltingung, Schriften der Gessellschaft fur Wirtschafts und Sozialwissenschaften des Landbaues e.V. Band 44.*

von Braun, J. and M. Torero (2009), "Implementing Physical and Virtual Food Reserves to Protect the Poor and Prevent Market Failure", *IFPRI Policy Brief*, No. 10, February.

WTO (2008), *Revised Draft Modalities for Agriculture*, TN/AG/W/4/Rev.3, 10 July 2008, WTO, Geneva.

OECD-FAO Agricultural Outlook 2010-2019

Chapter 3

Macroeconomic and Policy Assumptions

Current situation

As this year's *Agricultural Outlook* is being prepared, the macroeconomic situation has improved considerably compared with the same period last year. At that time, the world was undergoing its greatest economic downturn since World War II with persistent and frequent downward revisions to the short term economic Outlook from continuing turmoil in the financial sector that spilled to the real economy. However, with policy interventions in major economies through massive fiscal stimulus to shore up domestic demand, loose monetary policies that kept interest rates at very low levels and massive infusion of liquidity to help shore up the balance sheets of the banking and manufacturing sectors, the worst fears from that time were averted.

This policy intervention has come with a cost. Government budgets in OECD countries have deteriorated significantly as a result of the unprecedented government expenditure while falling incomes reduced tax revenues. The OECD area-wide deficit is projected to peak at a post-war high of 8.25% of GDP in 2010. In 2011, underlying deficits are projected to remain at record high levels of 8% of GDP. For the OECD as a whole, gross government debt is projected to exceed GDP in 2011. The resulting ratio of gross debt to GDP will be about 30% higher than in 2007, the year before the crisis. The crisis also had a human dimension as 27 million people lost their jobs in 2009, taking the global unemployment rate to its highest level ever and the number of jobless to more than 200 million people.

Overall, 2009 was a dramatic year with negative growth in all major OECD countries, save Australia, which generated modest growth. Average income in OECD countries was some 4% below 2008 levels, while worldwide, average income was more than 1% below 2008 levels.

The interdependence of economies around the world through financial and trade linkages was manifested in 2009 as many non-OECD countries also experienced turmoil in 2009. For some economies such as China and India, the upheaval resulted in reductions in their respective growth rates whereas for others such as Argentina, the Russian Federation and South Africa, their incomes contracted.

The global downturn in economic activity had coincided with a precipitous decline in international trade. World trade is estimated to have contracted by 12% in 2009, returning to its 2006 level. This decline has been the sharpest since World War II. This contraction in world trade has been broad based and has affected all regions. Estimates of the impact of the crisis on agricultural trade in 2009 are not yet available. However, if developments in the United States, a large trader of agricultural products are any indication, then agricultural trade also fell by a similar order of magnitude.

The end of 2009 has seen global trade resume with the start of the recovery in growth. World trade has now begun to recover, led by a marked rebound in trade volumes in the non-OECD Asian economies. This helped trade in those OECD countries with strong trading links with this region. The trade rebound has now spread to all regions, even if global trade volumes remain considerably weaker than before the crisis. It is now estimated that trade will grow some 9% in 2010.

While many governments introduced fiscal stimulus in an attempt to revise their economies, monetary policy has also been accommodating. Policy interest rates in most OECD countries are at historically low levels in order to increase the money supply and stimulate the economy. Countries with fragile financial sectors have attempted to recapitalise the banks and add liquidity through different venues. These actions mitigated the downturn and returned many countries to growth in the later part of 2009. The issue for 2010 and beyond is to devise exit strategies as to when and how to reduce the liquidity and fiscal stimulus without reversing the nascent recovery.

Projection highlights

- Presently, the Outlook for most OECD countries is more benign with all major economies expected to grow, albeit modestly, in 2010 and for many OECD countries, the medium-term prospects are somewhat muted compared to previous expectations.
- The recovery however, is fragile with reduced activity and high unemployment levels generating much uncertainty about the sustainability and vigor of the recovery as it faces substantial headwinds.
- The relatively tepid recovery is expected to prolong relatively high unemployment rates and there is uncertainty as to whether the recovery can be sustained with growing private demand once government support and liquidity is withdrawn.
- Prospects in the non-OECD area, however, are more positive as many major economies are projected to grow at least as fast as previous expectations continuing to fuel world income growth. Emerging and developing economies were generally less affected by the financial turmoil and are further ahead on the road to recovery, led by resurgence in Asia.

Macroeconomic developments

In what follows, the macroeconomic assumptions that underlie the *Agricultural Outlook* baseline are described along with a brief overview of the agricultural policy setting to give the reader a broad sense of the macroeconomic and policy environment influencing agricultural markets.

The Outlook baseline incorporates the macroeconomic projections from the latest available information. For OECD countries, these assumptions are from the *OECD Economic Outlook*, No. 86 (December 2009), while the World Bank's *Global Economic Prospects* (November 2009) is the source for non-OECD countries. For the countries within each grouping, these projections present a unified and consistent set of data on income growth and inflationary expectations.

Real gross domestic product (GDP)

In 2009, the financial and economic downturn affected almost all economies in the OECD area and much more than in the developing world. Within the OECD area, Mexico, Japan and Turkey are expected to have suffered, the largest declines with a decrease in GDP of 6.5%, 6.0% and 6.0%, respectively. As illustrated in Figure 3.1, Australia's economy is expected to grow in 2009.

Figure 3.1. **Income growth resumes in OECD area**

Annual percentage growth (real GDP)

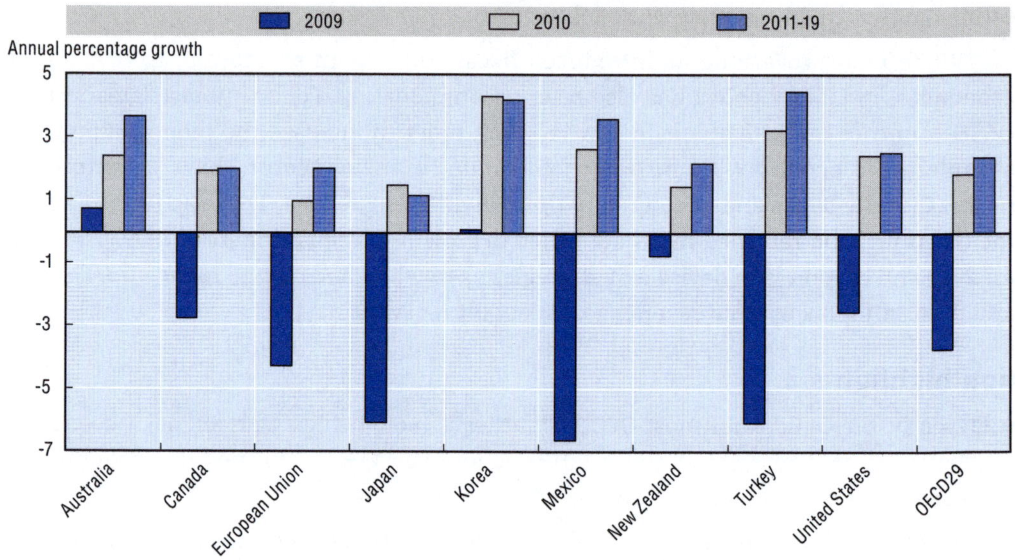

Note: Average annual growth is the least-squares growth rate (see glossary).

Source: OECD Economic Outlook, No. 86 (December 2009).

StatLink http://dx.doi.org/10.1787/888932286019

Economic growth is expected to resume in the OECD area in 2010 continuing the trend that started in the third quarter of 2009 reflecting the demand growth of major non-OECD countries especially in Asia, the policy stimulus and improving financial conditions. For most countries, this growth is expected to continue or even accelerate for the remainder of the projections period to 2019 (Figure 3.1). However, growth rates are somewhat muted compared to past trends as there are substantial headwinds to the recovery as households, businesses and eventually governments have to repair their balance sheets.

In the US, labour was shed very quickly during the downturn resulting in high unemployment rate. But, employment is expected to respond quickly to economic activity and unemployment may peak in the first half of 2010. Employment in the euro area did not fall as quickly as in the US but the schemes in place that cushioned employment in the downturn may also weaken the employment intensity of growth going forward. Unemployment is not set to peak before the end of 2010 or the beginning of 2011 damping consumer confidence and sapping the strength of the recovery. Although Japan is well positioned to benefit from strong growth in the rest of Asia, continued domestic weakness will remain a drag on growth which is not expected to be sufficiently strong to materially reduce unemployment.

With some exceptions, the impact of the financial and economic crisis for most non-OECD countries was shallower and of shorter duration. Among the major non-OECD countries covered in this report, the Russian Federation suffered the biggest drop in GDP, with a 9.1% decline in 2009. The upturn in the major non-OECD countries, especially in Asia reflects both the limited direct exposure to the financial origins of the crisis and the strong policy stimulus these countries were in a position to apply.

China and India have recovered quickly with growth of 8.1% and 5.8%, respectively in 2009. Over the next ten years, China and India should continue to grow at an impressive rate of 7.9% and 6.6% per annum, respectively. These countries remain the leaders in the growth stakes, one based on continued growth in manufacturing and the other on services. In Argentina and Brazil, GDP growth is expected at 4.5% and 3.0% p.a. respectively over the medium term.

Nonetheless, some aftermath to the financial crisis is expected. According to the World Bank's *Global Economic Prospects*, "... the crisis and the regulatory reaction to the financial excesses of the preceding several years may have lasting impacts on financial markets, raising borrowing costs and lowering levels of credit and international capital flows. As a result, the rate of growth of potential output in developing countries may be reduced by between 0.2 and 0.7 percentage points annually over the next five to seven years as economies adjust to tighter financial conditions. Overall, the level of potential output in developing countries could be reduced by between 3.4 and 8% over the long run, compared with its pre-crisis path."

Figure 3.2. **Stronger growth prospects in non-OECD area**

Annual percentage growth (real GDP)

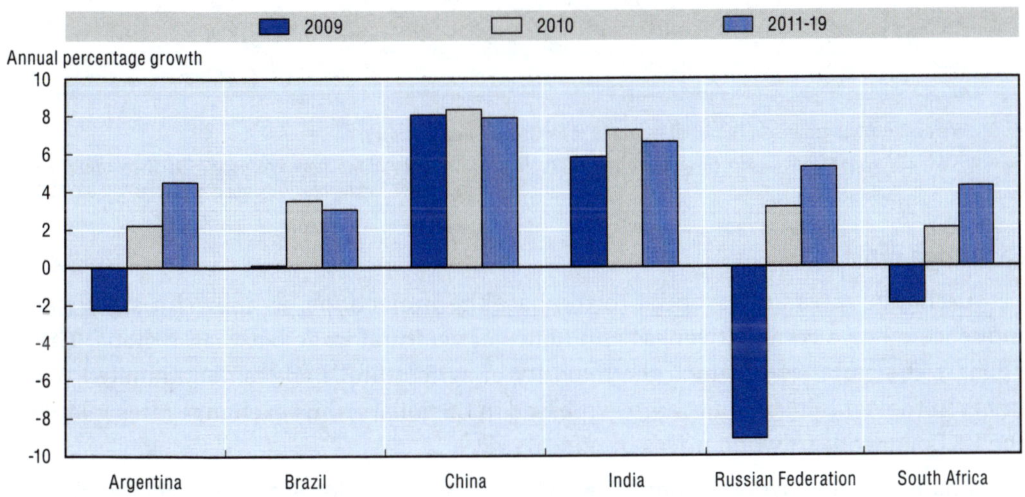

Note: Average annual growth is the least-squares growth rate (see glossary).

Source: OECD Economic Outlook, No. 86 (December 2009); World Bank, *Global Economic Prospects 2010* (November 2009).

StatLink http://dx.doi.org/10.1787/888932286038

Inflation

With relatively high unemployment rates, pressure for wage increases is mitigated. Coupled with elevated excess capacity, inflation is expected to remain at low levels in most OECD countries enabling monetary authorities to continue accommodating monetary policy to support economic activity in the short term. Over the medium term, monetary authorities are expected to keep inflation under control (Figure 3.3).

Inflation, as measured by the Consumer Price Index (CPI) in most cases, for the OECD area as a whole over the next ten years is assumed to average 2% p.a.[1] In contrast, inflation in Turkey is expected to average almost 6% a year. Inflation is expected to be problematic in many of the high growth emerging economies. In Argentina, India and South Africa, inflation is expected to be around 6% p.a. while in the Russian Federation inflation is expected to average more than 7% *per annum*. In contrast, China is expected to contain inflation to about 2% p.a.

Figure 3.3. **Inflation expected to remain under control**

Annual percentage growth (CPI)

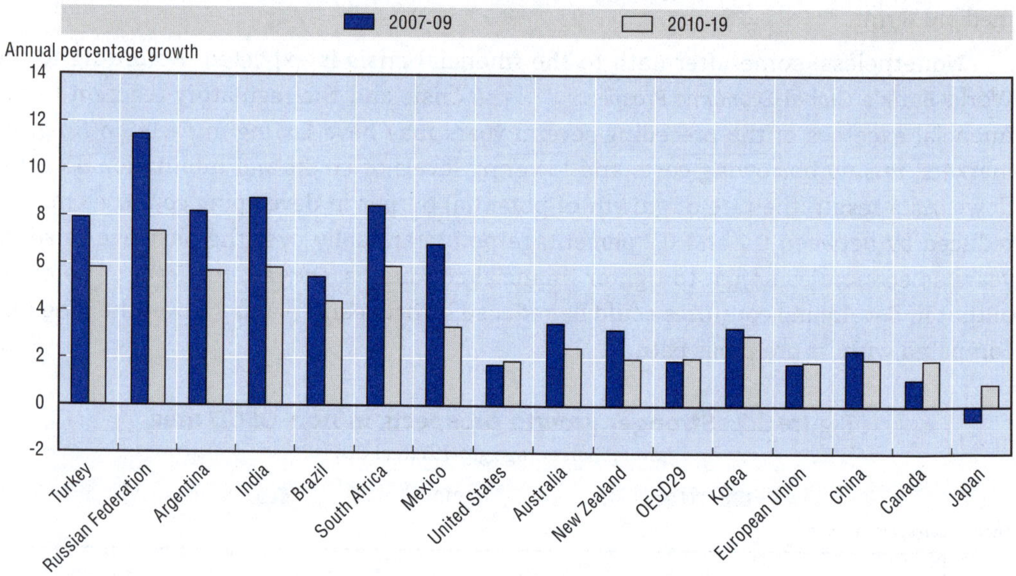

Note: Average annual growth is the least-squares growth rate (see glossary).

Source: OECD *Economic Outlook*, No. 86 (December 2009); World Bank, *Global Economic Prospects 2010* (November 2009).

StatLink http://dx.doi.org/10.1787/888932286057

Exchange rates

Exchange rates are essential to the baseline projections, as they can significantly influence relative competitiveness and hence agricultural trade between regions. The US dollar is the currency in which the majority of agricultural trade is denominated. World prices in the *Agricultural Outlook* are expressed in US dollars, thus exchange rates relative to the US dollar convert world prices into local prices.

Changes in the average nominal exchange rates of major economies included in the *Agricultural Outlook* relative to the US dollar are shown in Figure 3.4. For the 2006 to 2008 period, the US dollar depreciated against most currencies, while in 2009, the dollar appreciated against most currencies with the Japanese yen the notable exception.

Going forward for the projection period, the nominal exchange rates are adjusted under the assumption of constant real exchange rates. Thus, inflation differentials as measured by the GDP deflator of each country *vis-à-vis* the United States is the main determinant of the projections for exchange rates relative to the US dollar over the Outlook period.[2] This implies a strengthening of the US dollar against most currencies. The currencies of high inflation countries will depreciate most relative to the US dollar over the medium term, improving their prospects for agricultural exports while raising the cost of imports.

World oil price

The oil price is an important factor in the *Agricultural Outlook* as it impacts energy, transport and fertiliser costs and on the competitiveness of biofuels as an alternative source of energy. Oil prices have increased dramatically since the start of this decade culminating in a price of around USD 150 per barrel in the summer of 2008 (Figure 3.5).

Figure 3.4. **US dollar appreciated against most currencies**

Local currency *versus* US dollar

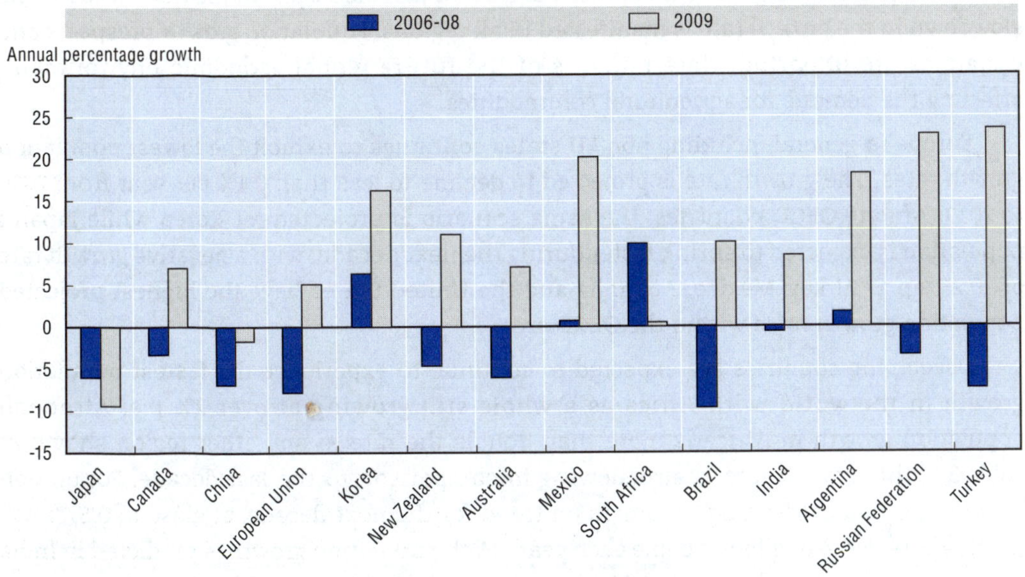

Note: Average annual growth is the least-squares growth rate (see glossary).

Source: *OECD Economic Outlook*, No. 86 (December 2009); World Bank, *Global Economic Prospects 2010* (November 2009).

StatLink http://dx.doi.org/10.1787/888932286076

Figure 3.5. **Oil price resumes upward trend**

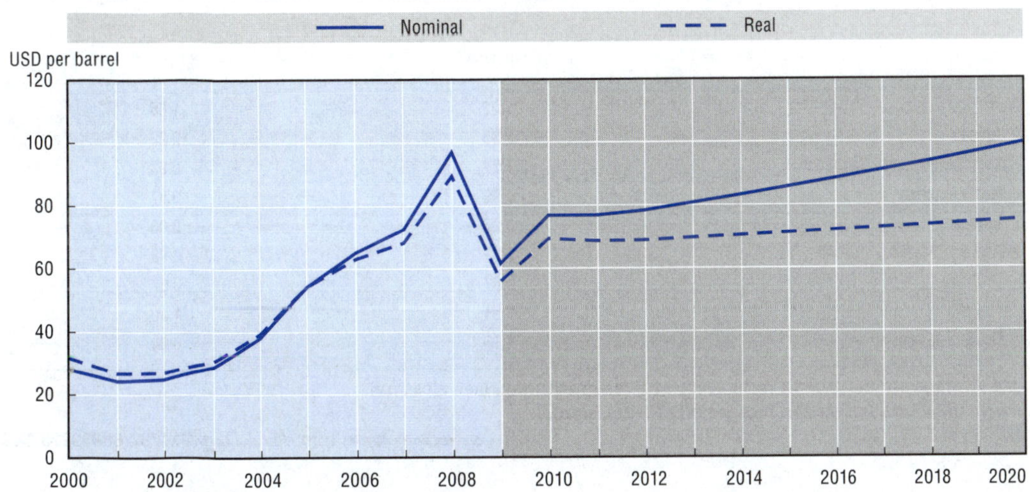

Source: *OECD Economic Outlook*, No. 86 (December 2009).

StatLink http://dx.doi.org/10.1787/888932286095

According to the *OECD Economic Outlook*, this was due to a much stronger than expected demand and a weaker than anticipated oil supply response to rising prices. Current projections have the oil price resuming an upward trend and by the end of the period will equal the previous peak in nominal terms. In real terms (deflated by the GDP deflator for the United States), however, the price rise is more modest and although substantially above historical levels, at the end of the period, it remains below the previous highs. Still, this suggests that the world oil price, even in real terms, will remain well above historical levels.

Population growth

World population growth is expected to slow to just 1.06% p.a. in the next decade. The slowdown in the growth rate is manifested in all regions. Population growth prospects and dynamics are important determinants of the future global economic environment, affecting the demand for agricultural commodities.

Europe in general including non EU states continues to exhibit the lowest population growth rates. The growth rate is projected to decline to less than 0.1% per year from 2010 to 2019. Among OECD countries, the same scenario is projected for Korea while Japan's population is expected to shrink faster during the next decade; with a negative growth rate of –0.25% p.a. Turkey, Mexico, Australia and the United States have the highest projected population growth rates within the OECD area.

Developing countries are expected to continue to experience the fastest population growth in the world, with Africa as a whole still growing at over 2% p.a. Although population growth in Africa is more than double the rate in any other region shown in Table 3.1, this growth rate is also slowing in comparison to the last decade. Population growth in China is assumed to remain on trend for the next decade, at close to 0.57% p.a. (adding between 7-8 million people each year). High population growth is predicted in India with an average growth rate of 1.2% p.a. adding some 14 to 16 million people per year. Overall, by 2019 some 77 million additional people per year will need to be fed.

Table 3.1. **Slowdown in population growth**

Annual percentage growth

	Population	
	2000-09	2010-19
World	1.22	1.06
Africa	2.35	2.22
Latin America and Caribbean	1.22	0.93
North America	0.99	0.87
Europe	0.27	0.06
Asia and Pacific	1.19	0.98
China	0.67	0.57
India	1.54	1.20
Oceania developed	1.16	0.94

Note: Average annual growth is the least-squares growth rate (see glossary).
Source: UN World Population Prospects (2008 Revision).

StatLink http://dx.doi.org/10.1787/888932287900

Agriculture policy developments

Agricultural and trade policies play an important role in both domestic and international markets for agricultural commodities and food products. OECD and emerging economies have gradually reformed their agriculture policies over the past two decades. Agricultural support levels in OECD countries have declined and governments have shifted to more decoupled programmes, although progress differs across countries. Key emerging economies have eliminated many policy distortions, but current low support levels disguise taxation of some sectors and protection of others. The recent food price fluctuations and economic crisis may become a challenge for further policy reform.

At the same time, non-agricultural policies, such as energy, environmental and rural development measures, have a growing impact on the agri-food sector. Policies change the composition and levels of both production and consumption, thereby creating (or sometimes correcting) market distortions and influencing prices.

The level of support to farmers in OECD countries has declined. As measured by the Producer Support Estimate (PSE), policy transfers fell from 37% of farmers' total receipts in 1986-88 on average to 23% in 2006-08. In the most recent years, the reduction of support levels resulted more from high world prices than from explicit policy changes decided by governments. There is a tendency towards increased price responsiveness on the supply side with ongoing policy reform in some OECD countries. Also, relatively elastic supply and demand in a growing number of developing countries, coupled with an increasing participation of these countries in world trade, has assisted adjustments in agricultural markets.

Sustained growth in the 2000s has strengthened agricultural prices and increased fiscal resources available to governments. The levels of agricultural support in emerging economies monitored by the OECD have tended to rise, but they remain far below those observed in the majority of OECD countries, varying from 4% of producer receipts in Chile to 14% in the Russian Federation. Nevertheless, these aggregate low support levels disguise in many cases negative support to some sub-sectors and high protection to others. Since the beginning of the current decade, the level of producer support has remained relatively stable in Chile, Brazil, and South Africa, increased in China and the Russian Federation and fluctuated in Ukraine.

For many non-OECD countries, the rising prices in 2007/08 led to tariff reductions, export restrictions, increased minimum prices and price controls, input subsidies sales of stocks and direct transfers to the most disadvantaged. Most of these measures were quickly removed once prices came back down. For the major emerging economies examined in the 2009 OECD Monitoring report (*i.e.* Brazil, Chile, China, India, the Russian Federation, South Africa and Ukraine), other common policy developments included expanded government-supported credit facilities and/or debt rescheduling, endeavours to improve the delivery and performance of agricultural policies, extended coverage of insurance programmes and further efforts in land reform.

Risks and uncertainties

The risks around the macroeconomic projections are substantial. A main risk on the downside relates to how rapidly consumers will choose to rebuild their balance sheets. Should households choose to increase their savings rates above those assumed and if the public sectors with large budget deficits implement exit strategies to restore fiscal balance and repair their balance sheets faster than assumed, then private demand may not be sufficient to sustain the recovery. Consumer expenditures will also reflect their confidence in the labour market and the recovery generating jobs. As mentioned above, the labour market conditions are expected to continue to weaken with unemployment rate that is projected to rise until end of 2010 or the beginning of 2011 but then job growth is expected.

International imbalances although narrowed somewhat from the crisis, are assumed to remain at levels unprecedented just a few years ago. The risk of disorderly exchange rate adjustment cannot be excluded. This underlines the importance of international efforts, recently given impetus in the context of the G20, to ensure a sustainable international growth pattern.

Although some countries have reacted to the economic crisis through increases in protectionist measures, such measures are not generally seen as the major driving force on plunging trade flows of 2009, rather it was primarily the result of falling demand. However, the WTO, the UN and the OECD continue to warn that protectionism may be a risk that could prevent trade from a sustainable rebound. Past experience shows that prolonged periods of job losses and unemployment are one of the main catalysts for more restrictive policymaking. The unemployment rates are expected to remain high as the recovery in its early stages is not expected to generate substantial job growth. Governments should remain vigilant and resist pressures for future protectionism and other measures with trade restrictive or distorting effects in response to the economic crisis.

No conjecture is included in the Outlook projections for the future outcome of negotiations in the WTO for the Doha Development Agenda. As a result, it is assumed that trade policies as agreed in the Uruguay Round Agreement on Agriculture (URAA) will hold for the entire Outlook period. Trade flows are increasingly influenced by policies that have been negotiated as part of regional trade agreements such as the North American Free Trade Agreement (NAFTA) and the Mercosur Agreement between Argentina, Brazil, Paraguay and Uruguay. The policy assumptions of the Outlook take into account the provisions of these agreements, as well as the European Union's EBA initiative and existing bilateral preferential trade provisions covering specific agricultural commodities.

The *Agricultural Outlook* assumes normal weather patterns and a stable policy environment. In fact, production and price volatility will increase if extreme events such as droughts and floods occur more frequently. Because of their unpredictability, such events are not included in the projections but, to the extent that they occur, will greatly influence agricultural markets at that time. An additional factor that can contribute to increased price volatility in the future is the closer link of agricultural prices to energy prices which have exhibited instability in recent years.

The Outlook also assumes a stable policy environment. No attempt is made to anticipate government policy reforms although there could be some significant policy developments in the coming decade. In February 2010, an Agriculture Ministerial meeting brought the 30 member countries of the OECD together with Argentina, Brazil, Chile, Estonia, Indonesia, Israel, Romania, the Russian Federation, Slovenia and South Africa to exchange views on the opportunities and discuss solutions to the challenges facing the farming and food system over the coming years. The resulting *Communiqué* acknowledged that the main priority is the need to provide an adequate supply of safe and nutritious food, on a sustainable basis, for the world's growing population and recognised a number of policy principles (the *Communiqué* is available at: *www.oecd.org/agriculture/ministerial*).

Notes

1. This is a broader measure of inflation as it includes all expenditures by households in contrast to the food price index discussed in the Overview chapter of this report.
2. See the methodology section for more details.

Chapter 4

Biofuels

Market situation

The economic downturn had direct consequences on biofuel markets. From mid-2008 to mid-2009, weaker energy prices caused a reduction in profit margins of biofuel plants which together with the financial crisis delayed private investments around the world. This associated with reduced policy support for biofuels in some countries and concerns regarding the sustainability of biofuel production has contributed to slow down the pace of expansion of the industrial capacity of the sector. When compared to their peak 2008 levels, ethanol and biodiesel prices decreased, respectively, by 6% and 26% in 2009.

Since the second half of 2009, the renewed firmness in mineral oil prices and the enforcement of higher biofuel blending obligations in some countries have triggered an acceleration in demand for ethanol and biodiesel and subsequently for feedstock used to produce biofuels. It contributed to the recent strengthening in world prices for ethanol and biodiesel.

Projection highlights

- The implementation of policies calling for growing ethanol use means that global ethanol production is projected to increase by more than 110% over the projection period and to reach some 159 bnl (Figure 4.1). Because of increasing mandates and consumption incentives, global biodiesel production is also expected to increase to almost 41 bnl by 2019 far above the average 2007-09 level (Figure 4.2).

Figure 4.1. **Global ethanol production to grow by more than 110%**

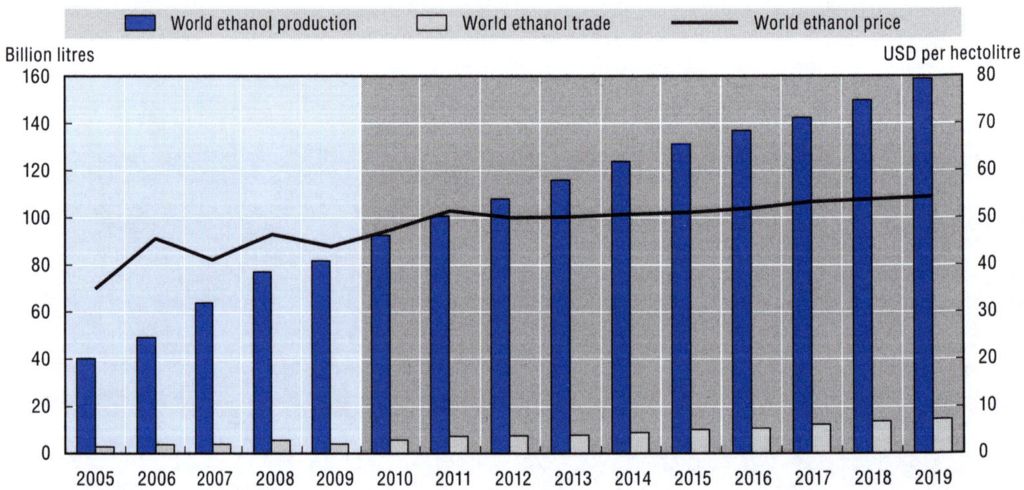

Source: OECD and FAO Secretariats.

StatLink http://dx.doi.org/10.1787/888932286114

Figure 4.2. **Global biodiesel markets to continue to expand**

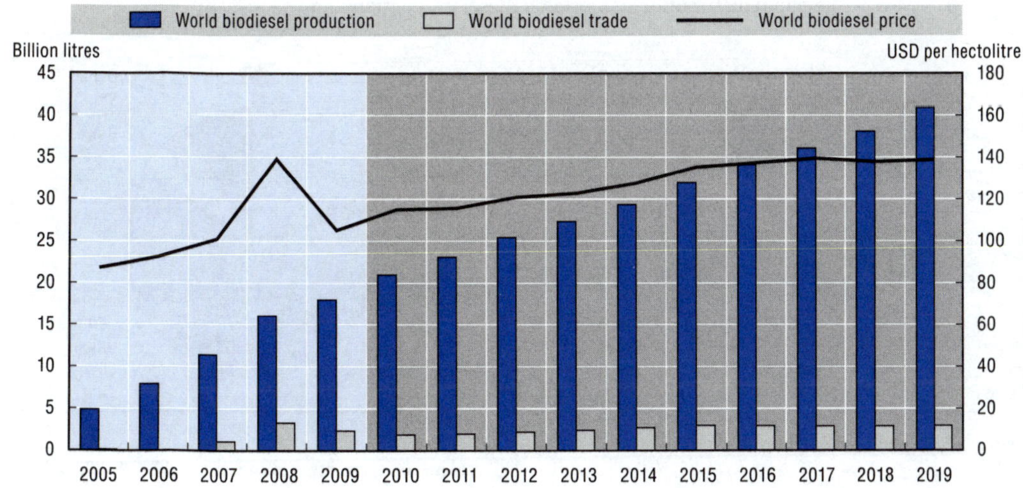

Source: OECD and FAO Secretariats.

StatLink http://dx.doi.org/10.1787/888932286133

- The US is expected to remain the largest ethanol producer and consumer throughout the Outlook (Figure 4.3). Brazil with its sugarcane based ethanol industry should be the main exporter. Part of Brazilian ethanol exports are expected to be channelled through Caribbean countries into the US to take advantage of preferential import conditions.
- With 60% of global biodiesel use in 2019 the EU remains the largest biodiesel market in the world. Other countries, notably India, the US and Brazil are set to reach substantial biodiesel use as well (Figure 4.4). Biodiesel is expected to be only little traded as most countries with consumption programmes produce their biodiesel domestically. Argentina should remain the major exporter with exports reaching a plateau after 2015.

Figure 4.3. **Ethanol markets dominated by the US, Brazil and the EU**

Regional distribution of world ethanol production and use in 2019

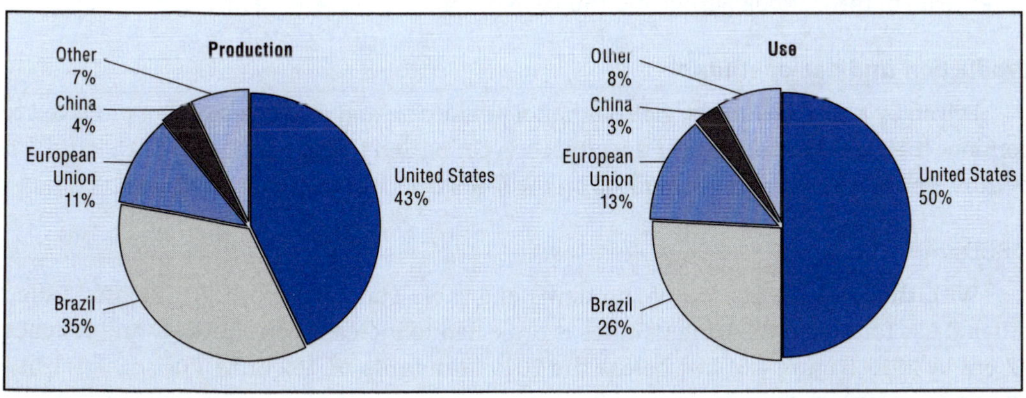

Source: OECD and FAO Secretariats.

StatLink http://dx.doi.org/10.1787/888932286152

Figure 4.4. **The European Union to dominate biodiesel supply and use**

Regional distribution of world biodiesel production and use in 2019

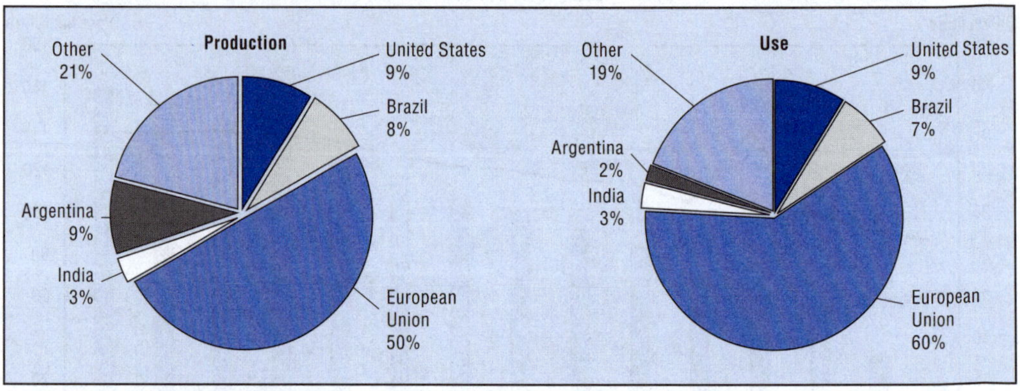

Source: OECD and FAO Secretariats.

StatLink http://dx.doi.org/10.1787/888932286171

Market trends and prospects

Prices

The *OECD-FAO Agricultural Outlook 2010-2019* assumes that crude oil prices will be strong and rising throughout the projection period, reaching almost USD 97/barrel by 2019. Biofuel markets are projected to be highly influenced by mandates and consumption incentives in countries all over the world, with the US and the EU playing major roles, respectively, on ethanol and biodiesel markets. The Outlook projects increasing world biofuel prices based on this situation.

The world ethanol price[1] should follow an increasing trend to reach USD 54.4 per hl in 2019 because of strong demand in the United States to meet the RFS2 standards and in Brazil due to the development of the flexi-fuel industry. The world biodiesel price[2] is projected to increase until 2017 and then to remain at a plateau of USD 144 per hl as it is assumed that second generation biodiesel will increasingly become available in the EU in the latter years of the Outlook period and thus will diminish the pressure on supply globally.

Production and use of ethanol

Driven by policy mandates, global ethanol production and consumption are projected to continue their rapid increases over the projection period and to reach some 159 bnl (Figure 4.1) by 2019. Box 4.1 presents a special focus on the feedstocks used to produce ethanol.

OECD

With the implementation of the new Renewable Fuels Standard (RFS2) Final Rule,[3] ethanol use for fuel in the United States is projected to increase continuously and to reach 77 bnl by 2019 (Figure 4.5) but below the 2019 standards of 102 bnl.[4] Fuel ethanol use should represent an average share of 8.4% in gasoline types for transport fuel by 2019.[5]

The current blending wall of 10%[6] of ethanol blended into regular gasoline is expected to be hit by 2015. However, the US Environment Protection Agency will provide a decision on expanding the blending permission to 15%[7] in the course of 2010. The 15% threshold is not projected to be met over the projection period as it is assumed that by 2019 the average share of ethanol use in gasoline types for transport fuels should be of 12.1% expressed in volume terms.

Box 4.1. **Agricultural commodities used for the production of biofuels at the global level**

This box provides another angle on the projections presented for ethanol and biodiesel markets, by linking the cereal, oilseeds and sugar market projections to biofuel markets.

The evolution of ethanol production by feedstock over the projection period is presented below in Figure 4.5. It shows that the major feedstock for ethanol production should remain coarse grains all over the projection period. The use of coarse grains for ethanol production should grow relatively less after 2015 when the mandate for Conventional Renewable Fuels reaches its maximum in the US.

Figure 4.5. **Ethanol to be produced mainly from coarse grains and sugarcane**

Global ethanol production by feedstock

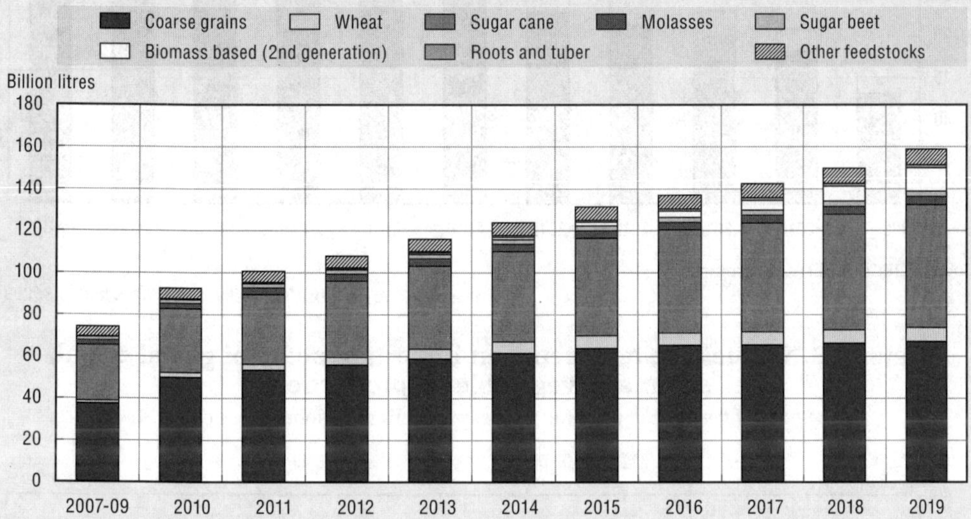

Source: OECD and FAO Secretariats.

StatLink http://dx.doi.org/10.1787/888932286190

Almost 40% of the increase in global ethanol production should be due to the increase in the production of ethanol based on sugarcane, mainly from Brazil, to meet both domestic and US demands. Biomass based second generation ethanol is only expected to develop in the latter years of the projection period, representing about 7% of total ethanol production. Roots and tubers and molasses are expected to be used as feedstocks for ethanol production in developing countries. Wheat, coarse grains and sugar beet should be used in the European Union to produce ethanol.

Figure 4.6 presents the evolution of biodiesel production by feedstock over the projection period. It shows that edible vegetable oil is expected to remain the major feedstock used to produce biodiesel. However its share in total biodiesel production should decrease from almost 90% over the base to about 75% by 2019. This is due to the development of the production of biodiesel based on *jatropha* mainly in India, to the increasing use of animal fats to produce biodiesel in the US and to the availability of biomass based second generation biodiesel in the latter years of the projection period. Biomass based biodiesel should represent almost 6.5% of total biodiesel production by 2019.

The importance of the biofuel use in agricultural commodity markets is illustrated in Figure 4.7. The shares of the different agricultural commodities used for ethanol and biodiesel production (molasses, vegetable oil, coarse grains, sugar beet, wheat and sugarcane) in global production is expected to increase strongly over the projection period.

Box 4.1. **Agricultural commodities used for the production of biofuels at the global level** *(cont.)*

Figure 4.6. **Vegetable oil is the major feedstock to biodiesel production**

Global biodiesel production by feedstock

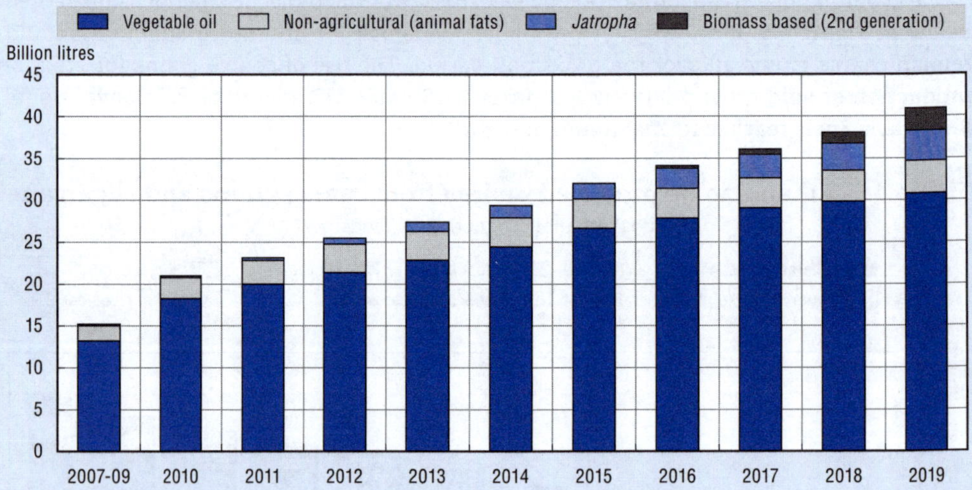

Source: OECD and FAO Secretariats.

StatLink http://dx.doi.org/10.1787/888932286209

Figure 4.7. **Biofuel use represents an important share of global cereal, sugar and vegetable oil production**

Share of feedstocks used for biofuel production in global production

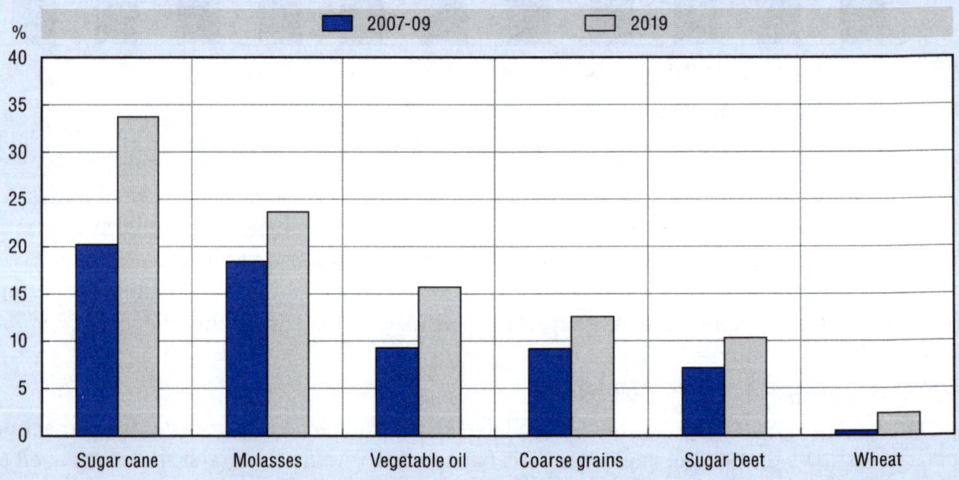

Source: OECD and FAO Secretariats.

StatLink http://dx.doi.org/10.1787/888932286228

By 2019, about 13% of the global production of coarse grains will be used to produce ethanol compared to 9% over the base. 16% of the global production of vegetable oil will be used to produce biodiesel compared to 9% over the base. The share of sugarcane to be used for ethanol production at the worldwide level is expected to reach almost 35% in 2019.

The US is expected to remain the major ethanol producer and consumer throughout the projection period. Following the mandate for conventional renewable fuels fixed in the RFS2, ethanol production derived from corn should increase strongly until 2015 and is projected to stabilise around the mandate level of 57 bnl thereafter. Cellulosic ethanol production mainly from crop residues, notably corn stover is only projected to expand significantly in the latter years of the projection period to reach almost 9 bnl in 2019 and to remain far from meeting the RFS2 32.2 bnl standard.

Figure 4.8. **US ethanol market to expand because of RFS2**

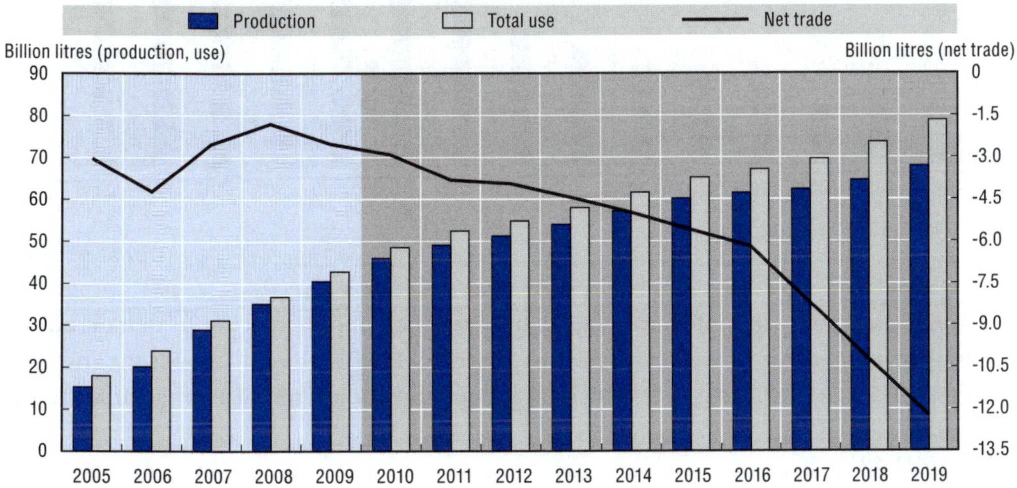

Source: OECD and FAO Secretariats.

StatLink http://dx.doi.org/10.1787/888932286247

The Renewable Energy Directive (RED) implemented by the EU states that the share of renewable energy sources (including non-liquids) should increase to 10% of total transport fuel use by 2020. This should foster the increase in domestic ethanol production. Complementing wheat, coarse grains and sugar beet based ethanol, the production of second-generation ethanol is assumed to take-off in 2016 and to reach 1.8 bnl by 2019, *i.e.* 10% of domestic ethanol production. The contribution of second generation biofuels will be counted twice toward EU RED mitigation targets. Fuel ethanol use is projected to increase to reach 18.7 bnl in 2019 (Figure 4.9) representing an average share of 8.5% in gasoline types for transport fuels by 2019. The EU is expected to confirm its position as the third largest ethanol market behind the US and Brazil.

In Canada, a requirement for 5% (3.4% in energy terms) renewable content based on the volume of the gasoline supply is filled by 2010 and maintained throughout the projection period. Canadian ethanol consumption is thus projected to grow in line with fuel consumption. Domestic production is expected to rise over the projection period at a rate of 3.2% per year to reach almost 1.9 bnl in 2019. This growth will be fostered by direct support to ethanol producers until planned termination in 2016.

Over the early years of the projection period, domestic use of ethanol in Japan is expected to stay at around 0.6 bnl with most of the domestic use to remain for non-fuel purposes. Domestic production should remain marginal. From 2015, Japan is projected to start producing cellulosic ethanol to be used for fuel purposes. Domestic consumption should rise to reach 1.1 bnl by 2019.

Figure 4.9. **EU ethanol production and use to develop**

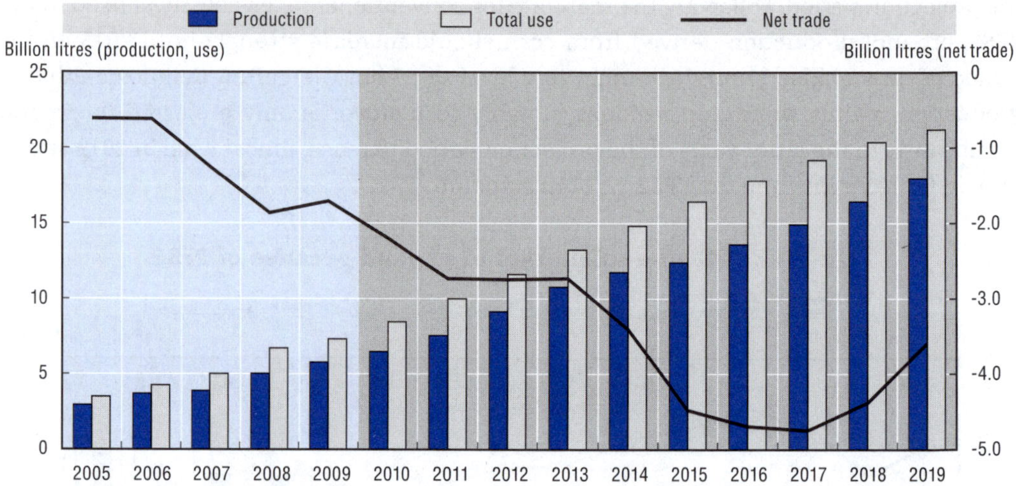

Source: OECD and FAO Secretariats.

StatLink http://dx.doi.org/10.1787/888932286266

In Australia, the ethanol share in gasoline type fuel use is expected to remain almost unchanged over the projection period at close to 1.3% after an important increase between 2006 and 2010. Most of the ethanol production should be based on coarse grains.

Latin America

Ethanol production derived from sugarcane in Brazil is expected to rapidly expand, growing by almost 7.5% per year over the projection period to meet both domestic and international demand. Brazil is projected to be the second largest ethanol producer, with a share of 35% of global production in 2019 (Figure 4.10). Due to relatively lower sugar prices relative to ethanol, a decreasing share of the growing sugarcane output is expected to be channelled to sugar production. Ethanol domestic use is expected to increase over the projection period to reach 41.7 bnl in 2019. This growth is mainly driven by the growing fleet of flexi-fuel vehicles.

Colombia achieved a sugarcane based ethanol production of over 400 million litres in 2009 mainly due to efficiency gains supported by pricing and fiscal incentives which is projected to grow on average at 7% per year to reach 880 million litres in 2019. This will allow domestic blending requirements of 10% in volume terms to be met by 2013 as gasoline consumption is expected to continue a declining trend. The subsequent expansion in export, reaching about half of the production by 2019, will be directed mainly towards the United States.

Argentine ethanol use is driven by mandates, supported by tax breaks. In 2007, Argentina decided to blend 5% of biofuels into gasoline from 2010 in volume terms, and this Outlook assumes that this would result in E5 blends. This should drive ethanol use to 0.5 bnl in 2019. It is assumed that the largely sugarcane-based ethanol industry will be able to marginally outpace domestic consumption over the course of the projection period.

Peruvian sugarcane yields per hectare are among the highest in the world. Domestic ethanol consumption is driven by a 7.8% blending mandate in volume terms effective in 2010. Based on a projected growth rate of 14% per year, the country's ethanol production is expected to grow from a currently low base to 190 million litres in 2019. The surplus is

Figure 4.10. **The flexi-fuel industry to drive increasing ethanol use in Brazil**

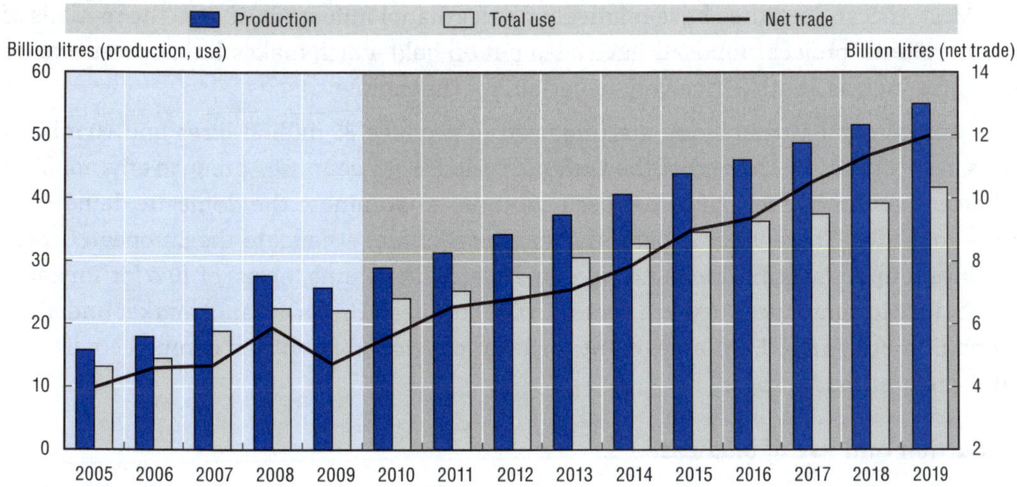

Source: OECD and FAO Secretariats.

StatLink http://dx.doi.org/10.1787/888932286285

expected to be exported to the United States and the European Union. Mexican domestic production, driven by domestic non-fuel use, is projected to grow to reach slightly more than 90 million litres in 2019.

Asia

In China, ethanol production should increase by 50% over the course of the projection period when compared to the 2007-09 base to reach almost 6 bnl in 2019. Most of the increase is assumed to be based on sweet sorghum, other low-quality cereals and sweet potatoes as official restrictions on the use of food commodities to produce biofuels apply. Ethanol share in gasoline type fuels is expected to reach 2.1% by 2019.

Ethanol production in India, from molasses, had a slight drop from 1.7 bnl in 2008 to 1.6 bnl in 2009 after a disappointing sugar harvest. The country had to temporarily import to fulfil the 5% blending mandate in volume terms and non-fuel applications. Output is expected to recover and grow by 4.5% per year to reach almost 3 bnl by 2019.

In Thailand, the government promotes gasohol[8] consumption through providing price cuts, tax incentives and marketing channels rather than compulsory phasing out of unblended gasoline sales. Ethanol production based on sugarcane, molasses and cassava is projected to grow at 11% per year and is closely matched by domestic consumption.

In the Philippines, the Biofuels Act passed in 2007 mandated a 5% ethanol blend in volume terms for gasoline within 2 years and 10% by 2011. To achieve the mandate, production based on sugarcane and molasses is expected to grow at an average rate of 15% per year over the projection period to reach 920 million litres in 2019.

Viet Nam currently produces ethanol from sugarcane, molasses and cassava, generally for non-fuel purposes. Growth in production is projected to be significant at 11% per year as the government induces an E-5 blending mandate by 2016.

Africa

Most African countries have relatively small ethanol industries, if any. The myriads of ethanol plants' project proposals have been put on hold which makes future prospects for a country such as South Africa very uncertain for the moment.

Tanzania and Mozambique are projected to produce 85 million litres and 40 million litres, respectively, by the end of the Outlook period. Both countries produce ethanol from sugarcane and molasses. Non-fuel uses continue to dominate the domestic demand in these countries. There is some potential for export of fuel ethanol to the European Union, contingent upon meeting domestic target mandates (in volume terms) of 10% for Tanzania and 5% for Mozambique. Duty free access to the European Union ethanol market under the Everything But Arms (EBA) agreement for least developed countries competes with the same preferences for sugar.

Production and use of biodiesel

Because of increasing mandates, supplemented by tax concessions and other forms of support, global biodiesel use is expected to increase to almost 41 bnl by 2019 far above the average 2007-09 level (Figure 4.2). Box 4.1 presents a special focus on the feedstocks used to produce biodiesel.

OECD

In the European Union, the major player on the biodiesel markets, total biodiesel use is expected to reach almost 24.4 bnl by 2019 given mandates and tax reductions by member states and the RED (Figure 4.11). The share of biodiesel in diesel type fuels is projected to grow to 8% (almost 10% in volume terms) on average[9] by 2019.

Domestic biodiesel production in the EU should increase to supply a large share of demand. From 2016, second generation biodiesel production is assumed to take-off reaching 2.6 bnl in 2019, *i.e.* almost 13% of domestic biodiesel production. As sustained profits thanks mainly to continued public support are expected, biodiesel production based

Figure 4.11. **EU biodiesel imports requirements to stabilise after 2016**

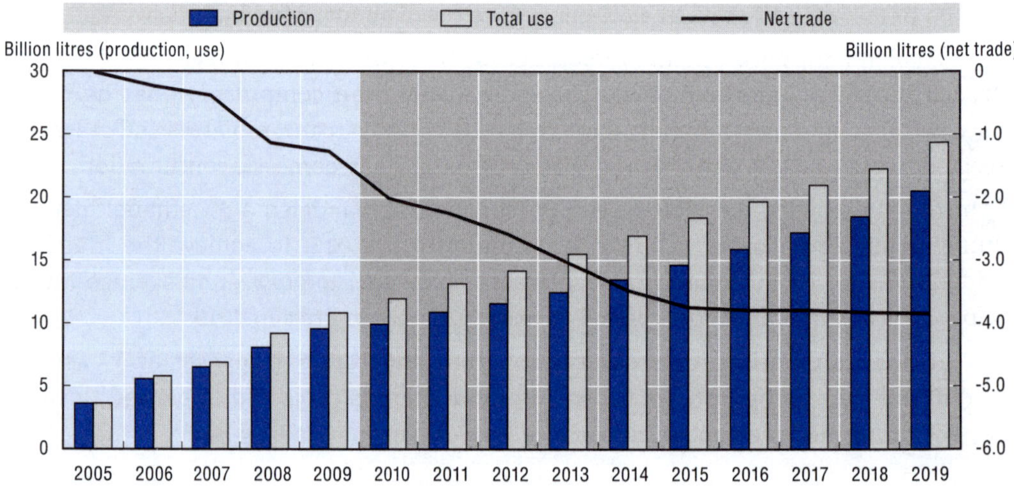

Source: OECD and FAO Secretariats.

StatLink http://dx.doi.org/10.1787/888932286304

on vegetable oil is expected to increase by 6.7% p.a. on average. This rate is lower in the latter years of the projection period when world biodiesel prices reach a plateau due to the anticipated increasing availability of second generation biodiesel.

In the US, the mandate defined in the RFS2 calls for 3.8 bnl of biodiesel to be used by 2012, driving the initial growth in US biodiesel use. The Outlook assumes biodiesel use to be held constant over the remaining years although no explicit mandate for biodiesel is legislated thereafter, the subsequent mandates referring only to advanced biofuel (Figure 4.12).

Figure 4.12. **US biodiesel use to increase up to 2012 to meet RFS2 mandate**

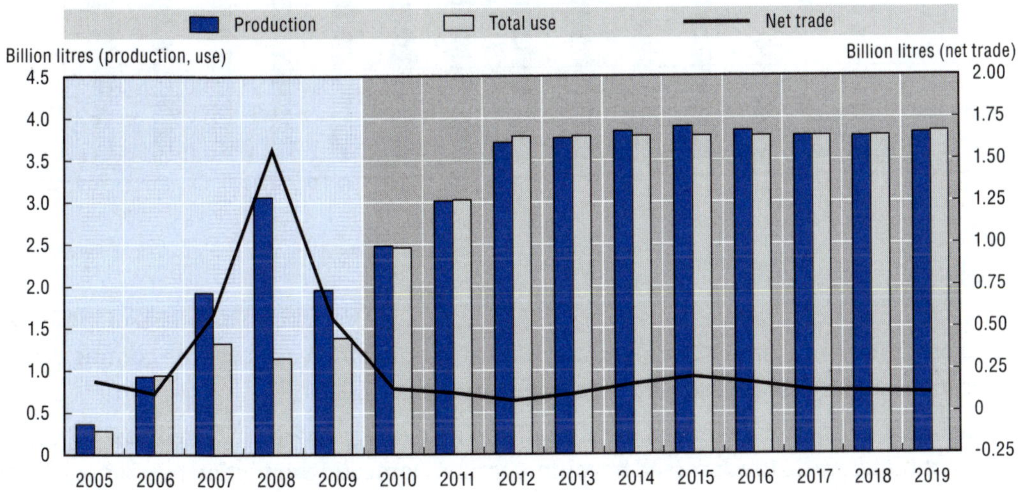

Source: OECD and FAO Secretariats.

StatLink http://dx.doi.org/10.1787/888932286323

It is projected that US biodiesel production will target to meet but not to exceed domestic use. US biodiesel production from tallow or other animal fat is expected to develop to reach 2.6 bnl in 2019 or two-thirds of total output. Increasing rate of returns means that biodiesel production based on vegetable oil is expected to grow until 2015. Afterwards, international biodiesel prices will reach a plateau as second generation biodiesel becomes available; hence, the price incentive to produce biodiesel from vegetable oil diminishes subsequently.

In Canada, biodiesel use is projected to grow in line with the biodiesel blending mandate of 1.6% (2% in volume terms) for all transport diesel as well as heating oil by 2011, or earlier, subject to technical feasibility. Biodiesel use is projected to reach 765 million litres by 2019. Domestic biodiesel production continues to be mainly based on other feedstocks such as tallow and other animal fat. It is projected to grow continuously over the projection period.

In Australia, the biodiesel share in diesel type fuel use to remain close to 2.5% all over the projection period. Most of biodiesel production should be based on animal tallow.

Latin America

Biodiesel production in Argentina is expected to increase rapidly until 2015 because of growing demand from international markets, especially in the European Union. The take-off of second generation biodiesel production in the EU lowers the need for soya-based biodiesel imports in the latter years of the projection period. Biodiesel production is thus expected to stabilise at 3.6 bnl in 2019 (Figure 4.13). Mandates and tax

Figure 4.13. **Argentine to export most of its biodiesel production**

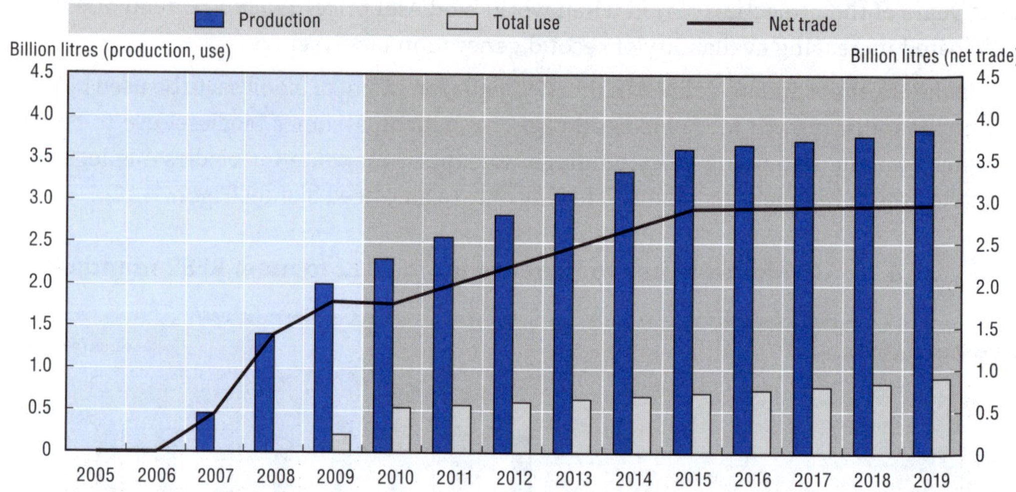

Source: OECD and FAO Secretariats.

StatLink http://dx.doi.org/10.1787/888932286342

breaks drive domestic biodiesel use. In 2010, Argentina is expected to blend 5% of biodiesel into diesel in volume terms. The blending mandate should increase over the coming years. Domestic biodiesel use is assumed to follow mandated levels.

In Brazil, the government's mandate calls for an average share of 5% in diesel types for transport fuel from 2010 expressed in volume terms. Domestic production based mainly on soybean oil should meet domestic mandate use. It should reach 3 bnl biodiesel in 2019.

In Colombia, palm oil based biodiesel production is expected to increase from currently 330 million litres to 875 million litres by 2019, an average annual growth rate of 10%. The country currently has a blending requirement of 5% in volume terms. Based on the mandate, domestic demand remains relatively flat over the Outlook period leaving an increasing exportable surplus.

Asia

The government of India promotes the use of non-food sources for production of biofuels. Edible oilseeds are not eligible for the production of biodiesel. Plans to boost the cultivation of *jatropha* from current experimental stage are a key component behind the biodiesel Outlook. Based on ambitious government targets, biodiesel production could reach 3 billion litres by 2019 all of which is destined to domestic usage.

Indonesian production and consumption of palm oil based biodiesel is projected to grow rapidly to comply with the 3% blending mandate expressed in volume terms which is to rise to 7% in 2015. Production should reach over 1 bnl in 2019. Malaysian output is expected to reach close to 1 bnl up from a current level of 600 million litres with an average annual growth of just 4.6%. Currently, about half of the production is exported mainly to the EU and North America, and the share moving into exports should remain unchanged during the Outlook period. The two countries continue to allocate a low proportion of palm oil production for biodiesel production as they attract higher premiums in the palm oil export markets.

Africa

Africa is a net importer of vegetable oils making their use as feedstock for biodiesel unlikely. Modest expansion of biodiesel production in Africa will derive mainly from *jatropha curcas* production. Numerous projects have been announced, but actual commercial production has been very small. Mozambique and Tanzania are projected to collectively produce 100 million litres of biodiesel in 2019. More than half of the production is destined for export to European Union markets.

Trade in ethanol and biodiesel

Ethanol

Trade in ethanol[10] is expected to almost triple over the projection period to reach 9% of global production. It is driven, in particular, by large US import demand. To keep pace with demand, net imports from the US should reach 10.9 bnl in 2019 (Figure 4.8). Part of this are expected to be Brazilian ethanol dehydrated in the Caribbean, imported with duty-free access under the Caribbean Basin Initiative. At the global level, growth in trade comes almost entirely by expanding export supplies from Brazil. They are expected to reach 13.3 bnl by 2019 (Figure 4.9).

In the EU, ethanol imports should initially grow to meet increasing ethanol demand to reach 4.3 bnl in 2017. With the increasing availability of second-generation in the latter years of the Outlook period, ethanol imports are expected to decrease to 3.2 bnl in 2019 (Figure 4.10). In Canada, net imports should remain close to 0.7 bnl over the projection period, while 0.8 bnl of ethanol are expected to be exported to the United States. Ethanol net imports by Japan are expected to stay around 0.5 bnl.

Biodiesel

Biodiesel is expected to be only little traded as most countries with binding mandates tend to produce biodiesel domestically. Biodiesel exports from Argentina are projected to increase rapidly up to 2015 because of growing demand in the European Union. The anticipated take-off of second generation biodiesel production in the EU lowers the need for biodiesel imports in the latter years of the projection period, which should stabilise at 3.8 bnl.

A countervailing measure has been put in place in spring 2009 by the EU to prevent the US imports of biodiesel benefitting from a blenders' tax credit. All exports from the US to the EU are now subject to an import tariff. As a consequence, the Outlook assumes that the US will not develop its exports of biodiesel. Canada's biodiesel imports should level off around 300 million litres once domestic mandate is met in 2011, or earlier, provided technical feasibility is demonstrated.

Key issues and uncertainties

Since the early 2000s, biofuel markets have experienced a considerable expansion following the implementation of policy packages aiming to foster the use of renewable fuels in transportation fuel in a number of countries. Future developments are subject to many uncertainties which are described in this section: the evolution of policies and of the economic environment, the set-up of environmental criteria, the development of second-generation biofuels and the competition between producing food or fuels from agricultural resources.

Evolution of policies and macroeconomic environment

Policies promoting the use of bioenergy are obviously a key factor behind the development of biofuels markets. However, they are not the only driver. The recent economic downturn has shown how crude oil prices and private investors' involvement interplay to shape the biofuel economy and, as a result, the economy of feedstocks used for biofuel production. Any change in the economic environment can potentially have a strong impact on biofuel markets as this may lead to changes in policy packages, raw material prices and modify the profitability of the industry and thus delay investors' decisions or spending on R&D.

National biofuel policies may well change over time and neither the future course of crude oil prices nor the response of private investors to price and policy changes can be foretold. In the Outlook, the assumption is made that the current policy set remains in place. In reality, over the medium term, national biofuel policies will likely be subject to continuous reviews and modifications. National targets for bioenergy utilisation are revised frequently and may be subject to further rescheduling, while government subsidies and other budget-sensitive measures in support of biofuel production or consumption could be subject to gradual cuts as the industry matures.

Greenhouse gas (GHG) emissions

The RFS2 Final Rule in the United States requires specific GHG emission reductions for the various biofuels included in the policy. Conventional renewable fuels must reduce GHG by 20% when compared to gasoline, advanced biomass-based diesel and non-cellulosic advanced biofuels by 50% and cellulosic biofuels by 60%. Existing conventional ethanol production facilities are exempt from this requirement, but new plants will have to comply. Similarly, imported biofuels will need to meet the requirement in the coming years.

The RED in the European Union specifies that a given biofuel has to achieve a saving of at least 35% in GHG. This 35% threshold should rise to 50% in 2017 for existing plants and 60% for new production facilities. Indirect effects of biofuel production commonly referred to as indirect land use changes (ILUC) effects have been discussed in OECD (2008). The European Commission is required by December 2010 to submit a report that addresses ways to minimize the impact of ILUC on GHG.

The measures taken and to be taken by the major countries consuming biofuels might affect international trade in biofuels and feedstocks. Internationally recognised standards are in the process of being developed. Their potential impacts on trade flows are not yet known.

Development of second generation biofuels and other sources of renewable energy

Considerable uncertainty remains regarding the future transition to second generation biofuels as bioenergy market developments are expected to increasingly depend on technological advances. The take-off of second generation biofuels produced from lignocellulosic biomass in the form of wood or miscanthus grass for example, waste material from food crops or residues from forestry or other non-food plants is expected in the latter years of the projection period. This potential development will obviously depend on capital availability and the advancement of R&D over the next few years.

Research is leading to new ways to make use of biomass for the production of liquid fuels are currently on the way, but production facilities are not yet operational at any commercial scale. However as most of current biofuel policy packages have set up ambitious mandates for the production of second-generation biofuels, commercial production is not dependant on full economic viability, at least in the United States and in the European Union.

Other sources of renewable energy could play a larger role in future years. The RED explicitly allows for renewable electricity used in the transport sector to count towards the 10% energy share in transport fuels. The development of the electrical or hybrid vehicles remains unknown for the time being but could potentially reduce the need for biofuels derived from agricultural products to meet the mandate set up by European States.

For more information on the development of second generation biofuels, the IEA (2010) has issued a report describing the potential and perspectives of sustainable production of second generation biofuels in major economies and developing countries.

Food versus fuel

The continued growth of the biofuel sector as a demand factor in agricultural commodity markets over the previous years has given rise to concerns about their impact on food security. Analysing the nexus between fuel and food is complex. Although there is a consensus that the demand for biofuel feedstocks affects food prices, no consensus exists on the magnitude of the impact, which may also vary across countries. The degree of competition among food, feed and fuel uses of biomass will hinge on a variety of factors, including crop selection, farming practices, agricultural yields and the pace at which next-generation biofuel technologies develop.

Higher food price levels as well as increased volatility can cause substantial problems to low income net food consumers including agricultural labourers, the urban poor and the large proportion of rural poor without sufficient productive assets. Competition for resource inputs places upward pressure on food prices, even if the feedstock itself is a non-food crop or is grown on previously unused land. On the other hand, farmers who are net food suppliers are likely to benefit from higher prices. Bioenergy growth can boost incomes by revitalizing agriculture, providing new employment opportunities and increasing access to modern energy, which can stimulate rural development. The food supply may be positively affected if the market for biofuel feedstock leads to new investments in agricultural research, infrastructure development and increased production.

Notes

1. Brazil, Sao Paolo (ex-distillery).
2. Producer price Germany net of biodiesel tariff.
3. More information can be found on the RFS2 Final Rule on the following website: *www.epa.gov/otaq/renewablefuels/420f10007.htm*.
4. The 102 bnl figure represents the sum of the Conventional Renewable fuels mandate in 2019 (15 bn gallons, *i.e.* 57 bnl) and of the mandate for total advanced biofuels except biomass-based diesel (12 bn gallons, *i.e.* 45 bnl).
5. All biofuel use shares are expressed on the basis of energy contained unless otherwise specified.
6. Expressed in volume terms.
7. Expressed in volume terms.
8. Mixture of ethanol and regular gasoline. The Thai Government provides price incentives through price subsidies and excise tax cuts for blends from E10 (mixture of 10% ethanol and 90% gasoline) through E85 (mixture of 85% ethanol and 15% gasoline), accordingly.

9. All biofuel use shares are expressed on the basis of energy contained unless otherwise specified.
10. Note that trade projections for ethanol, in addition to pure fuel alcohol, also include ethanol for other purposes as well as the ethanol share in gasoline blends.

References

OECD (2008), *Biofuel Support Policies – An Economic Assessment*, Paris.

IEA (2010), *Sustainable Production of Second-Generation Biofuels*, Paris.

Chapter 5

Cereals

Market situation*

The year 2009 marked a slow return to a more balanced situation in cereal markets. International prices, which in the previous year climbed to unusually high levels, driven by tight market conditions, fell following a significant improvement in global supply. The above-average cereal production in 2009 and large world inventories, especially among major exporters, weighed to depress international cereal prices, a trend that is continuing in 2010. The FAO Cereal Price Index averaged 174 points in 2009, down 27% from the all-time high average (in nominal terms) of 239 points in 2008.

Global cereal production in 2009 fell slightly below the previous year's record. Wheat production remained unchanged, but outputs of coarse grains and rice were smaller. Cereal production rose sharply in North Africa, following a strong recovery from the previous season's drought. In Asia, aggregate cereal production remained unchanged, with declines in China and India offsetting significant increases in several other countries. In India, a sharp drop in rice output, caused by an erratic monsoon and a recurring *El Niño*, more than offset a bumper wheat crop. However, in most other regions, cereal production declined in 2009. In Latin America, it fell significantly, driven largely by a poor wheat crop in Argentina, as a result of reductions in planted area and drought-driven low yields. In Europe, production dropped below the record in 2008 but was still high compared to previous seasons.

In spite of generally satisfactory production results in 2009, growth in world cereal utilisation remained sluggish for the second consecutive season. The main reasons were weaker growth in demand for feed grains, resulting from limited increases in 2009 world meat production, and a relatively slow expansion in industrial utilisation of cereals, especially of maize for production of ethanol. Against this background, world cereal stocks expanded for the third consecutive year, lifting the ratio of world cereal stocks-to-utilisation, an important indicator of global food security, above its five-year average.

Another major feature has been the sharp contraction of over 5% in world cereal trade in the 2009/10 marketing season, from the previous season's record. A fall in wheat trade, resulting from lower imports by several countries in Africa and Asia that harvested larger crops, accounted for most of the reduction in the international trade of cereals. Slightly lower imports were also anticipated for coarse grains against a modest rise in imports of rice.

Projection highlights

- Over the next decade, international prices of major cereals are projected to remain steady or decline slightly, but still exceed the levels of the previous decade. A noteworthy feature is the drop in wheat to maize price ratio, which is projected at between 1.1 and 1.2, well below 1.3-1.6 registered in the previous decade. This apparent decline in the

* All dates are on a marketing year basis (*e.g.* 2009 represents the 2009/10 marketing year and 2019 represents 2019/20) unless stated otherwise.

ratio is mainly explained by the anticipated expansion in the supply of lower (cheaper) quality wheat and the increasing new demand for maize from the expanding ethanol sector. Rice prices are projected to decline significantly compared to 2009, falling to USD 422 per tonne by 2019. However, in contrast with other cereals, international rice prices in 2009 were still particularly high, only 16% less than the record reached in 2008.

- World cereal production is likely to increase by 1.3% p.a. over the projection period to reach 2 579 Mt by 2019. This compares to 2.3% annual growth observed in the previous decade.
- World cereal utilisation is projected to expand at 1.4% p.a. between 2009 and 2019, to reach 2 564 Mt. This compares to 1.8% growth per year observed in the previous decade. The share of biofuel use of coarse grains will continue to increase until about 2015, reaching 13.1% of the total use of coarse grains, before declining slightly thereafter, dropping to 12.6% by 2019. World feed use is projected to increase by about 1.4% annually. Total food consumption of cereals is projected to reach 1 194 Mt by 2019, up 154 Mt, or 15%, from the base period. Food consumption of wheat is set to reach 514 Mt, growing at about 1.1% annually. Rice food consumption is projected at 449 Mt, growing by roughly 1.3% a year, which is slightly faster than observed in the past decade.
- World cereal stocks are projected to increase to 582 Mt by 2019, exceeding the very low levels in 2006 and 2007 by 153 Mt, or 26%. The world cereal stock-to-use ratio is projected at around 23% which is close to the comfortable levels observed in 2009 but well below the levels of the previous decade.
- World trade in cereals is projected to reach 313 Mt by 2019, up some 50 Mt, or 21% from the 2009/10 marketing season. Imports by the developing countries are projected to rise by around 2.3% p.a., or at about the same rate as in the previous decade, to 256 Mt. Largest increases in imports are projected for the Near East, followed by Africa, Latin America and the Caribbean and Asia.

Market trends and prospects: Wheat and coarse grains

Production

World wheat output is expected to rise by 1.1% p.a. during the projected period, to 746 Mt by 2019, up 90 Mt, or 14% from the base period (2007-09 average) (Figure 5.1). After several years of relatively fast growth, driven by high prices in international markets, the annual world wheat production is likely to expand at a slower rate; although sufficient to match the projected annual growth in world consumption. World wheat output is likely to grow at a slower pace than in the previous decade, marked by a slow-down in area expansion, which is projected to increase by 0.3% annually against 0.4% in the previous decade, and more subdued average yield growth, which is projected at only 0.8% compared to 1.3% in the previous decade. Having already reached high yields in many of the major wheat producing countries, the potential of further advancements in yields in those countries is more limited.

In the OECD countries, the rate of growth in wheat production is expected to remain muted, at only 0.8% per year, and its rebound restrained by equally sluggish growth in overall demand. Although the growth rate is higher in developing countries as potential for productivity gains are much stronger, wheat production in non-OECD countries may still not rise more than 1.3% per year, considerably below the rate in the previous decade.

Figure 5.1. **Wheat production and price**[a]

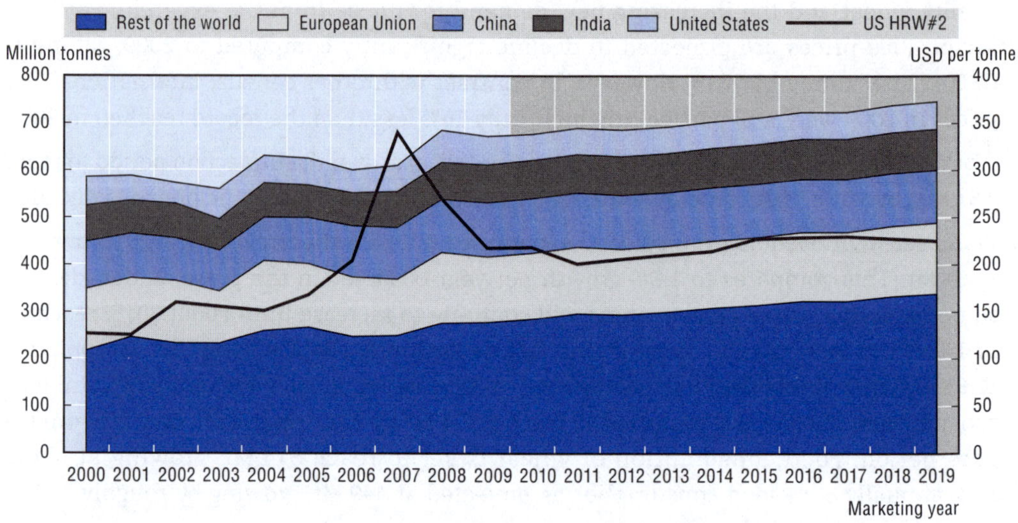

a) No. 2 hard red winter wheat, ordinary protein, USA f.o.b. Gulf Ports (June/May).

Source: OECD and FAO Secretariats.

StatLink http://dx.doi.org/10.1787/888932286361

Wheat production is expected to increase most significantly in those countries which suffered from exceptionally low yields in the base period, such as Argentina and Australia, while fast growth is also projected for a number of countries such as Pakistan and the Russian Federation. By contrast, production is likely to contract further in Saudi Arabia where the Government decided to phase out support to wheat farmers in the face of severe water scarcity problems. In China, the world's second largest wheat producer after the EU, a standstill is projected driven mostly by a diminishing consumption on annual per capita basis. In India, another leading producer, growth in wheat production is expected to decelerate while continuing to keep pace with the projected increase in demand. Among the OECD exporting countries, slower growth is projected for Canada and even a slight decline for the United States.

World production of coarse grains is projected to reach 1 311 Mt by 2019, some 19% higher than the base period (Figure 5.2). In aggregate, global coarse grains production is expected to expand by 1.6% annually during the projection period, which would be almost half as fast as in the previous decade. This slower growth is largely in response to weaker feed demand, reflecting the continued fallout from a recession during the early years of the projection period, and a slower rate of expansion in maize-based ethanol production, primarily in the United States. Overall, the deceleration in yields is more significant than the declining rate of expansion in planting area. Average yields are projected to increase by just 1.0% annually during the projection period, well below a relatively fast growth rate of 2.1% per year observed in the previous decade. In the case of area, the projected expansion is put at 0.6% per year, which is slightly below the rate recorded in the previous decade. A considerable part is on account of the expected coarse grain production development in the United States which accounts in 2019 for 30% of the world total; the area is expected to remain unchanged during the projection period compared to a growth by 0.9% annually the decade before, and yield to increase annually by 1.1% compared to 2.4% in the decade before.

Figure 5.2. **Coarse grains production and price**[a]

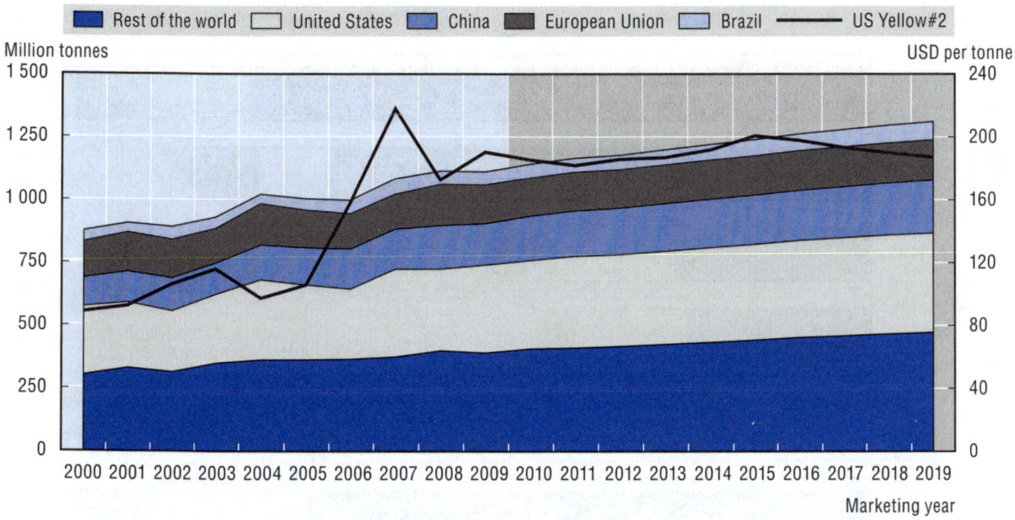

a) No. 2 yellow corn, US f.o.b. Gulf Ports (September/August).
Source: OECD and FAO Secretariats.

StatLink http://dx.doi.org/10.1787/888932286380

The bulk of the increase in world production of coarse grains is expected to come from non-OECD countries and that mostly because of the more rapid expansion in their planting area compared to the OECD countries. Productivity gains in non-OECD countries are also projected higher than in the OECD countries, especially in Asia where the increase in the land base is more restrained because of mounting water constraints and continuing urbanisation.

Among leading producers, maize production in the United States is projected to exceed the base period by roughly 54 Mt and that of coarse grains by 45 Mt in China. Substantial increases are also projected for Argentina, reflecting not only a recovery in its maize production from the drought-reduced yields witnessed in the base period, but also higher plantings as compared to less profitable wheat. In addition, Brazil, the EU, the Ukraine and the Russian Federation are projected to produce notably higher volumes of coarse grains (maize and barley) than in the past. However, an emerging new feature during the projection period is likely to be in Africa where the production of coarse grains is expected to continue increasing rapidly in several countries, especially in Nigeria (maize, millet and sorghum) and in Ethiopia (maize and sorghum), driven by good potential for both land expansion as well as further gains in yields.

Utilisation

World wheat utilisation is projected to rise by around 1.2% annually in the next decade, to reach 740 Mt (Figure 5.3). The increase is mostly on account of the rise in direct food consumption in the developing countries, driven by growing population, income and continued urbanisation.

World food consumption of wheat is projected to reach 514 Mt by 2019, representing 69% of the total projected utilisation. This translates into a world average of roughly 68 kg per person per annum, very close to the base period. Algeria (210 kg), Egypt (202 kg), Turkey (199 kg), and the Islamic Republic of Iran (167 kg) will continue to top the list of the highest consumers of wheat. On the other hand, the average per capita in sub-Saharan Africa is projected to reach 19 kg by 2019; remaining the lowest compared to other regions in spite

Figure 5.3. **Wheat consumption in OECD and developing countries**

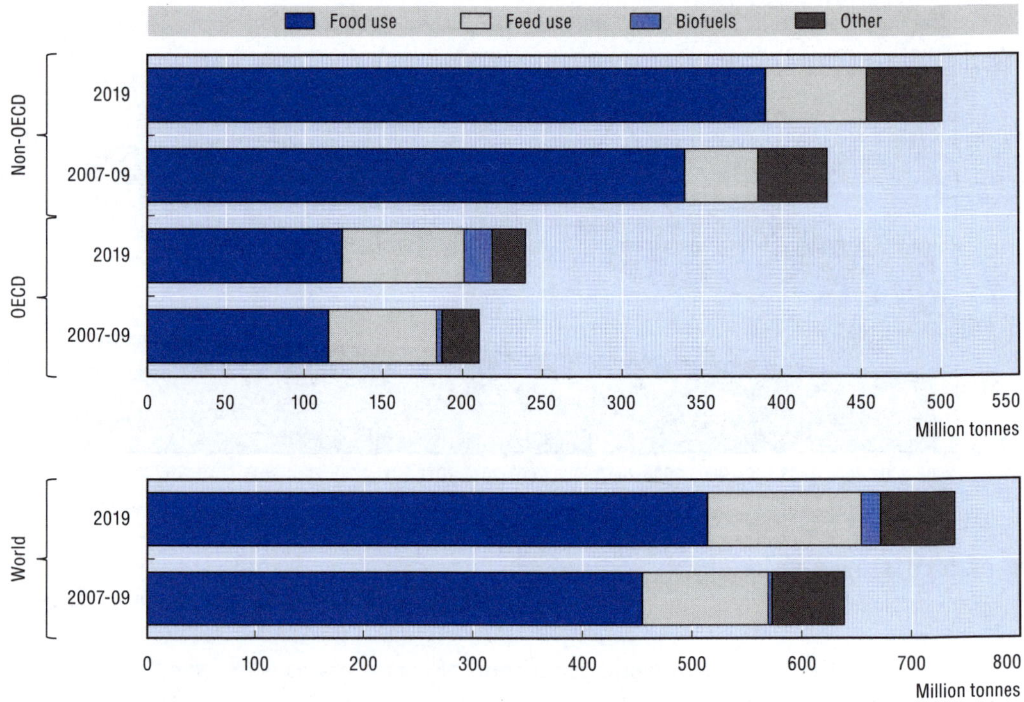

Source: OECD and FAO Secretariats.

StatLink http://dx.doi.org/10.1787/888932286399

of relatively significant increases projected for several countries in the region, in particular for Tanzania and Mozambique. Annual per capita wheat consumption levels in Brazil and South Africa are projected to increase the fastest, to 64 kg and 67 kg respectively.

The other major use of wheat is for animal feed, which is expected to grow by 1.2% annually, or more than twice as fast as in the previous decade, to 140 Mt. Especially, the tight season 2007/08 reduces the average of the reference period considerably. In addition, more ample wheat supplies and more favourable prices than in recent years are among the main factors behind this increase. In the EU, the largest user of wheat for animal feed in the world (41% of global feed wheat use in 2019), total feed use of wheat is projected to show a slight increase during the projection period, after a contraction in the previous decade, to finally round out at 57 Mt.

Among the other main utilisation categories, the industrial usage of wheat for production of biofuels is projected to increase the fastest, mainly driven by the anticipated increase in production of ethanol in the EU, which accounts for about 90% of the total use of wheat for biofuels. The Renewable Energy Directive (RED) requires to reach, by 2020, a share of 10% of renewable fuels in EU transport fuel use.

World utilisation of coarse grains is projected to increase by 1.5% p.a., to 1 303 Mt by 2019 (Figure 5.4). The projected expansion is below the pace observed over the past decade and that mostly on account of much slower growth in the use of coarse grains for production of biofuels in the United States, the world's largest producer of maize-based ethanol. The share of biofuels in total use of coarse grains is projected to increase until about 2015, reaching 13%, before declining during the latter years of the projection period. This development is largely on account of a slower increase in the use of ethanol for fuel

Figure 5.4. **Coarse grains consumption in OECD and developing countries**

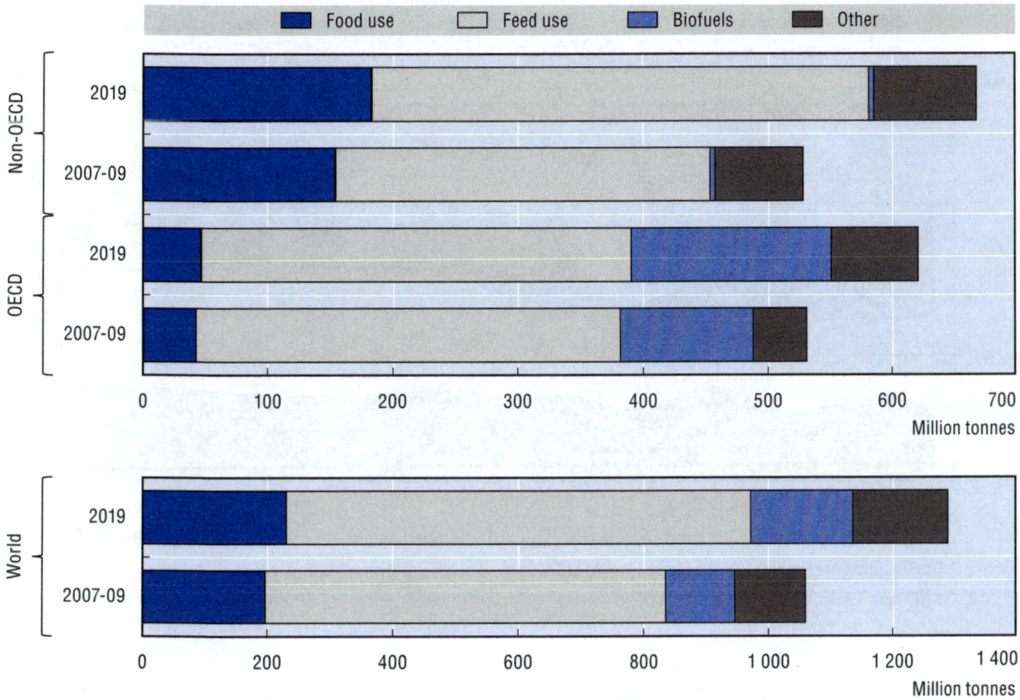

Source: OECD and FAO Secretariats.

StatLink http://dx.doi.org/10.1787/888932286418

(to 82 bnl by 2019 and thus below the mandate of 102 bnl set by the new Renewable Fuels Standard (RFS2) in the United States) and increasing production of second generation cellulosic ethanol towards the end of the projection period.

As in the previous decade, growth in the consumption of coarse grains as food is likely to slow, mostly keeping pace with population growth. Globally, direct food consumption of coarse grains accounts for only 17% of its total utilisation and this share is expected to remain largely unchanged during the next decade.

Total use of coarse grains for animal feed is projected to increase at a slightly faster pace than in the previous decade, by 1.5% p.a., to 741 Mt by 2019. A continuing rapid increase in the developing countries, driven by strong economic growth and changes in dietary preferences toward protein-based diets, is responsible for most of the projected expansion in world feed utilisation of coarse grains. China alone accounts for 40% of the world increase in coarse grains feed use over the projection period. In the developed countries, despite a reversal from a small contraction observed in the previous decade, the recovery will be small, reflecting a slow anticipated growth from the livestock sector and large supplies, hence utilisation, of distillers dried grains (DDGs), co-products from grain-based ethanol production used in feed rations.

Stocks and prices

The gradual recovery in the level of world wheat stocks which started in 2008 is projected to continue throughout the first years of the projection period before inventories stabilise at around 242 Mt by 2019, up 49 Mt, or 25%, from the base period (Figure 5.5).

Figure 5.5. **Wheat stocks and real price[a] development**

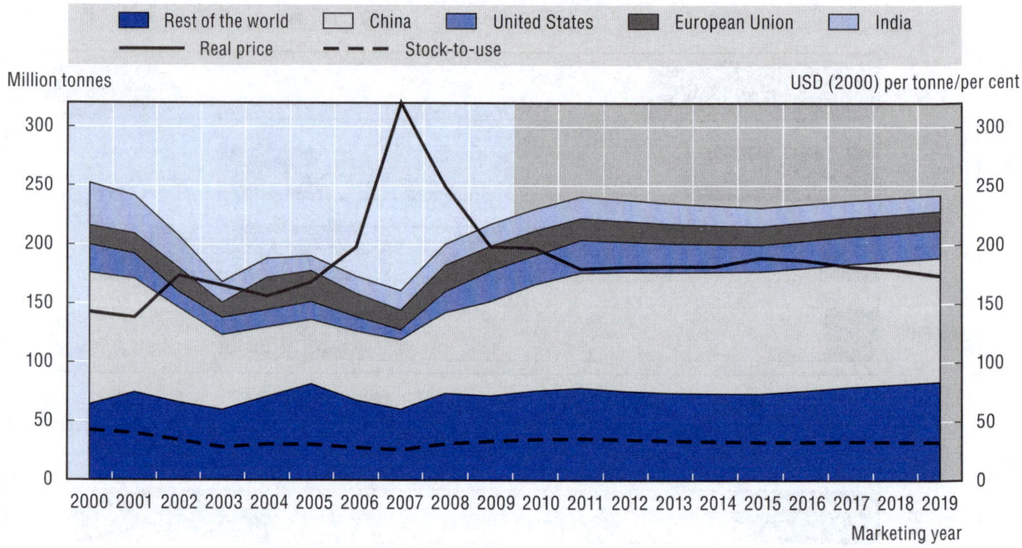

a) No. 2 hard red winter wheat, ordinary protein, USA f.o.b. Gulf Ports (June/May).

Source: OECD and FAO Secretariats.

StatLink http://dx.doi.org/10.1787/888932286437

Most of the increase in world wheat stocks would be in non-OECD countries, in particular in China where inventories could increase by as much as 36 Mt compared to the base period. Besides China, significantly higher stocks in the Russian Federation and Ukraine are also expected to boost world stock levels.

Aggregate stocks held by the major exporters, traditionally considered as the main buffer against unexpected production shortfalls, are projected to reach 54 Mt, which would be only 5 Mt above the relatively low average registered in the base period. This development makes world price stability more conditioned on the size and stock holding policies in other countries. As wheat production in many of those countries are often prone to sudden, or unexpected, changes because of erratic yields, this may give rise to even higher price variability compared to the historical trend. This, coupled with the fact that the global stocks-to-use ratio for wheat is projected to reach only 33%, which would be below the ratios registered in the 1990s (although higher than the low of 2006), suggests that prices may not only become more unstable in the future but also higher than in the past, at least in nominal terms. The price of the benchmark US wheat (No. 2 Hard Red Winter, f.o.b. Gulf) is projected to increase to USD 225 per tonne by 2019, up 3% from the average in 2009. In real terms, however, wheat prices are likely to continue their long-term decline, albeit falling less rapidly and from higher levels.

World stocks of coarse grains are also anticipated to increase during the projection period but the build-up is likely to be smaller than that of wheat and, contrary to the wheat situation, most of the increase is projected to occur during the latter years when the growth in utilisation is expected to slowdown. By 2019, world stocks of coarse grains are projected to reach 223 Mt, some 25 Mt, or 13%, above the relatively low level of the base period (Figure 5.6). However, as in the case of wheat, higher inventories in China account for most of the increase (some 70%) of the projected rise in world stocks. In addition, the stocks-to-use ratio may not exceed 17% by the end of the projection period. This could be

Figure 5.6. **Coarse grains stocks and real price[a] development**

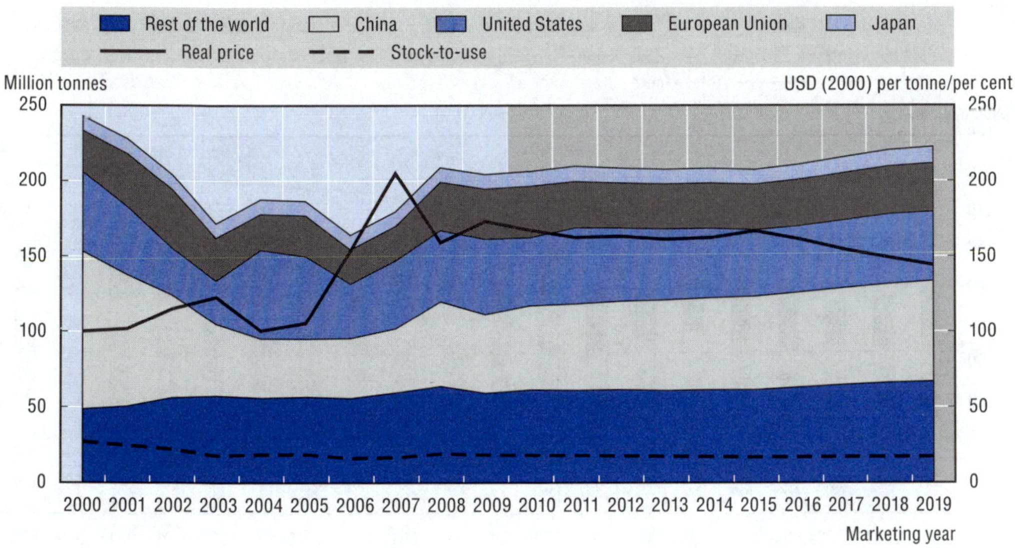

a) No. 2 yellow corn, US f.o.b. Gulf Ports (September/August).

Source: OECD and FAO Secretariats.

StatLink http://dx.doi.org/10.1787/888932286456

considered as relatively low when compared to the previous decade, especially in light of the fact that aggregate stocks held by the traditional exporters are projected to reach 86 Mt, only 2% above the base period.

Current projections for stocks and utilisation point to a somewhat tighter supply and demand balance during the early years of the projection period resulting in prices rising the fastest until 2016. The situation is likely to improve thereafter with utilisation growing at a slower pace and keeping up with production growth. By 2019, the price of the benchmark US maize (No. 2 Yellow, Gulf) is projected to reach USD 187 per tonne, almost unchanged from the average in 2009. However, even at this price, the wheat to maize price ratio would drop to a notably low ratio of 1.1-1.2, compared to 1.3-1.6 in the past. One reason for this apparent decline in the ratio is the anticipated expansion in the supply of lower (cheaper) quality wheat, increasingly taking larger share of world wheat markets and hence putting downward pressure on world average prices of wheat. Another reason is the increasing new demand for maize from the expanding ethanol sector, which is supportive to maize markets and push up world prices of maize.

Trade

World wheat trade (including wheat flour in grain equivalent) is projected to reach 142 Mt by 2019, expanding at a slower pace than in the previous decade, especially when compared to the base period (Figure 5.7). Global trade in wheat peaked in 2008/09 marketing season when it registered an all time high of 135 Mt primarily due to significantly larger imports by several drought-hit countries in Asia and Africa. This sudden sharp jump was followed by an equally significant dive in the following season, reflecting a recovery in production in many countries and a gradual return of confidence to markets as inventories replenished.

Trade in wheat is projected to resume an upward trend from 2010/11, increasing by about 1.8% annually, almost entirely on projected expansions in imports by the developing countries. Demand for wheat continues to be driven mostly by population growth, especially in several

Figure 5.7. **Main wheat traders**

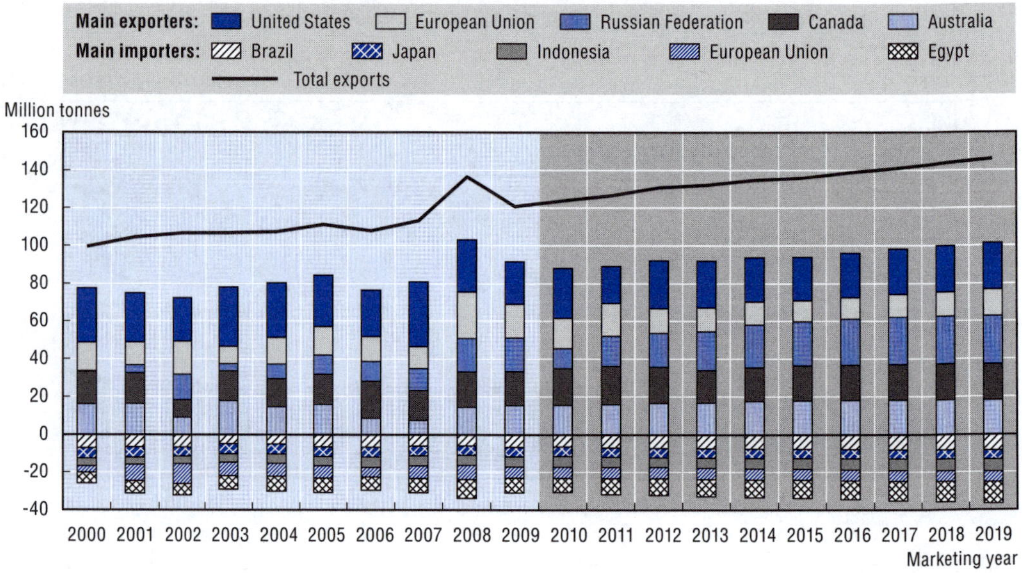

Source: OECD and FAO Secretariats.

StatLink http://dx.doi.org/10.1787/888932286475

countries in Asia and North Africa where wheat is a leading staple. Aggregate wheat imports by the non-OECD countries are projected to reach 117 Mt in 2019, representing 82% of projected world trade. Compared to the base period, the biggest increases in terms of volume are projected for Egypt, which will retain its position as the world's largest importer, followed by Pakistan, Brazil, Nigeria, and Algeria. Imports by Saudi Arabia will continue to increase following the government's policy to phase out support to domestic producers. By 2019 Saudi Arabia is likely to source almost 90% of its needs through imports. Wheat imports are also projected to increase with relative strength in India, Mexico, South Africa, and Viet Nam. Wheat purchase by the Islamic Republic of Iran are likely to fall below the exceptionally high imports seen during the base period when the country was hit by a severe drought, but still remain large in spite of the country's drive for self-sufficiency.

On the export side, a select few exporters are expected to dominate the wheat market. Aside from the traditional five major exporters, namely Argentina, Australia, Canada, the EU, and the United States, shipments from the Russian Federation could surge by 63% from the base period to reach around 26 Mt by 2019. In fact, based on the projection, the Russian Federation will be slightly ahead of the United States, currently the world's largest wheat exporter, by 2019.

Global trade in coarse grains is projected to reach 132 Mt, up 12% from the base period largely on account of a continuing expansion in feed demand especially among faster growing developing countries (Figure 5.8). Similar to the wheat situation, trade in coarse grains experienced a significant increase during the base period, particularly in 2007/08 when it climbed to a record 131 Mt. The surge in imports (maize and sorghum) by the EU, due to shortages of feed wheat, was one of the main drivers for the sudden jump in world trade in 2007/08 but, in the subsequent seasons, world trade returned to more normal levels as the situation in the EU improved. Given the fact that the increase foreseen during the Outlook period starts from a relatively high volume, trade expansion is likely to remain subdued during the early years before accelerating towards the latter years.

Figure 5.8. **Main coarse grain traders**

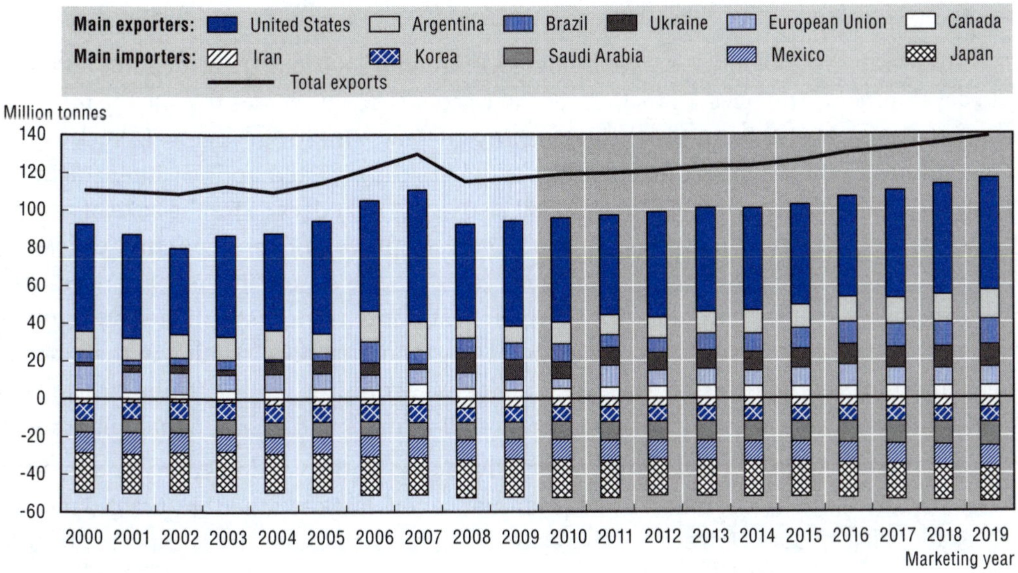

Source: OECD and FAO Secretariats.

StatLink http://dx.doi.org/10.1787/888932286494

Aggregate imports of coarse grains by non-OECD countries are projected to increase by 2.7% p.a. to 79 Mt; representing 60% of the world's total imports by 2019. In terms of volume and compared to the base period as well, the fastest increases are projected for Saudi Arabia, Egypt, Mexico, Chile, most countries in North Africa, Ethiopia, the Islamic Republic of Iran, Peru, Malaysia and the Philippines. While Mexico will retain its position as the world's largest importer of sorghum and second largest maize importer, Saudi Arabia will remain the leading importer of barley. China is expected to become a net maize importer from the early years of the projection period, driven by fast growth in domestic demand, but imports by China will not exceed even one half of its tariff-rate-quota (TRQ) of 7.2 Mt. Among the OECD countries, lower imports are expected in Japan, where higher meat imports dampen demand for coarse grains, and in Canada, mostly on lower feed demand and larger use of DDGs.

The traditional major exporters are likely to increase their overall market share slightly during the projection period with total exports of nearly 98 Mt by 2019, or 71%. The United States is expected to retain its position as the world largest exporter, and be able to ship more maize and sorghum, compared to the base period, despite its own growing domestic utilisation. Having suffered from the severe drought during the base period, Argentina is projected to return to the export market as a major exporter starting in 2010 given the good prospects for a strong recovery in its maize production. Australia, Canada, and the EU will also increase their exports of barley, meeting most of the demand in North Africa and countries in the Middle East. In addition, the projection points to larger exportable supplies of maize and barley in Ukraine. While Paraguay and India are projected to increase their maize exports, exports from South Africa, a traditional supplier of maize in southern Africa, may change little compared to the base period mainly due to its own growing domestic demand.

Market trends and prospects: Rice

Production

Following the 2008 price surge, many governments revisited their medium and long run development strategies. Fears that resources could become scarce in Asia, the principal source of rice trade, fostered the launching of new ambitious rice self-sufficiency programmes in major importing countries or, where they existed, to their intensification. This supportive stance is expected to prevail also in the next decade for the sake of gaining more control over food supplies, as food security prevails over economic efficiency concerns. Much of the resources necessary to boost the sector in the least developed countries are expected to come from large investments from institutional and private investors, who have been actively buying or leasing land for rice cultivation in their own or foreign countries in recent years. Besides their economic implications, these new tendencies may have substantial impacts on small farmers and international relations.

As a result, a brisk expansion of rice cultivation is anticipated in Africa and in a few Asian countries, such as Cambodia and Myanmar, where large swathes of suitable land and water are still available. However, at the world level, the area under rice is expected to grow little, given an expected retreat in various Asian countries, notwithstanding existing regulations that limit the conversion of agricultural land into other end-uses. Emblematic is the case of China, where the size of rice fields is projected to shrink by 1.3% per year over the next decade, or 3 Mha between 2009 and 2019. At the world level, the area under rice is projected to remain around 163 Mha over the next ten years. Nonetheless, a process of consolidation of small farms into large concerns is likely to prevail alongside.

Thus, yield increases are likely to remain the principal factor sustaining production in the coming decade. However, on average, the pace of their growth is expected to slow down from 1.3% between 2000 and 2009 to 1.0% per year in the coming decade, to reach 3.2 tonnes per hectare by 2019, in milled rice equivalent. Indeed, technological and varietal improvements, for instance through the adoption of hybrids or transgenic rice, or improved agronomics, are expected to focus less on yield enhancement and target instead other traits, such as nutritional characteristics (Golden Rice), or tolerance to abiotic stresses, such as drought. Despite fast progress, yields in Africa are projected to remain some 40% short of the world average and well below those reaped in the other regions. On the other hand, continued brisk productivity gains are anticipated in OECD countries.

At the world level, rice production is projected to expand by about 1.0% per year to reach 522 Mt in 2019 (Figure 5.9). Production by OECD countries, which followed a declining trend in recent years, is set to rebound somewhat by 2019. The recovery is projected to be sustained by a recovery in Australia, on the assumption of a return to a favourable rainfall pattern, and continued gains in the United States. However, the OECD share in the total will remain marginal, at less than 5%. Virtually all of the anticipated growth in production is expected to materialise in non-OECD countries, especially in Asia and Africa. However, China stands as an exception; as the country is foreseen to cut production in the next ten years, amid falling domestic demand and shrinking land and water resources.

Consumption

Rice continues to be predominantly a food commodity, as only 14% of overall utilisation is used for feed and other ends, the latter consisting mainly of post-harvest losses. Over the next decade, the status of rice as major staple food is not expected to be

Figure 5.9. **Rice production and price**[a]

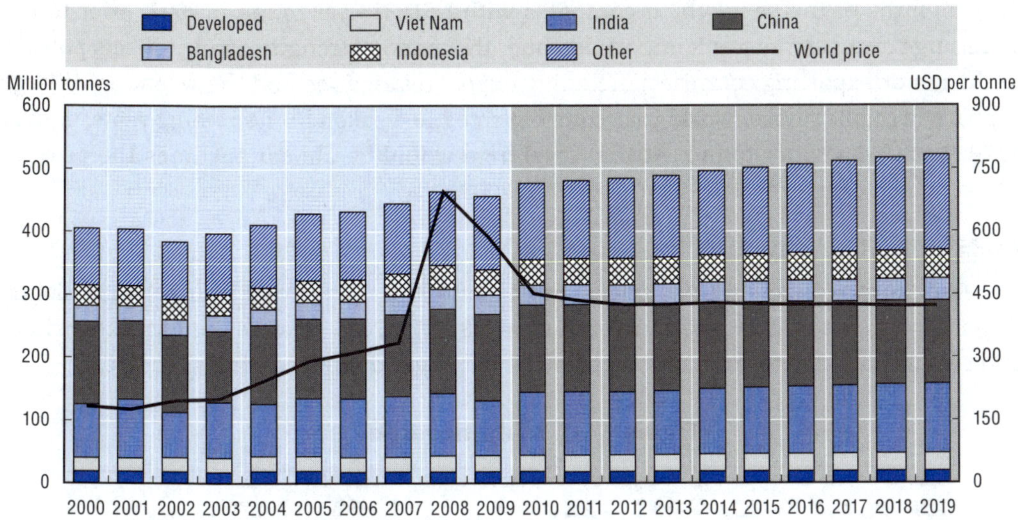

a) Milled, 100%, grade b, Nominal Price Quote (NPQ), f.o.b. Bangkok (August/July).

Source: OECD and FAO Secretariats.

StatLink http://dx.doi.org/10.1787/888932286513

challenged, although the utilisation of rice as feed will continue to be promoted and expand in countries such as Japan or the Republic of Korea. Utilisation of rice as feedstock for biofuel still does not appear a plausible alternative.

World overall rice utilisation is projected to reach 522 Mt in 2019, rising at the annual rate of 1.1% over the ten year period, virtually matching the increase in population. Indeed, on a per capita basis, consumption (all uses included) is set to expand at a marginal rate to 59.2 kg per year in 2019. However, per capita consumption is expected to follow diverging directions, either stabilising or even falling in Asia, while rising in Africa and part of Latin America and the Caribbean, where rice is increasingly vying with wheat to become the main staple. The fall of rice consumption in China, on the other hand, appears to replicate the historical pattern already witnessed in Japan and the Republic of Korea, whereby growingly affluent populations cut rice consumption to shift to higher protein-based diets. Nonetheless, Asia is to remain the major growth market for rice, given the sheer size of its population.

Because important countries, such as China, India, the Philippines, Bangladesh or the Islamic Republic of Iran are soon to engage in the commercial production of genetically modified rice, the questions of product segregation and labelling are anticipated to gain prominence in the commercialisation of rice in domestic and export markets.

Stocks and prices

Following the poor 2009 season production results, the size of world rice reserves had to be reduced in 2010, a situation that is foreseen to be redressed in the next decade. By 2019, world rice inventories are projected to reach 117 Mt, up from some 100 Mt in 2009. Much of the build-up is foreseen to be concentrated in China, where stocks would come to represent 44% of domestic requirements, a level of security much higher than the 30-34% coverage in recent years. At the world level, rice inventories would cover 22.5% of global rice consumption, slightly up from a stock-to-use ratio of 21.6% in 2009.

World rice prices continue to be high in 2010, averaging USD 570 per tonne in the first three months of the year, which compares with USD 335 per tonne in 2007, prior to the price surge event. Amid weak import demand, the market strength largely reflects policies in the major exporting countries, including export restrictions and the retention of large supplies in public stocks. World prices are expected to weaken in the coming years, ending in 2019 at USD 422 per tonne. At that level, rice would be almost 1.9 times the price of wheat, which compares with a ratio of 2.7 in 2009.

Trade

Trade in rice (Figure 5.10) is forecast to expand at 2.1% per year over 2010-19, slightly slower than in the past decade, and reach some 38 Mt by 2019. Despite the 7 Mt expansions, trade would continue to account for a small share of world production of about 7%.

Figure 5.10. **Main rice traders**

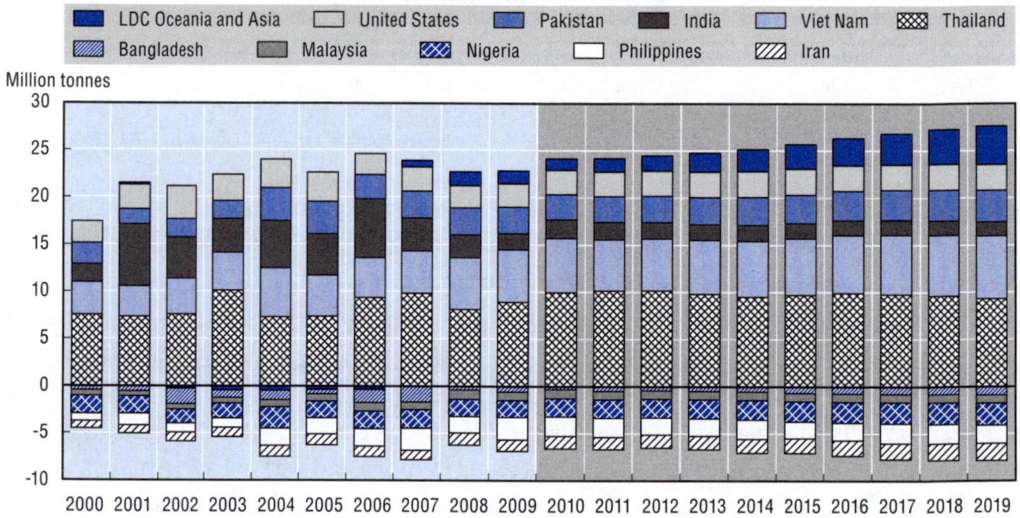

Source: OECD and FAO Secretariats.

StatLink http://dx.doi.org/10.1787/888932286532

Within the OECD group, the EU is projected to import substantially more in ten years time, but sizeable increases are also anticipated for the United States and South Africa. Imports by Japan and the Republic of Korea are foreseen to change little, as they are assumed to depend largely on the WTO minimum quota provisions. However, non-OECD countries are expected to continue driving import growth in the next decade. In particular, Near East countries are expected to boost their purchases, including Egypt, which has passed more stringent limitation on rice cultivation, the Islamic Republic of Iran and Saudi Arabia. While still rising, deliveries to African countries are projected to grow at half the rate of the past ten years, a consequence of the expansionary production strategy currently pursued in the region. However, Africa is not foreseen to emerge as a major source of supplies for the rest of the world, as much of the region's production gains will be needed to cover expanding domestic consumption.

Among traditional rice exporters, Thailand, Viet Nam and Pakistan are anticipated to increase deliveries over the decade. By contrast, India is now forecast to limit sales abroad to the high quality basmati rice, while Egypt may turn into a net importer position under its restrictive production policy. China's exports may also shrink. On the other hand, new countries appear ready to emerge as important suppliers to the world market, namely Cambodia, Laos and Myanmar, where large investments to expand rice cultivation are undergoing. The process would likely be facilitated by the newly implemented ASEAN free trade agreement, which has abated most restrictions on rice movements within the block. Among OECD countries, the United States is expected to export substantially more in 2019.

As already experienced in the past few years, trade in rice may well undergo frequent upheavals associated with changes in government policies. The presence of genetically modified rice in a number of markets could further expose trade to potential disruptions. Based on the current status of research and development, about 15 varieties of GM rice could be grown commercially by 2015, compared with none as of today [Stein and Rodriguez-Cerezo (2009)].

Key issues and uncertainties

Links to energy markets

Cereals are one of the main feedstocks used in the production of ethanol (*e.g.* maize in the United States and wheat, maize, barley and others in the European Union) but the shift towards second generation biofuels might curb or even reduced the demand for cereals as ethanol feedstock. The biofuel demand of cereals have been included in this Outlook to the available knowledge but the introduction of new biofuel policies or the changing of existing ones can have a considerable impact on the world demand for cereals. In addition, some biofuel policies use tax incentives rather than mandates and thus, link the production of cereal-based ethanol on the price difference to mineral fuels. This increases the price link between cereal and energy. This adds to the existing link via input costs, especially energy use in agricultural production and fertiliser costs.

A new development which could develop rapidly over the projection period is the use of biomass for energy purposes. This could be the production of biofuels (second generation biofuels) but also the use for the production of biogas, heat and electricity. The question arises how the biomass is produced, especially in the case of dedicated crops like switch-grass, short-rotation crop rice or green maize it may reduce the availability of land for the production of cereals.

Production uncertainties

The values presented here assume normal weather conditions and average yields. Recent experience and expectation result in the question: "What are normal weather conditions?" The high number of droughts in large parts of Australia does question whether this needs to be expected on more regular terms in the future. The impacts of climate change on agriculture (Box 1.4) could change also the medium term Outlook, but how and especially in which magnitude remains largely uncertain.

The spreading of plant diseases may have adverse effects on cereal production. In recent time the spreading of the stem rust Ug99 from East Africa towards Asia affecting wheat causes concern. It is uncertain, how large the effect on world wheat will be as most of the currently planted wheat has little or no protection against Ug99. On the other hand, plant breeding and plant protection could limit the effects before it has a global impact.

The development in plant breeding (Box 5.1) can positively influence the evolution of yields in the coming decade but the propagation of new varieties to a wide spread use in agricultural production takes some years. Through closer interaction and faster

Box 5.1. **New crop varieties and high quality seed – keys to enhanced productivity**

In September 2009, the FAO, OECD Seed Scheme, International Union for the Protection of New Varieties of Plants (UPOV), International Seed Association (ISA) and International Seed Federation (ISF) jointly organised the second World Seed Conference in Rome with the key focus on the role of new plant varieties and high quality seed in responding to the challenges facing agriculture in a changing world. The main message emanating from the Conference was that "urgent government measures and increased public and private investment in the seed sector are required if agriculture is to meet the challenge of food security in the context of population growth and climate change".

Plant breeding is a major factor contributing to increased food security whilst reducing input costs, greenhouse gas emissions and deforestation. Long run data for various crops across different countries suggest that about half the yield increase of around 2% per year has come from plant improvements and half from improved husbandry practices such as better irrigation, use of fertilisers and crop protection. In some cases, the contribution of plant breeding to production growth has been higher. For example, a UK study found that half the yield increase in wheat, barley and oats over the 1947-86 period was due to new varieties, and attributed 90% of the yield increase over the 1986-2006 period to new varieties.

Future developments in new plant varieties are expected to maintain and enhance this trend in yield improvements, thereby helping to mitigate the effects of population growth, climate change and other social and physical challenges. For example, maize was not grown in temperate climates above the 46th parallel before the 1960s; however, with the development of early maturing varieties it is now grown in the south of Sweden. Temperate crops can also be adapted to tropical climates. Sugar beets, which have lower water requirements, higher drought tolerance and shorter growing periods compared with cane sugar, can now be grown in tropical regions such as India.

Concerns have been expressed that global plant breeding capacity has been declining. A recent study by the FAO in co-operation with CGIAR found that plant breeding capacity in most developing countries is not adequate, or sufficiently integrated, to achieve the stated national production goals, leading to a lack of effective access to germplasm and technologies. Participants at the 2009 World Seed Conference recognised the importance of research on genetic enhancement while noting that other technologies, such as quality seed production and seed treatments, contribute substantially to improved yields and capacity building in all these areas is urgently needed in developing countries. It was also noted that plant breeding depends heavily on the conservation of genetic resources and that a multilateral system of access and benefit sharing of plant genetic resources for food and agriculture is necessary to support the global plant breeding process. Moreover, the development of new varieties alone is not enough to enhance agricultural output. Governments need to develop and maintain a regulatory environment that encourages the production and marketing of high quality seed. In developing countries, most threatened by food insecurity, farmers do not have access to improved varieties. In Eastern and Southern Africa, for example, this was attributed to a number of bottlenecks including a lack of market infrastructure and information, poor extension services, seed supply shortages and inadequate credit markets. Several institutional weaknesses were also identified. Few developing countries have well-defined seed development guidelines, the variety release process can be too long, state-controlled seed markets can hinder trade, and a failure to establish accreditation can limit access to international markets.

Notwithstanding, the global seed market continues to grow rapidly and is currently estimated at about USD 37 billion, a threefold increase over the last 30 years. There have been even higher growth rates in many developing countries such as China (with seven fold increase over 30 years), Argentina, Turkey and India. The development of reliable and internationally acceptable certification standards, contributes substantially to the strong growth in international trade and development of seed markets. For 2007, international seed trade was valued at USD 6.4 billion, with Europe, North America and Asia accounting for almost 80% of the total.

More specifically, the adoption of international standards has encouraged the growth of the seed trade by reducing technical barriers to trade, increasing transparency, lowering transaction costs and increasing access to high quality seed. The OECD Seed Scheme is the most widely used global certification system for the export and import of high quality seed, with some 58 participating countries from Europe, North and South America, Africa, the Middle East, Asia and Oceania.

Source: www.worldseedconference.org/en/worldseedconference/home.html.

information transfer this time could be more reduced in the coming decade. Other technical developments like precision farming can enhance the side-specific optimisation of agricultural production but impacts on world-wide cereal production depend on the speed of the knowledge transfer and availability of investments.

Governments, especially in those nations that rely heavily on food imports, reacted to the 2008 price hikes by launching or intensifying food self-sufficiency programmes, a move that was especially manifest in rice-dependent countries. Efforts to promote rice cultivation were facilitated in 2008 and 2009 by the permanence of relatively high world prices. However, there is concern that many rice self-sufficiency programmes could be put in jeopardy in the longer run should world prices fall.

The recent price hike also stirred interest of domestic or foreign investors in agriculture in land and water rich countries including Brazil, Cambodia, Ethiopia, Indonesia, Guyana, Madagascar, Myanmar, Nigeria, Sudan, etc. While potentially positive for agricultural production, it may be disruptive for the local economies and harm small farmers and their historical rights on land. Tensions could also arise between the host country and the foreign investors, should the produce be exported, depriving indigenous populations from locally produced food supplies.

Market integration

Asynchronous approval of genetically modified (GM) crops and other sanitary and phyto-sanitary border measures could result in trade diversions between regions and also substitutable crops. Especially on the short term this could have large effects although on the medium term new trade equilibriums would establish. Being such an important food staple, the advent of GM rice on markets will pose a series of critical consumer-related issues, but may have also potentially disruptive effects on world trade. Marketing of GM rice may require the segregation by the industry of traditional rice *versus* GM rice and clear label indications on their GM content.

Several important bilateral trade agreements are under negotiation, *e.g.* European Union-India, Canada-European Union. These agreements could enhance trade in cereals between the parties but also divert trade flows. The situation regarding an agreement in the World Trade Organisation (WTO) negotiation remains uncertain and an agreement will affect also world trade in cereals.

Reference

Stein, Alexander, J. and Rodriguez-Cerezo, Emilio (2009), "The Global Pipeline of New GM Crops – Implications of Asynchronous Approval for International Trade", *JRC Scientific and Technical Reports*, Institute for Prospective Technological Studies, European Commission Joint Research Centre.

Chapter 6

Oilseeds and Oilseed Products

Market situation

In the last few years, considerable movement has interrupted the historically flat price trend for oilseeds, oils and meals, especially when prices for all three product groups rose dramatically in mid 2008. The price surge was caused by an unusually tight supply and demand situation, spill-over effects from related grain markets as well as a growing importance of biofuel demand and of macroeconomic factors. Even though the rise was followed by a sharp fall in prices towards the end of 2008, values for oilseeds and derived products have since moved at levels above those prevailing prior to the general increase.

In early 2009, prices embarked on a new upward trend, mainly reflecting concerns about the progressive tightening of global supplies following the weather related decimation of South America's soybean crop. Furthermore, demand growth had returned to normal levels (compared to the slowdown at the onset of the global economic crisis), with main importing countries showing robust buying interest. Eventually, the shortage of supplies relative to demand led to record-low stock levels.

The ensuing strengthening in oilseed prices lasted until mid-2009, when ample crops from the Northern hemisphere (notably soybeans) started entering the market. With global oilseed production poised to resume growth, prices for oilseeds started to weaken gradually, although those of oils and fats and in particular of oilmeals remained firm. Sustained oilmeal values are explained by sustained Asian and European import demand and the market's strong reliance on a single type of meal (soymeal) and only few suppliers. By December 2009, meal prices almost matched the level recorded during the 2008 price peak. More recently, however, due to the prospect of a full recovery in South American soybean production in 2010, meal prices started to ease again.

Also, the market for edible oils and fats was characterised by a tight supply and demand situation: in the second half of 2009, limited export supplies and reduced output of high oil-yielding oilseeds coincided with revived growth in global demand for oils from both the food as well as the energy sector. The renewed firmness in crude oil prices contributed to the gradual strengthening in oils and fats values. Unlike meal values, oil and fat prices continued to be firm in early 2010, reflecting the prospect of limited growth in the production and export of palm oil during the remainder of the year, while oil utilisation for food purposes was poised to rise further and demand from the biodiesel industry started accelerating in countries enforcing higher domestic biofuel blending rates.

The medium-term market projections for oilseeds and derived products presented in this chapter are subject to assumptions of average weather conditions, the continuation of current policy packages (concerning agriculture and bioenergy) and an evolution of the macroeconomic environment that has been described in Chapter 3.

Projection highlights

- Given steady demand growth and strong crude oil prices, prices for oilseeds and oilseed products should remain above long term levels in both nominal and real terms. Given biofuel mandates and sustained food demand in developing countries, vegetable oil prices are expected to increase in nominal terms and be fairly flat in real terms over the Outlook period (Figure 6.1).
- Compared to the 2007-09 average, world oilseeds acreage and production are expected to be 10% and almost 30% higher, respectively, by 2019, based on sustained profitability and yield improvements (Figure 6.2). Global oilseed area expansion is limited compared to the previous decade due to higher marginal costs of area expansion and new constraints coming from environmental regulations in developing countries, as well as sustained profitability of competing crops, mainly maize, in the US.
- At the world level, vegetable oil production should increase by almost 40% over the projection period when compared to the base. The share of vegetable oil consumption used for biodiesel production is expected to increase from 9% for 2007-09 to 15% in 2019, driven by biofuel mandates and consumption incentives in many countries. In non-OECD economies, because of increasing population and income, food use accounts for 90% of the 3% annual growth in vegetable oil demand (Figure 6.3).
- During the Outlook period, annual growth in protein meal consumption is projected at 1% in OECD economies, compared with 3.2% in non-OECD economies based on sustained growth and intensification of livestock production around the world.

Figure 6.1. **Oilseeds and oilseed products prices to remain well above long term levels**

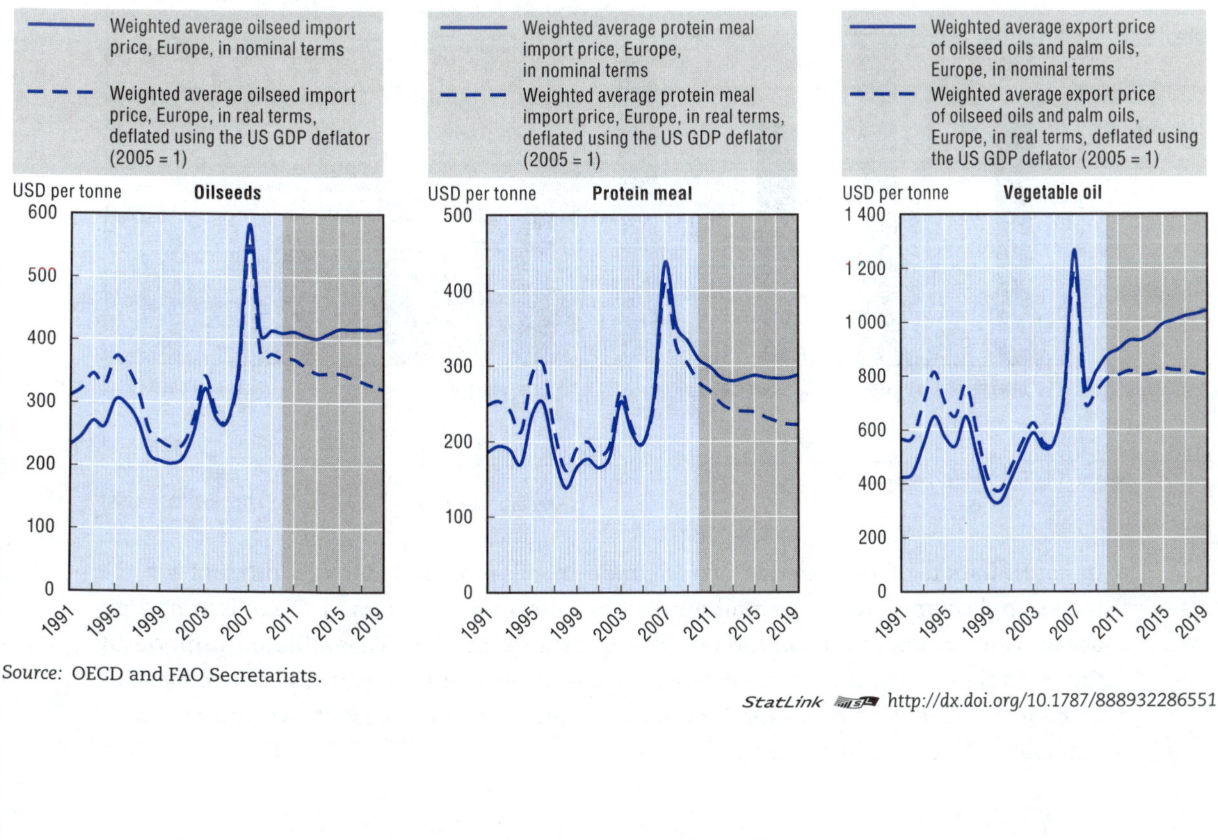

Source: OECD and FAO Secretariats.

StatLink http://dx.doi.org/10.1787/888932286551

Figure 6.2. **Oilseed production to expand**

Evolution of global oilseed production over the projection period

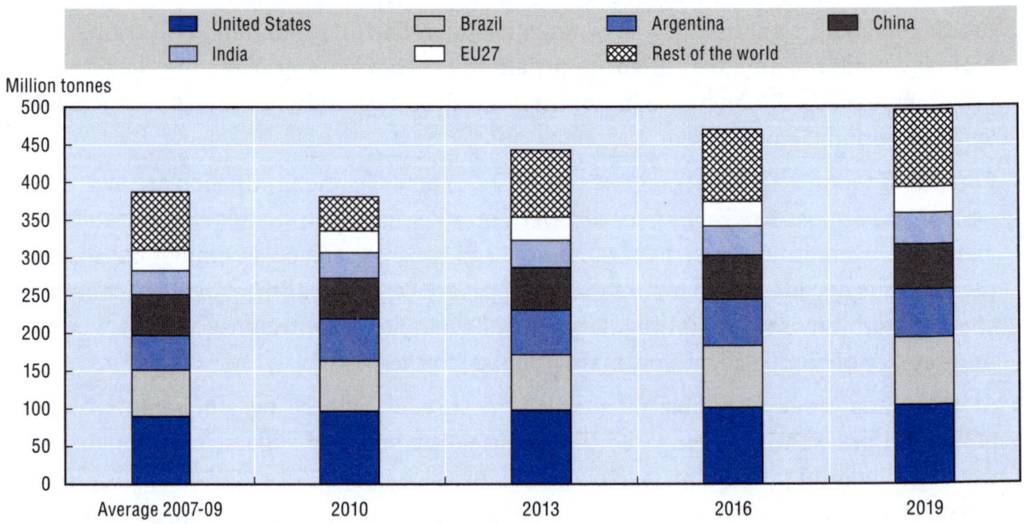

Source: OECD and FAO Secretariats.

StatLink http://dx.doi.org/10.1787/888932286570

Figure 6.3. **Food use to drive vegetable oil consumption in non-OECD economies, biofuel use to drive consumption in OECD economies**

Comparison of average annual growth rates of vegetable oil consumption

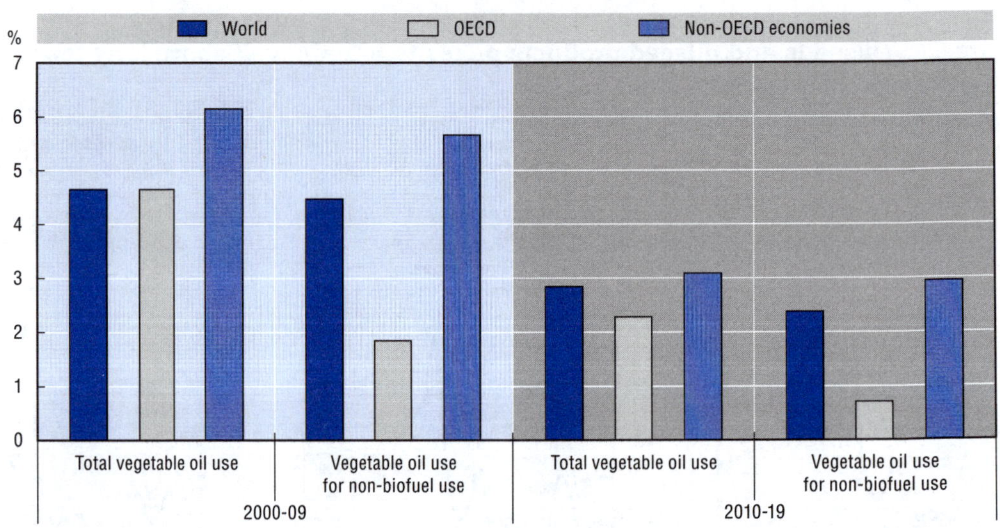

Source: OECD and FAO Secretariats.

StatLink http://dx.doi.org/10.1787/888932286589

- In continuation of past trends, world trade in oilseeds and derived product will be increasingly dominated by a small number of countries, with no more than four nations accounting for about 85% of global oilseed exports in 2019. Vegetable oil imports by China, India and the EU are expected to represent 55% of total world imports. By 2019, palm oil producers Malaysia and Indonesia are projected to represent two-thirds of total vegetable oil exports (Figure 6.4).

Figure 6.4. **Vegetable oil exports to be concentrated**

Evolution of vegetable oil trade over the projection period

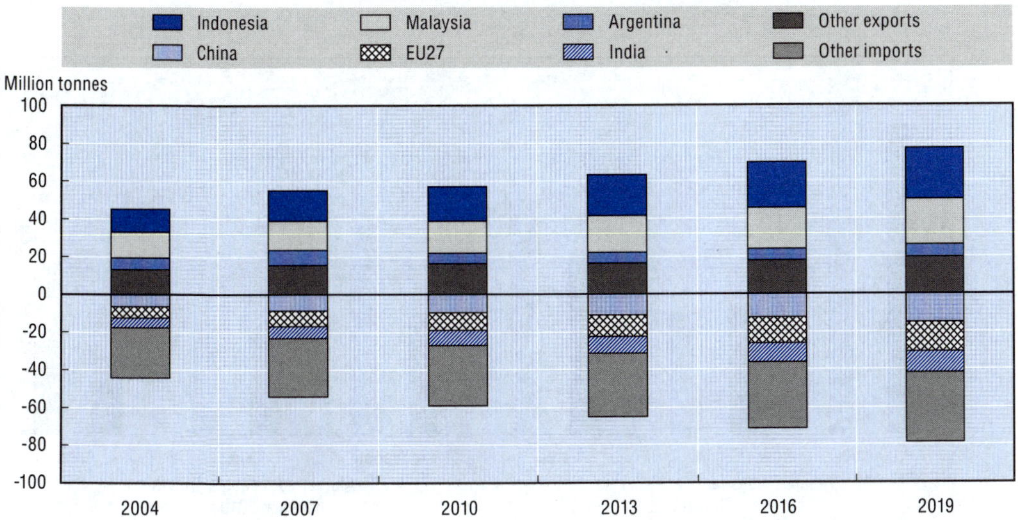

Source: OECD and FAO Secretariats.

StatLink http://dx.doi.org/10.1787/888932286608

Market trends and prospects

Prices

World prices for vegetable oils should remain well above levels prior to the food crisis in both nominal and real terms (Figure 6.1). Sustained food demand growth in developing countries and ambitious targets for biodiesel consumption, together with strong and increasing crude oil prices throughout the Outlook period are expected to lead to a sustained increase of prices in nominal terms. The pace of increase should be slower in the later years of the Outlook period as the pressure on biodiesel markets becomes lower because of the assumption of second generation biofuels slowly becoming available in the European Union. World vegetable oil prices expressed in real terms should remain fairly flat over the projection period.

Compared to the levels prior to the food crisis, also oilseeds and protein meal prices are projected to remain strong over the projection period as global stock-to-use ratios are expected to stay at low levels. In a context of sustained crush demand and increasing production, oilseeds prices are expected to remain about unchanged in nominal terms and to decrease in real terms all over the course of the projection period. In line with other feed commodities, protein meal prices are expected to decrease in the early years of the Outlook period before marginally increasing over the rest of the projection period.

Oilseed production and crush

Compared to the 2007-09 average, world oilseeds acreage and production are expected to be 10% and almost 30% higher, respectively, by 2019, based on sustained profitability and yield improvements (Figures 6.2 and 6.5). Global oilseed area expansion is limited compared to the previous decade due to higher marginal costs of area expansion and new constraints coming from environmental regulations in developing countries, as well as sustained profitability of competing crops, mainly maize, in the US. The yield gap between OECD and non-OECD countries is expected to diminish over the Outlook period. The US

Figure 6.5. **Potential for oilseed area expansion is limited**

Comparison of average annual growth rates of oilseed production, acreage and yield

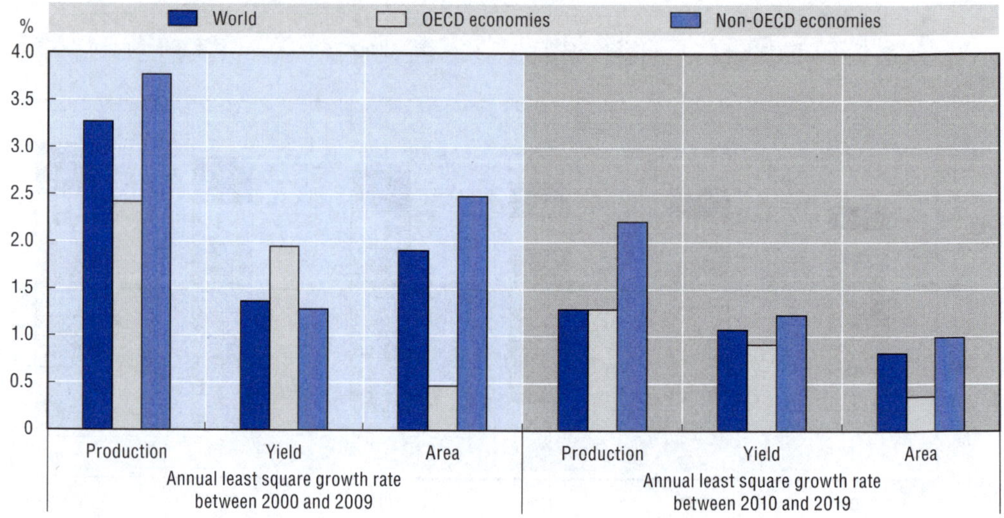

Source: OECD and FAO Secretariats.

StatLink http://dx.doi.org/10.1787/888932286627

should remain the major oilseed producer over the projection period (Figure 6.2) with oilseed acreage remaining constant at about 35.2 Mha as US domestic producer prices for soybean and maize are expected to move together over the projection period.

Brazil and Argentina should strengthen their role in oilseeds supply with a combined share of global oilseed production of more than 30% in 2019, compared to about 27% in the base period. Brazil is projected to be the main contributor to area expansion in non-OECD economies although due to increasing environmental limitations plantings should expand at a much slower pace than over the previous decade. Paraguay has been emerging as the third South American oilseeds producer. Driven by sustained high profitability, the country's area under soybeans is expected to expand to more than 3.2 Mha producing over 8.3 Mt by 2019.

China's agricultural policy doctrine is expected to continue to focus on domestic production of coarse grains. As a result, oilseeds production growth is limited as yields and area are not projected to improve significantly. India is expected to continue its expansion of edible oilseeds production (*i.e.* this does not take into account the expansion in *jatropha* cultivation) by adding about 2.7 Mha over the projection period to reach 36 Mha in 2019. Improved yields also contribute to production growth. However, the anticipated annual growth rate of 2.2% over the projection period compares to almost 6% in the past decade, reflecting a comparable slow down in both, yield improvements and area expansion, indicating that investments and public support for and returns from oil crops tend to remain behind those of alternative crops. Oilseed production in other parts of Asia is expected to expand by less than 2% per year.

Oilseeds production should increase by almost 30% in the EU over the projection period with yield gains contributing strongly to this increase. Driven by strong prices, oilseed area – mainly rapeseed – is projected to continue to grow over the projection period at a rate of 0.5% per year. The growth should however be slower than over the previous decade where the rapeseed sector developed quickly in response to strong demand from the expanding biodiesel sector. Within the OECD, oilseed production expansion is expected to also occur in Canada and Australia to satisfy both domestic crushing industries and export markets.

Driven by strong demand for protein meals from the livestock industry, oilseed plantings and production in the Russian Federation should expand, respectively, at a rate of 1% and 2.7% per year. As export opportunities within Europe have developed over recent years, Ukraine has become a significant sunflower seed producer with compound annual growth close to 15% during 2000-09 and total output exceeding 10 Mt in 2008. Further consolidation and modernisation in oilseed production should allow production to continue growing at more than 5% per year reaching more than 15 Mt by 2019.

With growing demand for oilseeds products, global oilseed crush is projected to continue expanding. It should grow at an average growth rate of 2% compared to 3.8% over the previous decade. This is due to the slowdown in protein meal and vegetable oil demand growth rates compared to the previous decade. Countries oriented towards the export of oilseed products or of biodiesel like Brazil and Argentina are expected to continue to develop their crushing industries with annual growth rates close to 2.5% over the projection period. Oilseed processing in Paraguay has been expanding along with seed production and is expected to grow further, with about 30% of domestic production being crushed in the country.

China is expected to continue to favour the imports of oilseeds to capture the value added from processing oilseeds into protein meals and vegetable oils. Its crushing capacity is expected to continue to grow at a rate of 2.2% per year which, however, is significantly below the previous decade. In India, about 80% of the domestic production is crushed; the remainder is consumed directly as food. Direct consumption is expected to fall slightly as the mix of oil crops shifts slowly towards seeds with higher crushing rates. The crushing sector in Ukraine has not quite kept up with seed production and is expected to maintain a growth rate just below seed production, processing 8.2 Mt in 2019.

In the US, strong soybean oil prices should be behind higher average annual growth rates of oilseed crush over the projection period than over the previous decade. In the EU, after the strong increase in crushing facilities between 2003 and 2009 due to the emerging biodiesel industry, crush is expected to continue to develop but at a slower pace of 1.3% per year.

Products production and consumption

Vegetable oil

At the world level, vegetable oil production should increase by almost 40% over the projection period. Global production of palm oil remains very concentrated. Malaysia and Indonesia are expected to produce almost 70 Mt in 2019, *i.e.* about 40% of global vegetable oil production. Because of environmental concerns and area restrictions the growth rates are projected below recent trends.

In line with the development of their crush industries, the main non-OECD oilseed oil producers, Argentina, Brazil and China, are expected to reduce the increase in their ouputs to about 2.5% per year. When compared to the base, the increasing EU oilseeds production and crush capacity are expected to lead to a rise in oilseed oil production of 25% over the Outlook period. Vegetable oil is also produced from other tropical oilseeds as described in Box 6.1.

Box 6.1. **Tropical oilseeds and oilseed products**

The 2010-19 Outlook covers a wider range of oilseeds and products than previous editions. Traditionally, temperate climate oilseeds, soybean, rapeseed, and sunflower plus tropical palm oil were covered in the Aglink-Cosimo model. Starting with the current edition, the oilseeds complex also includes groundnut, cottonseed, coconut and palm kernel. While the quantities of all of these products are small in a global context they have significance for certain regions, especially tropical island nations. The inclusion of these additional products enhances the coverage of the OECD-FAO commodity aggregate for oil bearing commodities and their processed products especially for tropical countries and regions. Groundnuts are mainly grown in the US, Argentina, West Africa and East Asia. The oil is of high value, mostly used for special cooking applications. Confectionary groundnuts are an important source of fat and protein in many parts of Africa and Asia.

Cottonseeds are a by-product of the cotton harvest, where lint represents the main product. Cotton cultivation is concentrated in developing countries, notably in the higher latitudes of the Asian continent, but is also present in North and South America and Africa. Cottonseeds are crushed as any other oilseed, primarily to obtain oil.

Coconut palms are grown throughout the tropics. The dried flesh of the coconut (copra) is processed for oil and meal. The oil has industrial and human consumption uses and, together with palm kernel oil, stands out for its high melting point. The meal is fed to animals. Coconuts play an important role in the culture and economy of coastal areas in many tropical countries.

The Philippines are by far the largest producer of copra and its products. It is projected to produce 1.6 Mt in 2019, up from 1.4 Mt currently. Coconut oil accounts for more than 90% of the country's vegetable oil production. Furthermore, the oil is expected to play an increasingly important role as biofuel feedstock. Other notable producers are Indonesia and India. Coconuts are processed all around the tropics where they play an important role in the rural economy. In the least developed countries of the Oceania region, coconut oil accounts for about one third of local oil production, copra meal for over 80% of total protein meal production. At the global level, the share of global coconut oil production in total vegetable oil production is currently of about 2.4% and the share of copra meal in total protein meals production stands at 0.8%.

Palm oil is one of the world's leading vegetable oils, produced from the fruits of the oil palm. These fruits contain a kernel which is also processed into an oil and a protein-rich meal. The oil is mostly used for industrial applications; the meal is fed to animals. Indonesia and Malaysia are the dominant producers. Palm kernel oil and meal are exported to Europe and Asia. Malaysia also maintains a specialised chemical industry for further processing of palm kernel oil.

Palm kernel meal and oil are co-products of palm oil, Malaysia and Indonesia are the leading producers. Combined they currently produce about 4.5 Mt of palm kernel oil, which is expected to grow to 6.3 Mt. Palm kernel meal production amounts to currently about 5.6 Mt, projected to grow to around 8 Mt. At the world level, palm kernel oil production currently represents almost 4% of total vegetable oil production. Palm kernel meal has also currently a share of 4% in total protein meals production.

Rising per capita income and population growth combined with continued growth in non-food uses of vegetable oil are behind the solid vegetable oil demand expected worldwide over the projection period. Per capita food consumption is expected to level off in more and more countries due to saturation effects. Overall, the average annual growth rate of global consumption is expected to weaken compared to the past decade, reflecting relatively slower growth in demand from the biofuel industry.

The share of vegetable oil consumption used for biodiesel production is expected to increase from 9% during the base to 15% in 2019, driven by biofuel mandates and consumption incentives in many countries, *i.e.* almost 16 Mt of vegetable oil will be additionally used to produce biodiesel over the Outlook period, a third of the global use increase. The biofuel chapter describes in details the evolution of biodiesel markets projected over the Outlook.

In the EU, vegetable oil used for biodiesel production is expected to represent 9% of worldwide and 53% of domestic vegetable oil consumption by 2019 (Figure 6.6). Vegetable oil use for biodiesel production is expected to increase by 130% over the course of the Outlook period compared to the base. The strong annual growth rate should slow-down a little toward the end of the Outlook period when second generation biofuels (based on feedstock other than edible crops) are assumed to take-off. The availability of second-generation biofuels is expected to calm down the pressure on edible feedstock-based biodiesel prices and thus on vegetable oil prices.

Figure 6.6. **Vegetable oil use for biodiesel production to increase because of mandates**

Share of vegetable oil consumption used for biodiesel production in selected countries

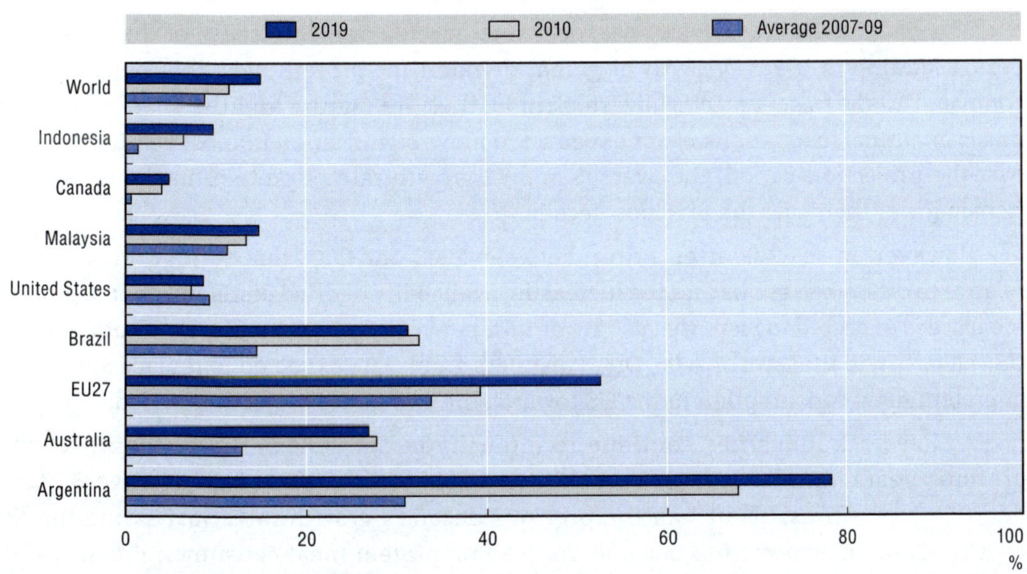

Source: OECD and FAO Secretariats.

StatLink http://dx.doi.org/10.1787/888932286646

Biodiesel consumption in the US should increase to the mandate set by the Renewable Fuels Standard in 2012. Vegetable oil use for biodiesel production should represent about 9% of total vegetable use after 2012. Argentina is expected to continue developing an export-oriented biodiesel industry. Vegetable oil for biodiesel production is expected to reach 3.5 Mt by 2019. Government targets for domestic biodiesel use are projected to encourage the use of vegetable oil for biodiesel production in Brazil which should account for more than 30% of domestic consumption by 2019. The soybean oil based biofuel industries in Argentina and Brazil account for 60% of the expansion in biofuel use of vegetable oils in non-OECD economies.

Malaysia and Indonesia have invested into flexible production structures that will allow channelling more palm oil towards biodiesel production if economically profitable. Currently, biodiesel output is very low, but production is expected to increase over the projection period, mainly to satisfy domestic consumption mandates. Actual outcomes will depend on economic and political conditions. Malaysia also maintains a flourishing oleo-chemical industry that is projected to utilise over 4 Mt per year by the end of the coming decade.

Because of increasing population and income, food use accounts for 90% of the 3.1% annual growth in vegetable oil demand in non-OECD economies (Figure 6.3). Growth is strongly concentrated in Asia, with China expected to remain the world leading vegetable oil consumer with an annual growth rate of 3.2% p.a. Vegetable oil market in India is dominated by the food demand for vegetable oils. Despite expansion efforts, vegetable oil production from locally grown crops covers only about 45% of domestic requirements with a declining tendency. In countries where production is strongly driven by the export market, such as the Ukraine or Paraguay, the weight of domestic vegetable oil consumption is expected to fall further in the future.

Protein meal

The global livestock industry is expected to develop over the course of the projection period leading to a 25% growth of global demand for protein over this period when compared to the base. As OECD livestock industries are mature and the share of protein meals in animal feed rations is not expected to move significantly in most OECD countries over the projection period, the average annual growth rate of protein meals use should remain lower than over the previous decade.

However, in the US, after a drop between 2006 and 2008 due to high meal prices relative to other feed grains and the increasing availability of dried distillers' grains, protein meal use is expected to increase at a rate of 1.6% per year. By 2016 meal consumption in the US will return to the level of 2006. Dried distillers grains, are expected to replace almost 8% of protein meal consumption in the US towards the end of the projection period.

As livestock industries continue to intensify but at a slower pace than over the previous years, annual growth in meal consumption is expected to average 3.2% in non-OECD countries, about half the previous decade's growth rate. Surpassing the EU in 2012, China is expected to become the leading protein meal consumer. Brazil is also projected to see an increase in protein meal use of more than 30% over the projection period compared to the base. In Paraguay, more than half of local meal output is exported; the cattle oriented domestic livestock industry consumes the remaining amount.

In India, protein meal output should be more than sufficient to cover the domestic feed market, leaving about 5.5 Mt for the more lucrative export market. Oilseeds' processing in Ukraine has turned into an export oriented industry. Local consumption of meal accounts for only about 25% of domestic output. With feed demand and production growing at similar rates, that share is expected to remain unchanged during the Outlook.

Trade in oilseeds and products

Oilseeds

Oilseed trade is expected to grow by more than 26.5 Mt over the Outlook. World oilseed trade should continue to be dominated by few major market players: the US, Brazil, Argentina, Canada and increasingly Paraguay. US exports are determined by the relative development of crushing facilities and oilseed production. Their comparable development is anticipated to lead to stable oilseed exports over the projection period at about 38 Mt on average. Despite the rapid development of its crushing industry, Brazil is expected to affirm its position as the global trade growth leader. Over the Outlook period, almost 70% of the total growth in oilseed exports is projected to originate from Brazil, growing from 26% of global exports over the base to 35% in 2019. Brazil should become the largest oilseeds exporter surpassing the US in 2018.

The differential export tax system in Argentina continues to favour domestic processing of seeds and exporting oilseed products rather than exporting oilseeds. It favours even more exports of biodiesel. As a consequence, total oilseed exports are expected to diminish over the projection period at a rate of 3.5% per year. Paraguay is an emerging exporter of oilseeds and products. Its seed exports are projected to reach over 5.5 Mt in 2019. The processing industry is mostly export oriented, but is small in size and is not yet a major player on the global scale. Traditionally, India does not trade in oilseeds. Prohibitive tariffs prevent imports and the domestic market absorbs all domestic production.

Within the OECD, Canada remains the second largest oilseed exporter. Exports are projected to remain stable over the projection period because of the development of the Canadian crushing industry in line with increasing oilseed production. The fast growth of the Ukrainian seed production is projected to provide the country with a growing export potential of almost 5 Mt over the Outlook period.

China and the European Union are the major oilseed importers. Because of the expansion of the Chinese crushing industry over the projection period, Chinese oilseed imports are expected to increase by 13 Mt accounting for most of the global increase in imports over the projection period when compared to the base. Despite the expansion of domestic oilseed production, mainly rapeseed, the European Union should remain a strong and stable importer of oilseeds over the Outlook period.

Vegetable oil

World vegetable oil trade is projected to increase over the projection period as production in the main consumption regions is not anticipated to keep up with demand. Trade is projected to increase to reach 77 Mt in 2019, 40% higher than the base level. World vegetable oil exports are very concentrated with Indonesia and Malaysia providing two-thirds of shipments. Strong demand in India, China and Europe support their expansion in trade to over 50 Mt by 2019. By 2019, Argentina is expected to produce almost 6% of the global vegetable oil and to be the third largest exporter with a share of 8.5% of total exports. Its differential export tax system favours exports of biodiesel and oilseed products rather than oilseeds exports. In Ukraine, vegetable oil production continues to grow leading to exports of 3 Mt over the coming decade.

Due to biofuel consumption incentives and mandates being in place in many countries, vegetable oil trade will be increasingly influenced by biodiesel markets on both the exporting and the importing side. To meet both industrial and traditional vegetable oil demand, EU imports are expected to increase by almost 75%, becoming the leading vegetable oil importer after 2011. Income and population growth support a strongly expanding vegetable oils market in India which has to import more than 11 Mt in 2019. Imports by the three dominant trading countries (EU, China and India) are expected to represent 55% of total world imports.

Protein meal

Over the Outlook period, because of the slowing of global protein meal use, the protein meal trade growth rate is expected to fall to about 2% per year, significantly lower than over the previous decade. Argentina is expected to strengthen its position as the largest protein meal exporter in the world because of its well developed, efficient crushing industry and small domestic market. The differential export tax system affects the exports adversely in general, but provides a relative advantage to the exports of processed products. Argentina should account for more than 40% of global exports by 2019.

Brazil is expected to remain the second largest protein meal exporter. Exports originating from Brazil are expected to increase by almost 20% over the course of the Outlook period despite an increasing domestic consumption. As their domestic crush industry develops US protein meal exports are projected to expand over the projection period at a similar rate as over the previous decade.

The expansion of the Indian livestock industry will keep pace with the production of protein meals, limiting the country's annual meal exports to nearby destinations in South East Asia (where it enjoys a competitive advantage) to about 5.7 Mt. Annual protein meal exports from the Ukraine are projected to grow from 1.7 to 2.7 Mt by 2019 because the processing industry outpaces feed demand and the country is well placed as a supplier for the growing feed markets in the Middle East and Western Europe.

The European Union is projected to keep its position of principal protein meal importer over the Outlook period. Because of increasing domestic rapeseed production and the development of new crushing facilities, oilseed meal imports are expected to remain stable over the course of the projection period. In China, meal demand should be mostly satisfied by domestically produced meal, as well as through an expansion in meal imports. They should reach 7 Mt by 2019.

Key issues and uncertainties

A number of issues and uncertainties that are embedded in the projections presented in the *OECD-FAO Agricultural Outlook* are discussed in this section. Obviously all these uncertainties are inter-related and changes may also arise in light of developments in other sectors of the economy.

Development of the biodiesel sector

Since the early 2000s, vegetable oil-based biodiesel production experienced a considerable expansion. Increased demand for oils and fats from biodiesel producers became one of the driving forces of the global vegetable oil market – with repercussions for the oilseed complex as a whole as well as marked cross-sectoral linkages (notably in the form of competition for land with maize, another highly demanded biofuel feedstock).

Any deviation from current national bioenergy policies and investors decisions may affect the biofuel demand for vegetable oil and thus the global oilseeds and oilseed products market. This was evident from mid-2008 onward when weaker energy prices together with weaker policy support for biofuels and sluggish private investment, caused a slow-down in the expansion of biofuel production, which contributed to the sharp drop in world prices for vegetable oils and fats. Conversely, the renewed firmness in mineral oil prices during the second half of 2009, along with the implementation of higher biofuel blending rates in 2010 triggered acceleration in oil and fat demand from the biodiesel industry. Both factors were contributing to the recent strengthening in world prices for oils and fats. For a more extensive treatment of the major issues and uncertainties associated with the expansion of biofuel markets please refer to the "Biofuels" chapter of this publication.

Supply and demand concentration

The global market for oilseeds and derived products is characterised by a high degree of concentration. Today, soybeans account for well over half of global oilseed output, with production concentrated in just three countries. Similarly, palm oil occupies a key role in global vegetable oil production, with the bulk of output stemming from two countries. The potential for further expansion in the production of oilseeds and oil also remains strongly concentrated in a few regions: the global market will depend heavily on South America and Southeast Asia for additional supplies of respectively, soybean and palm oil. At the same time, the growth potential in these areas could be increasingly constrained by resource limitations and environmental concerns, a prospect that introduces considerable uncertainty into the global market.

Furthermore, as far as international trade is concerned, exports of the worlds' most widely shipped oilseeds, oils and meals tend to be dominated by no more than two or three nations. Given the size of production and exports by the countries in question, any weather anomalies, important shocks to their economies or radical policy decisions would have huge consequences on markets and world prices for oilseeds and oilseed products. Also on the demand side, consumption growth tends to be heavily concentrated in a few, import dependent countries, in particular China and India, which can again lead to increased market instability as any unexpected developments in these countries would strongly affect the global market.

Sustainability issues

Concerns about the environmental and/or social sustainability of current crop production methods are voiced by society in many countries and involve annual and perennial oilseeds in developing and developed nations. Large scale palm oil plantations in Asia and extensive soybean cultivation in South America, for example, are increasingly facing challenges in this regard.

Environmental issues also started playing a prominent role in the discussion on how agriculture should contribute to mitigating climate change and on ways of adapting to it. Policy makers and the private industry in many countries – including producers as well as consumers of oilseeds and derived products – started paying more attention to the problem and efforts to identify appropriate responses are increasing. An array of measures can be conceived: from bans on deforestation in high conservation value areas, to voluntary or compulsory production restrictions in environmentally sensitive areas, to general requirements regarding the application of good agricultural and social practices in production and processing, to the certification and preferential treatment of sustainably produced products. When introduced on a larger scale, such measures could lead to gradual shifts in oilseed production patterns as well as global trade flows.

In recent years, a number of initiatives have been launched by the private sector for application on a voluntary basis. In particular, commercial production and global marketing of palm oil produced in South East Asia following specific sustainability standards is slowly taking hold and the soybean industry in parts of South America is about to embark on a similar path. The speed and direction, and thus the likely market impact of ongoing and future initiatives in this field remain difficult to predict.

Rising demand for new oils with special properties

In recent years, the production of oilseeds and oils featuring particular traits required by the industry for special food and non-food uses started growing in a number of mainly developed countries – a development supported by new genetic engineering techniques that allow cost-effective modification of planting materials. The trend applies in particular to vegetable oils with special nutritional profiles, notably regarding the trans-fat content and fatty acid composition, which have become increasingly important to health conscious consumers. Relatively small but lucrative markets for such products are beginning to emerge, especially among developed countries.

The further evolvement of the market for specialty oils depends on a variety of country specific factors of a technical, economic, regulatory and social nature, which implies considerable uncertainty in any projections. A continuation or intensification of recently emerged trends could eventually lead to adjustments in the global pattern of oil and oilseed production and trade, especially if the health concerns currently present in developed countries should start spreading to the developing world.

OECD-FAO Agricultural Outlook 2010-2019

Chapter 7

Sugar

Market situation

The world sugar market is experiencing considerable turbulence and stress at the start of the Outlook. The sugar market is currently facing a second year of global deficit with a wide gap remaining between world consumption and production that has reduced stocks to very low level). World sugar prices have rallied strongly in response to tightening supplies, and rising imports, to reach 29 year highs in February 2010 (Figure 7.1). Prices have since fallen back to around the pre-peak levels of mid-2009 on expectations of improved supply prospects, particularly in Brazil. The past several months have also witnessed considerable volatility in the white sugar premium. The premium advanced rapidly in the second half of 2009 on fears of low stocks and tightness of export supplies of white sugar to reach over USD 146/t in January 2010, the highest nominal margin between raw and white sugar prices since July 1995.

Figure 7.1. **Consecutive sugar deficits set stage for price surge in 2009**

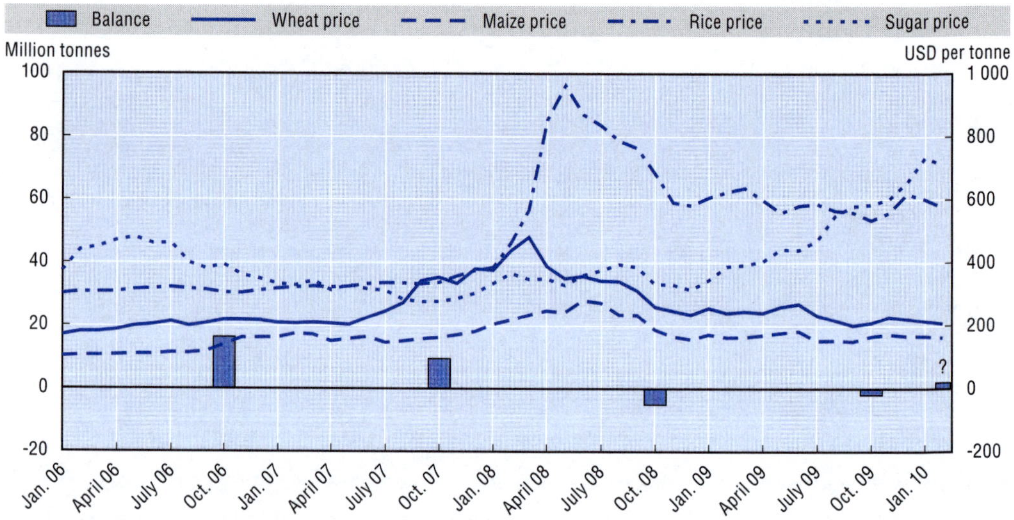

Note: Balance is annual surplus or deficit for crop year October/September.

Source: OECD and FAO Secretariats.

StatLink http://dx.doi.org/10.1787/888932286665

While part of the cause of the global supply shortage can be assigned to bad weather in key producing countries, such as Brazil and India, reduced incentives for producers were also partly to blame. World prices of sugar, historically one of the most volatile of all commodity markets, have been out of sync of late with those of the major temperate zone crop commodities. For instance, world sugar prices were low in 2007/08 when other commodity prices surged to historical peaks and then reached historical highs in December 2009, while the others declined. The fact that sugar prices did not rally with other crops in 2007/08 sent a clear signal to farmers at the time to switch out of sugar into other production. This contributed to a collapse in production and the world shifted to a global deficit between production and consumption in 2008/09. Initially stocks were drawn down to buffer the production shortfalls and this tightening of supplies set the stage for the eventual surge in prices. However, the rise in prices in late 2009 came too late to affect many planting decisions in the 2009/10 crop year (1 October to 30 September), slowing the global production response. As a result, it is now unlikely that production will be able to respond fully to the current high prices before 2010/11 (Figure 7.1). Consecutive sugar deficits over the past two years have lead to sharply lower global stocks at the commencement of the Outlook and a lower global stocks-to-use ratio, which should continue to be supportive of market prices, at least in the near term.

Projection highlights

- World sugar prices are expected to remain volatile over the coming decade and to be shaped by policy driven market and trade adjustments in Europe and North America, the continuing sugar production cycle in countries of Asia (primarily India and some neighbouring countries), and the extent to which Brazil's large sugarcane crop is allocated to ethanol production, influencing sugar export availabilities. World sugar consumption continues to grow but is dampened at the start of the Outlook by the current high prices and the lingering effects of the global recession. Demand then strengthens in following years with economic recovery leading to some rundown in sugar stocks by the end of the projection period that help to lift prices by 2019/20.
- Projected sugar prices to 2019/20 are, nevertheless, lower than the near record levels attained at the start of Outlook, but still exceed the levels of the previous decade. The world indicator price for raw sugar in nominal terms (Intercontinental Commodities Exchange Contract No. 11, spot, f.o.b. Caribbean ports) and white sugar (White Sugar Futures Contract No. 407, Euronext market, Liffe, London), are projected at USD 372 per tonne and USD 439 per tonne, respectively, in 2019/20, a decline of around 5-7%, when compared to the average for 2007-09 (base period) (Figure 7.2). World sugar prices in real terms (once adjusted for inflation) are expected to continue their long term decline, but at a slower pace (Figure 7.3).
- The margin between raw and white sugar prices, or the white sugar premium, should narrow in the near term, as export availabilities and the trade balance for white sugar improves. The premium is expected to average nearly USD 67/t over the projection period.
- Larger sugarcane production should account for the bulk of the growth in sugar output. The world sugarcane crop is projected to expand by 32%, whereas that for sugar beets actually contracts slightly by around 7% by 2019/20, when compared to the base period. World sugar production is projected to increase to by some 39 Mt or 24% by 2019 above the average level for the base period. The longstanding production cycle in some countries of Asia, an important dynamic in the world market, will influence the growth in, and pattern of global production in particular years.
- World consumption is expected to grow relatively fast, averaging 1.8% p.a., to 2019 to reach about 198 Mt in 2019/20. The fastest growth in use is expected in the developing countries, whereas OECD countries experience only slow growth in sugar disappearances to 2019/20.

Figure 7.2. **Nominal sugar prices remain variable and average higher than the last decade**

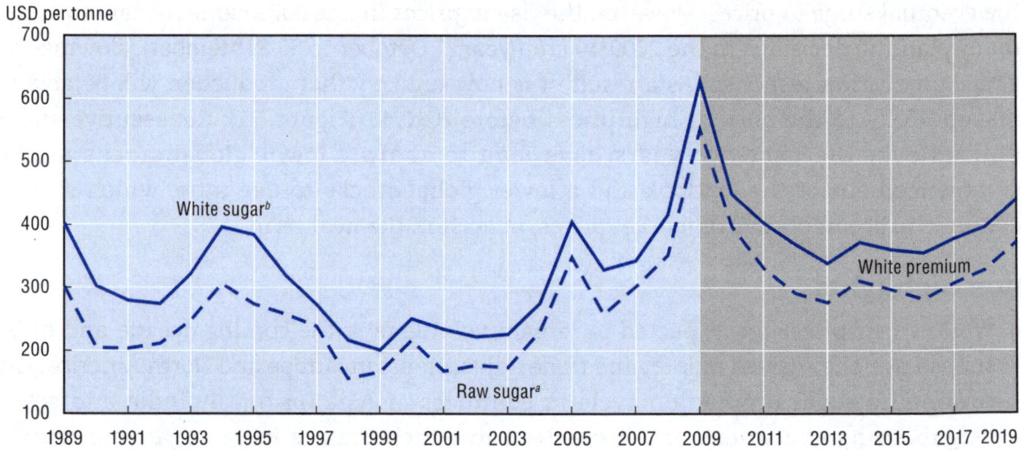

a) Raw sugar world price, ICE Inc. No. 11, f.o.b., bulk spot price, October/September.

b) White Sugar Futures Contract No. 407, Euronext market, Liffe, London.

Source: OECD and FAO Secretariats.

StatLink http://dx.doi.org/10.1787/888932286684

Figure 7.3. **World prices to trend lower in real terms**[c]

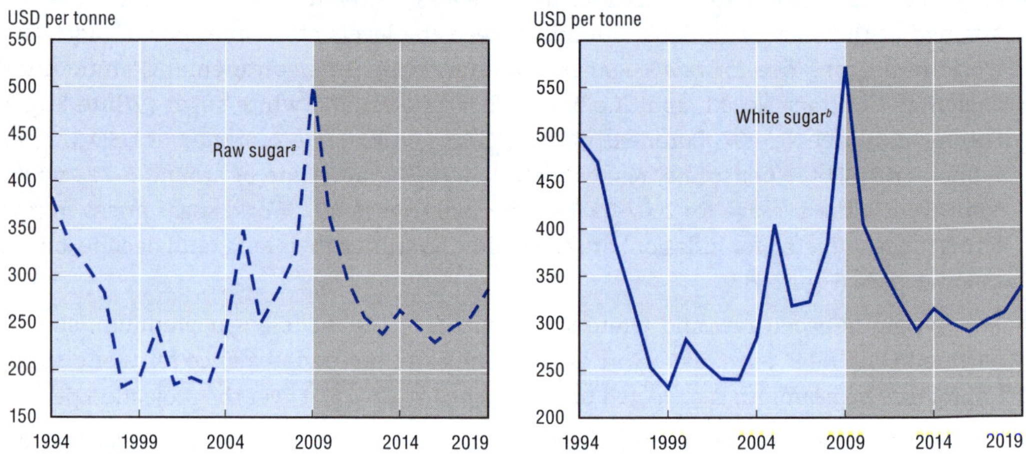

a) Raw sugar world price, ICE Inc. No. 11, f.o.b., bulk spot price, October/September.

b) White Sugar Futures Contract No. 407, Euronext market, Liffe, London.

c) World prices deflated by US GDP deflator (2005 = 1).

Source: OECD and FAO Secretariats.

StatLink http://dx.doi.org/10.1787/888932286703

- Low global stocks in 2009/10 are rebuilt in the early years of the Outlook period as world production outpaces the growth in consumption in response to the current higher prices. The global stocks-to-consumption ratio rises from a low 39% in 2009/10 to 44% by 2012/13, and then gradually declines to 38% by 2019/20 (and below the average of 45% for the last decade), as steady consumption growth eats into the surplus and as additional sugarcane (and beet in some countries) is allocated to ethanol production (Figure 7.4).

Figure 7.4. **The global stocks-to-use ratio to rise in near term and then decline**

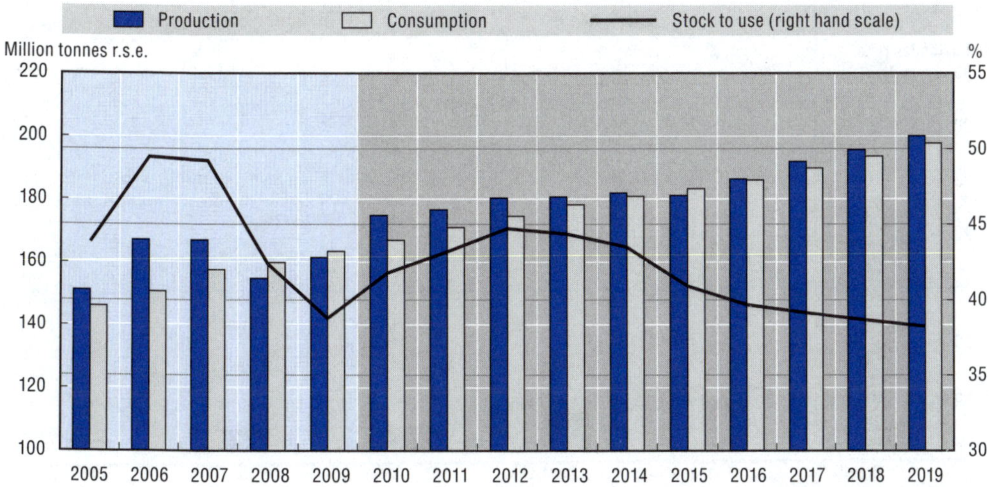

Source: OECD and FAO Secretariats.

StatLink http://dx.doi.org/10.1787/888932286722

- The non-OECD region's dominant share of global production, consumption, and exports increases further by 2019/20 to reach new records of 81%, 77% and 89%, respectively (Figure 7.5). Brazil is the world's largest producer and exporter with trade projected to increases by 41% to 2019/20, when compared to the base period, and accounting for 50% of global trade. Brazil together with Thailand, Australia and South Africa, collectively account for most of the growth in world exports (Figure 7.6). Imports remain more diversified over a larger group of countries, but with much of the growth accounted for by the EU, the United States and countries in Asia, particularly India and China. India's imports are the largest over the projection period, but remain periodic and are essentially driven by its longstanding production cycle. Imports by the Russian Federation continue to decline in favour of domestic production (Figure 7.7).

Figure 7.5. **The non-OECD countries are the leading sugar producers and consumers**

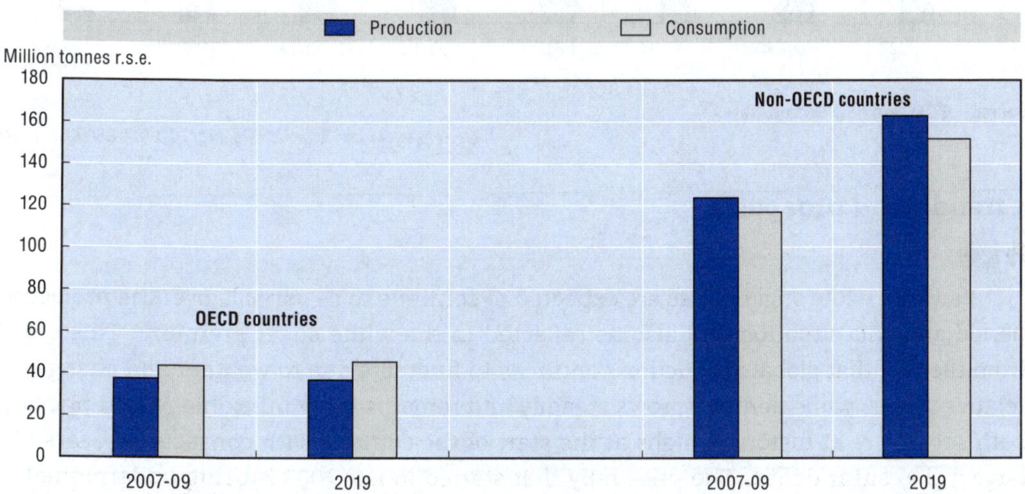

Source: OECD and FAO Secretariats.

StatLink http://dx.doi.org/10.1787/888932286741

Figure 7.6. **Sugar exports remain highly concentrated and dominated by Brazil**

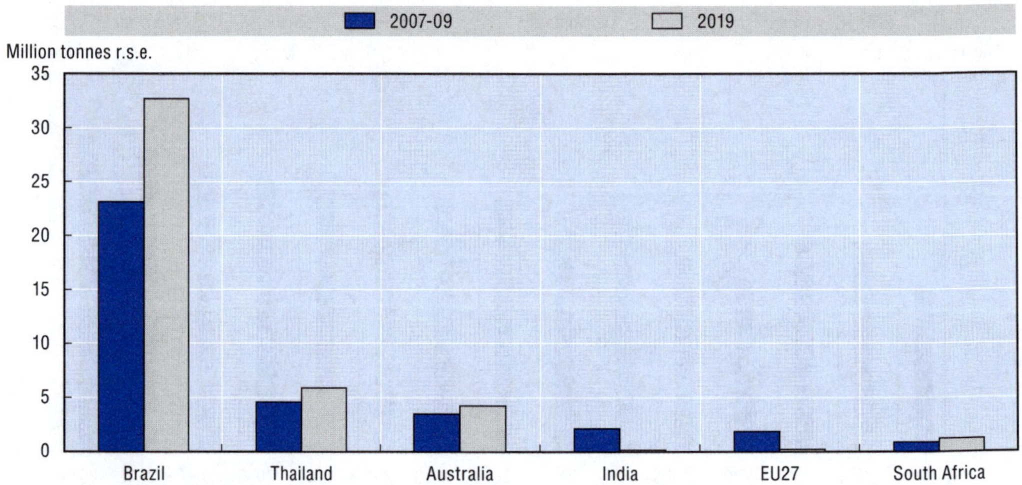

Source: OECD and FAO Secretariats.

StatLink http://dx.doi.org/10.1787/888932286760

Figure 7.7. **Imports are more diversified and lead by India, European Union and United States**

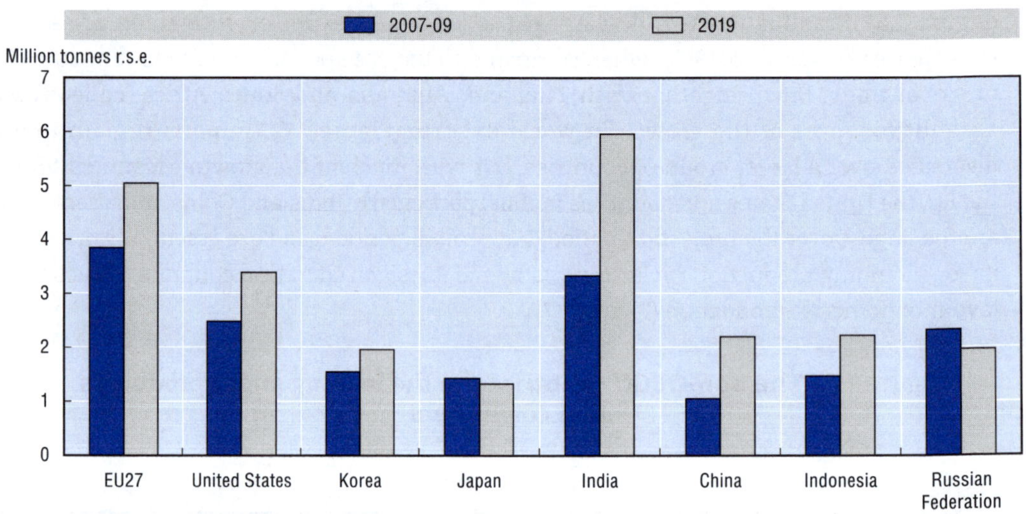

Source: OECD and FAO Secretariats.

StatLink http://dx.doi.org/10.1787/888932286779

Market trends and prospects

Prices

Raw and white sugar prices are expected to continue to be variable over the projection period and this situation will also be reflected in the white sugar premium. This arises from the fact that global production continues to fluctuate due to weather and changes in relative prices, while demand grows steadily and remains fairly inflexible. World prices of both sugars are at historical highs at the start of the Outlook with consecutive years of a large global sugar deficit. The price rally that started in late 2009 is being underpinned by strong import demand from India after a sharp fall in production and from a number of smaller importers as well in a general context of lower export availabilities from Brazil and

Thailand, the two largest exporters. The current high prices can be expected to lead to a supply response at some point. As the high prices occurred too late to allow most growers to adjust production for the 2009/10 crop year, a more complete supply response is not expected to occur before 2010/11. Once production adjusts, the world sugar market is likely to switch from a deficit to an overall surplus situation, increasing export availabilities and relieving some of the current price pressures.

Notwithstanding a sharp decline in world prices from the initial high starting levels, and their continuing variability in response to changes in the market balance and other influences, prices are expected to average higher over the next ten years than the depressed levels of much of the last decade. Steady growth in global sugar consumption along with expanding demand over time for alternative products of sugar crops such as biofuels along with rising costs in major supplying countries are expected to maintain sugar prices at elevated levels or on a higher plateau, in nominal terms, relative to those at the turn of this century.

The sugar price projections are also predicated on certain developments taking place in Brazil. Brazil's role in the world sugar economy has grown rapidly over the last twenty years and it has now attained the status of a "price setter" on the world market. Brazil's sugarcane crop is split between the co-production of sugar and ethanol based on sugarcane juice. As most sugar mills in Brazil can produce both products, this gives the country considerable flexibility to switch production between the two end uses depending on relative prices and mill profitability. The world price projections assume that Brazilian sugarcane production, sugar output and exports will all expand strongly over the medium term increasing Brazil's share and dominance of the world sugar market. However, it is also anticipated that relative profitability will favour ethanol production over sugar, on average, in the decade ahead. As a result, the share of the larger cane crop that is allocated to sugar production, currently less than 50%, is expected to continue to contract over the period as ethanol production increase to 55 bnl, for domestic use and export by 2019-20. However this development is not expected to constrain sugar production and exports in coming years.

The margin between raw and white sugar – the white premium – is expected to narrow over the near term as additional white sugar supplies become available from traditional exporters and from an increasing number of destination and toll refineries in North Africa and the Middle East. These refineries are filling the gap left by the withdrawal of large EU white sugar exports from the world market with the CAP sugar reforms of 2006, and can be expected to have a growing demand for raw sugar for further refining into white sugar for domestic use and export. The white sugar premium is expected to reflect mainly the cost of additional processing and refining of raw sugar and transport cost in moving sugar from origins to destinations.

Production

Sugar crops in most of the world's leading producing countries are projected to expand in 2010/11 in response to the historically high prices switching the world market balance from a deficit to a surplus and lowering world prices. Further growth in production to just over 200 Mt is projected for 2019/20 to meet the strong increase in demand for sugar and other uses. Nearly all the increase in sugar production will come from larger sugarcane harvests, rather than sugar beets, expanding the share of cane sugar in the world total (Figure 7.8). Amongst the world's sugar producers, developing countries will account for essentially all of the growth in output, and with a major share coming from Brazil. Brazil's dominant role in the world sugar economy is expected to increase with production

Figure 7.8. **Larger sugarcane production accounts for most of the higher sugar output**

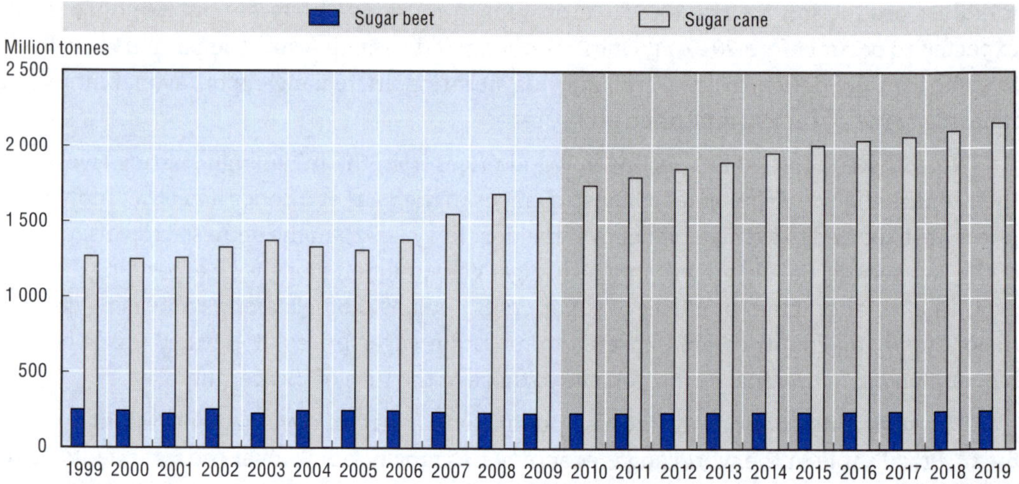

Source: OECD and FAO Secretariats.

StatLink http://dx.doi.org/10.1787/888932286798

expanding by an additional 11 Mt or 31% to 2019/20, when compared with the base period. Despite the higher output from Brazil and other leading producers such as Thailand, Australia and South Africa, the pattern of global growth and production in particular years will likely remain highly variable. The main reason for this is the production cycle in certain countries of Asia, particularly in India. Sugar production in India is subject to violent swings every 4-5 years as a result of cyclical factors and along with similar, albeit less pronounced, cycles in China and Pakistan, these contribute to periodic sugar market imbalances, trade adjustments and world price turbulence, as was the case in late 2009. These cycles are expected to continue to influence world markets and result in price variability over the coming decade. India's production, although variable, is projected to increase over time to reach nearly 32 Mt in 2019/20, or 52% higher when compared to the low production in the base period. Chinese sugar production has grown rapidly in recent years under the stimulus of high support prices and strong domestic demand. However, the rate of growth of sugar crops is expected to slow in coming years and to rely mainly on yield growth rather than area expansion. China's sugar production is expected to reach 18 Mt in 2019, some 28% above the average level for 2007-09. Sugar production in Pakistan is projected to increase by over 21% between 2007-09 and 2019/20, but to display continuing variability over the period.

In terms of other countries, production in the EU is projected to stabilise at around 14.4 Mt to 2019/20 as production quotas have been reduced by over 5 Mt to equilibrate the domestic market in a context of stable consumption, fixed exports and rising imports. Sugar production in Australia is also expected to show little change from current levels due to continuing pressure on land availability for cane production. Despite some increase in sugar support prices and a guaranteed share of the US market, US farmers are not expected to expand the area harvested of sugar crops to keep pace with the anticipated increase in domestic demand due to high returns from competitive crops. Rather, they should focus on maximising higher margins and production remains below the 85% minimum allotment level of domestic consumption in the FCE Act of 2008, over the projection period. Similarly, production in Mexico is expected to grow only slowly despite further investment in the sector and the prospects of higher and more certain returns from sales to the US market, following full market integration under NAFTA.

Production in Mexico is projected to rise by about 4% above the base period to 5.7 Mt in 2019/20. In the case of Thailand, sugar production has shown considerable volatility over the last decade, as a result of variable weather, changing support policies and competition for resources from other crops such as cassava, fruits and vegetables. Projected growth in the bioethanol is not expected to limit sugar output, as it is based on molasses, a by-product of sugar processing, as the main feedstock. Sugar production is projected to increase by 14% to 9 Mt in 2019/20, compared to the base period, in order to meet rising domestic and regional demand.

Consumption

While consumption growth has been slowed at the start of the Outlook, faster growth is expected in following years once the recession recedes and economic recovery becomes well established around the world. In contrast to more variable production, consumption has exhibited steady, year-on-year growth and remains the main driver of the world sugar market. Global sugar consumption is projected to increase by 23% or 38 Mt by 2019/20. The developing countries account for the major share and growth in world sugar use, due to a combination of faster population growth, changing diets and rising incomes; albeit with considerable variation between individual countries. The global share of sugar off take in this group of countries is projected to increase from 73% in 2007-09 to 77% in 2019/20 and with growth in use to average over 2.2% p.a. The fastest growth in consumption is expected in the growing sugar deficit region of Asia. In contrast to these demand trends, sugar consumption in the OECD area is expected to remain relatively stable to 2019/20 as population growth declines and with dietary shifts away from sugar due to increasing health awareness and concerns with obesity and related issues.

Trade

Exports

World sugar exports remain highly concentrated in a few countries and are increasingly dominated by a single emerging economy. Brazil's sugar exports which have grown markedly since the beginning of this century have outpaced those of all other exporters to become the leading world exporter of raw and white sugar. Brazil is expected to boost exports in the near term as it responds to the high prices at the beginning of the Outlook. In following years exports are projected to increase to 32 Mt in total, and comprising 20 Mt of raw sugar (up 32% from 2007-09) and 12 Mt of white or refined sugar (up 70% from 2007-09), respectively, by 2019/20 (Figure 7.9). At the same time Brazil's dominant share of world exports are projected to grow from 41%, on average, over the base period to 50% in 2019/20. The growing concentration of sugar exports is not without risks for sugar users and may be another factor contributing to future market volatility as world supplies depend increasingly on the growing conditions of a single country. One possible counterweight is that a majority of Brazil's sugarcane is now used for ethanol production and many mills have the capacity to produce both sugar and ethanol. This capacity to switch between end uses of sugarcane should help assure sugar production and export availabilities when relative prices favour sugar over ethanol.

Figure 7.9. **Both sugar and ethanol production to increase in Brazil**

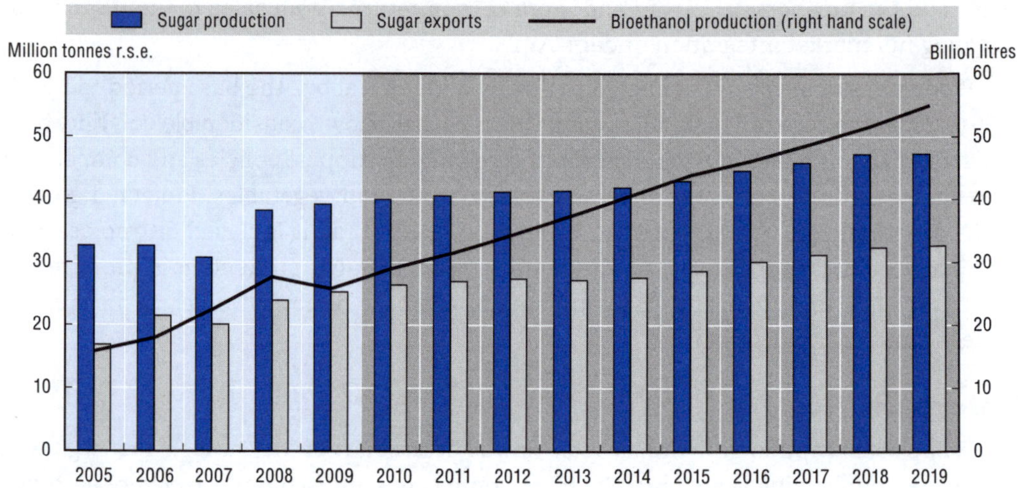

Source: OECD and FAO Secretariats.

StatLink http://dx.doi.org/10.1787/888932286817

The other major sugar exporters of Thailand, Australia and South Africa and several smaller suppliers such as Argentina, Guatemala are also expected to follow suite and expand their shipments to the world market over the medium term. Thailand's sugar export should reach nearly 6 Mt, mainly in the form of raw sugar, with the bulk of the shipments likely to be directed to higher priced regional markets. With average seasonal conditions, Australian sugar exports, also mainly of raw sugar, are projected at 4.2 Mt in 2019/20. Despite rising consumption, increased sugar production in Argentina leads to higher sugar exports that reach over 0.6 Mt in 2019/20. The EU sugar industry has been transformed since 2006 from a major into a minor player in the white sugar export trade, due to the imposed limit on subsidised exports of 1.3 Mt.

Following the reform of its sugar sector over four years from 2006, the EU's role in the world sugar economy has changed dramatically as it has emerged as a large importer of essentially raw sugar to meet domestic food requirements (Figure 7.10). Sugar imports may slow somewhat in 2009/10 in the context of high world prices that make the Community a less attractive destination for exports and with a bumper domestic harvest of sugar beets available to boost local production. Excess domestic production and high world prices encouraged the EU to increase exports of out-of-quota sugar in 2009/10. Once world prices retreat from current levels, the relatively high prices in the EU, even after a 36% price cut as part of the sugar reforms, will make it an attractive destination once more. EU imports from all origins, including under preferential access agreements, such as those with the EBA, EPA and the Balkan countries, and other arrangements are projected to reach 5 Mt by 2019/20, making the European Union a large sugar importer.

With the full unification of the United States and Mexican sugar markets under NAFTA and the adoption of the higher sugar price provisions of the FCE Farm Act in 2008, the US has become an attractive destination for Mexican sugar exports (Figure 7.11). At the same time, the US market remains heavily insulated from the world market by prohibitive over quota tariffs and safeguard measures with imports restricted to the minimum tariff rate quota (TRQ) under the WTO and Central American Free Trade Agreement (CAFTA) to around 1.2 Mt p.a. Total US sugar imports are projected to reach 3.4 Mt in 2019/20, but with

Figure 7.10. **EU sugar reforms lead to lower quota production, fixed exports and rising imports**

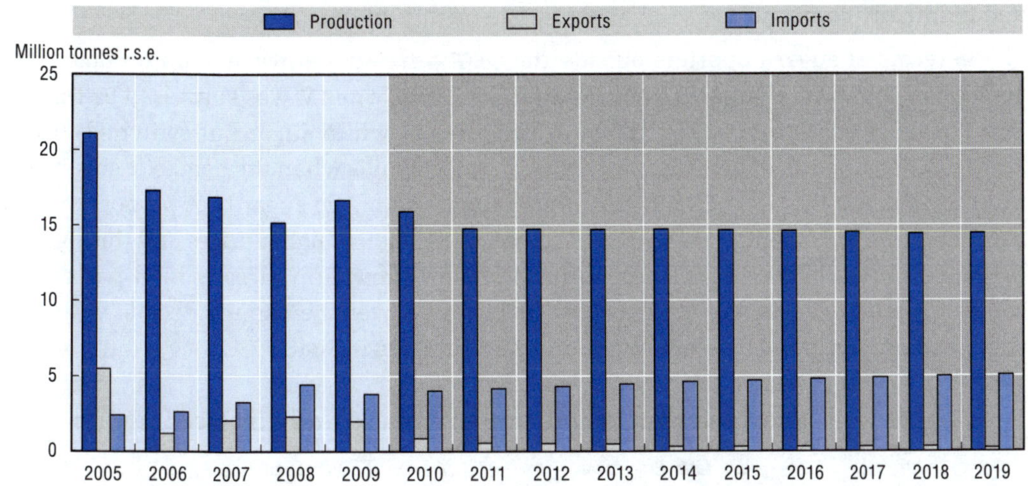

Source: OECD and FAO Secretariats.

StatLink http://dx.doi.org/10.1787/888932286836

Figure 7.11. **Rising sugar consumption in the United States by import from Mexico**

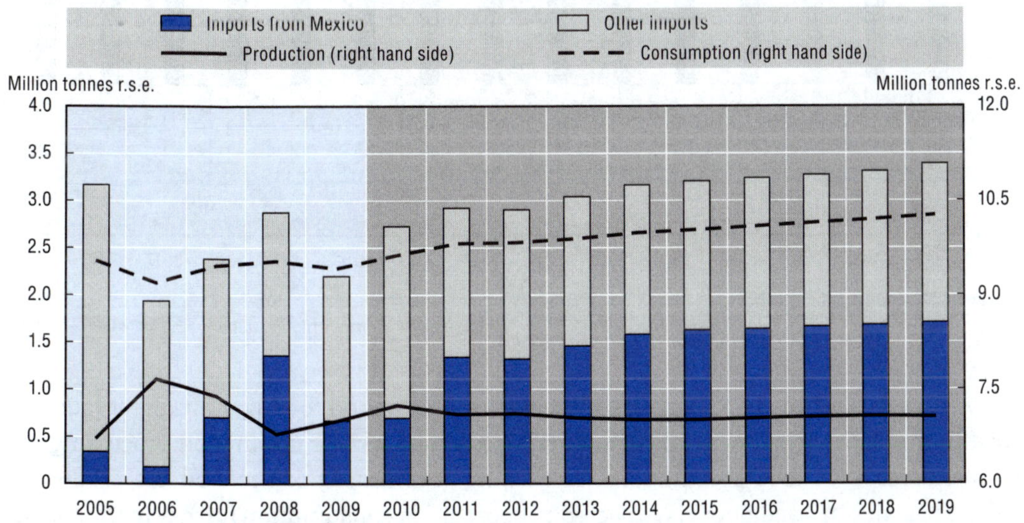

Source: OECD and FAO Secretariats.

StatLink http://dx.doi.org/10.1787/888932286855

little growth in domestic production, these imports do not trigger the Feedstock Flexibility Program (FFP – a sugar to ethanol programme) under the FCE Act. This programme is intended to remove surplus sugar imports from the market to ensure US market prices remain above higher support levels, without accumulating stocks. Within NAFTA, Mexico represents the logical source to supply of additional sugar to the US market. Mexican exports to the US market are projected to expand in response to the higher US market price and as high fructose syrups (HFCS), sourced from US imports and domestic production displace sugar in domestic beverage and food manufacture. Backfilling from the world market is used periodically by Mexico to assure domestic requirements in periods of reduced production. The other member of NAFTA, Canada remains open to raw sugar imports to supplement low domestic production for further processing into white sugar and for mainly domestic use.

In terms of sugar importers outside the OECD, the Russian Federation occupied the position of the world's largest sugar importer until 2007 when it was surpassed by the EU. The Russian Federation was initially an importer of white sugar but switched in the early 1990s to imports of raw sugar for refining by local mills when the domestic sugar beet processing season is over. The Russian Federation's domestic sugar beet production has been increasing in recent years with tariff protection ensuing higher prices and this has led to steady import replacement in a context of slow growth in already high per capita consumption. These trends are expected to continue over the coming decade with raw sugar imports declining to a level of around 2 Mt in 2019/20 (Figure 7.12).

Figure 7.12. **Russian sugar production to expand and displace imports**

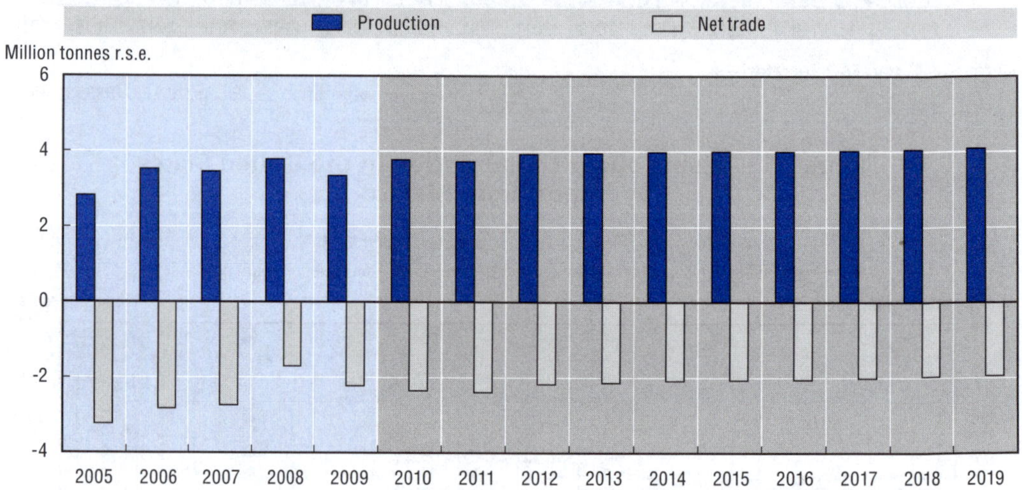

Source: OECD and FAO Secretariats.

StatLink http://dx.doi.org/10.1787/888932286874

Within Asia, China has been experiencing strong demand for sugar for direct food use, albiet from relatively low per capita consumption levels, and for use in food processing and the manufacture of food preparations. Tightening government controls on the production and use of artificial sweeteners is expected to boost sugar consumption further in China with sugar use growing by nearly 2% per year to reach over 20 Mt in 2019/20. This level of use would imply rising imports that will exceed the tariff rate quota of 1.95 Mt established when China joined the WTO. With substantial year to year swings in production as a result of its sugarcane production cycle, India is required to periodically turn to the world market to dispose of surplus sugar or to make large scale imports. The latter was the case for 2009/10, when annual production fell for a second year and stocks were unavailable to buffer the shortfall, leading to estimated imports of 6-7 Mt. With a continuation of the production cycle, large, but irregular, imports are anticipated on a periodic basis over the projection period. These are projected to reach around 6 Mt in 2019/20, making India the largest sugar importer. In Indonesia demand is expected to continue to outpace the growth in domestic production, leading to additional imports that rise to 2.2 Mt, or by over 51%, by 2019 when compared to the base period.

Key issues and uncertainties

The sugar projections discussed in this chapter are a conditional scenario based on a number of assumptions regarding the evolution of the macroeconomic environment and exchange rates, a continuation of existing agricultural policies, average weather conditions, longer term productivity trends and the absence of market shocks. Should any of these assumptions change, the resulting set of sugar projections would also be different. A number of major uncertainties remain. At the beginning of the Outlook, the lingering effects of the global financial market turmoil and the deep economic recession are intermingled with the onset of a strong recovery in the large developing countries and a more hesitant or fragile turnaround in much of the OECD area. At some stage, many OECD countries will need to implement exit strategies to address excessive budget deficits and restore fiscal balance and unless offset by growth in private demand this can be expected to impact the national economies and agriculture as well. These developments are likely to slow growth in consumption and trade, while other external factors such as oil and energy prices, freight rates, inflation, interest rates and access to credit and exchange rate changes can also have profound impacts on the competitive positions of national sugar industries and their trade performance. Among other uncertainties are future changes in sugar policies and the eventual outcome of the current Doha Round of international trade negotiations. All of these factors play an ever increasing role in sugar price formation. The sugar market, therefore, can be expected to face a period of continuing price instability and uncertainty given a growing number of price determinants in addition to traditional market fundamentals.

When and where will supply respond to high prices?

World sugar prices reached historical highs in later 2009. History suggests that a sustained period of high prices at the beginning of the Outlook can be expected to trigger a supply response at some point in time and this will force prices down from current levels. One question is, when and where will this take place? As noted, the high prices of late 2009 are thought to have come too late to affect many producers planting decisions for the 2009/10 season. Most commentators now expect sugar supplies to respond at the earliest in 2010/11 (Figure 7.1). Having answered "when?" the other part of the question is "where" will the response come from? In essence, the supply response should take place in those countries where local sugar prices have been influenced by the high prices on the world market. This suggests that a supply response will occur not only in exporting countries open to world markets, but also in the importing countries that operate behind trade barriers such as tariffs that allow some pass through of high world prices to domestic producers to increase production incentives. With sugar prices currently out performing most other crops, there will be a strong incentive to plant sugar crops in 2010/11.

In terms of the major exporters, Brazil can be expected to have additional production available, as in past years, although the credit crunch has slowed investment in new mills and plantings. On the other hand, if high sugar prices are sustained they will offer greater return to Brazilian mills than ethanol leading to some switching of cane use in favour of sugar production. Another exporting country with large scale expansion potential, although not of the same order of magnitude as Brazil is Thailand, provided returns from sugar remain more attractive than other crops. With both the Australian and South African sugar industries facing constraints on expansion, higher prices will provide some temporary relief and enhanced profitability. However, neither of these industries is likely to be a major source of additional global sugar supplies. Although the EU could expand production as was the case in 2009/10, exports are technically constrained by a WTO ruling.

On the importers side, a number of relatively small producers such as Indonesia and Viet Nam could also expand production to some extent in the face of high world sugar prices. Similarly China has seen a rapid increase in output in recent years and could see some rebound in production, along with Pakistan, if producers have the right price incentives. However, of all the importers, the country with the greatest potential to boost production without additional investment is India. It suffered a major downswing in production in 2008/09 to around 16 Mt and has shown in the past it has the capacity to produce around 30 Mt. However, whether Indian farmers exploit this potential to its fullest will depend on good weather and high relative returns for cane *versus* competing crops such as rice. If India's response is only average, overall supply response may be sufficient to tilt the world into a small global surplus which would not be sufficient to fully rebuild depleted stocks. In this situation, high prices will need to be maintained to bring demand into line with the available supply. Should the response be even lower, prices can be expected to remain at elevated levels until 2011.

Is there a counterweight to Asia's sugar production cycle?

There exists an important dynamic that lies at the heart of the world sugar economy: the longstanding production cycle in some countries of Asia (India, China and Pakistan), but predominately in India. These regular cycles are a leading cause of world market imbalances and along with widespread policy distortions, result in the price turbulence that is a periodic feature of the world sugar market, as witnessed again in 2009 (Figure 7.13). There are two main reasons for India's widely fluctuating production – weather conditions or the timing of the arrival of the monsoon season and government sugar pricing policies. While little can be done to influence the weather, government policies could be made more effective.

The Indian government intervenes heavily in the domestic sugar market in many ways, including the setting of sugarcane prices. Administered sugarcane prices set to guaranteed farmers' income are often in conflict with market determined sugar prices.

Figure 7.13. **India's production cycle influences world sugar prices**

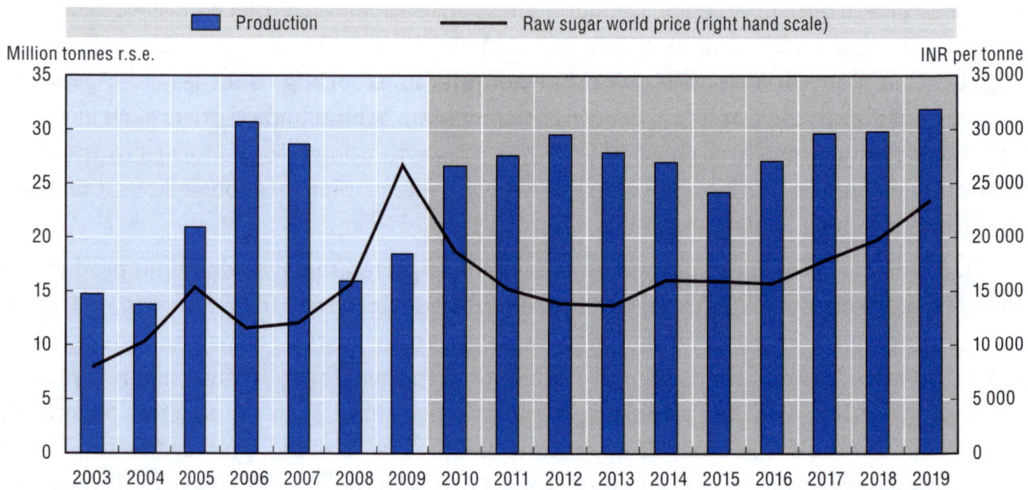

Source: OECD and FAO Secretariats.

StatLink http://dx.doi.org/10.1787/888932286893

This situation, in periods of large domestic crops and low sugar prices, can squeeze sugar mills profitability and their capacity to pay growers for the sugarcane they deliver (leading to large cane payment arrears or IOUs by the mills), that triggers an exodus from sugarcane production to more profitable alternative crops. This action then signals the next round of the sugar production cycle. Similar, although less pronounced, production cycles are a feature in China and Pakistan as well. Because of the large scale of India's production and consumption, there is a clear link between the sugar production cycle in Asia and the global supply and demand balance, with the former accounting for most of the global swings or market imbalances. These cycles, which appear to have increased in amplitude over the historic period, are projected to continue to influence trade volumes and world markets over the coming decade.

Initially importing countries can use accumulated stocks to limit their recourse to imports to address any emerging imbalance from a production shortfall, and exporters can do the same to increase export availabilities within a crop year. Brazil, unlike other exporters, has an additional weapon in its arsenal to draw on to increase its sugar output significantly in a short period of time. It can do this by changing the mix of sugar and ethanol production from sugarcane. The main requirement for this to occur is that world sugar prices must offer mills a more attractive return than ethanol. At present more than 50% of Brazil's enormous sugarcane crop is used for ethanol production and some of this cane could be redirected to sugar production within a crop year and, thus, become available for export. In theory, this flexibility can provide a counterbalance to the Asian production cycle and associated market volatility by allowing Brazil to align its export availabilities with the periodic changes in Asian import demand. While the extent of this flexibility is necessarily limited and depends on the number of mills with both sugar and ethanol production capacity, it can be expected to increase as the industry expands further with increased investment in mill processing capacity in future years. The potential exists in Brazil, therefore, to provide a counterweight to production and price volatility arising from regular production cycles in Asia.

What were the policy responses to high sugar prices?

With the price spikes of 2007/08 in the world cereal and rice markets, many governments responded with a range of trade measures in importing and exporting countries to either safeguard or encourage additional supplies and protect domestic consumers from the high prices. By and large, measures taken by exporters encouraged the hoarding of cereals, reducing available supplies and further inflaming already volatile markets. To some extent, similar emergency measures have also been implemented by governments in response to the sugar price surge to 29 year highs in 2009. However, there was one major difference in that the response was mainly by importing countries to reduce the level of import protection by cutting tariffs and duties to counter high prices and encourage import demand. In contrast to cereals, no major sugar exporter took measures to restrict exports to increase domestic supplies and lower local prices. It would appear that government interventions in domestic sugar markets may have learnt a lesson from the experience of the cereal price spikes by implementing measures that do not exacerbate price volatility on world markets. To what extent the measures that were taken have managed to insulate domestic consumers from the volatility of world sugar prices in this period is, however, less clear.

Chapter 8

Meat

Market situation

The economic downturn triggered by the financial crisis severely impacted the meat sector. Falling purchasing power and difficult access to credit affected both demand and supply. All meats were affected, although beef suffered the most compared to others, as consumers preferred cheap beef cuts and cheaper alternative sources of animal proteins. The economic crisis also accelerated structural changes in the meat industry, as evidenced by the numerous mergers and acquisitions recorded last year, of which the most notable was Perdigao and Sadia that formed the world's largest poultry processor by market value. The structure that results from this crisis, characterised by increasing economies of scale, is expected to speed the globalisation of meat trade, as large scale operations both in production and marketing are better suited to exploit global agribusiness growth opportunities, through a larger portfolio of meat origins and meat types. In addition, economies of scale are anticipated to increase the ability of the industry to manage risk through spatial and meat product diversification, as well as through hedging on futures markets.

Key meat markets are set to recover quickly in the first years of the projection period, and an understanding of this phenomenon requires a review of recent markets events. Over the past two years, meat market prices failed to show the exceptional developments of many other agricultural commodity prices. Their relative stability can be partially explained by the fact that meat plays a more limited role as a staple food, which combined with limited storage capacities makes panic buying for stockpiling unlikely. In addition, price movements were also buffered by changes in meat trade. On the supply side, falling profit margins resulting from inflated feed costs pushed beef farmers to liquidate herds, increasing meat supplies and putting a downward pressure on world market prices at the time when other agricultural commodity prices were increasing. Feed is a key input cost of production systems, not only in pig and poultry sectors but also in the bovine meat sectors in certain OECD countries where animals are finished to higher weights for slaughtering with a heavy use of processed feed concentrates: faced with sudden increases in feed costs producers responded as usual, through herd liquidation. After this episode of herd liquidation, supply became tighter but prices failed to increase because of the contraction in demand induced by falling incomes/purchasing power around the world. All meat sectors were affected, as consumers, confronted with declining budgets, opted for reducing their intake. Indeed, preliminary data for OECD countries in 2009 suggests that aggregate meat production was lower after decades of continuous growth.

The global meat market grew by 1% in 2009, *i.e.* a lower rate than the 2.4% observed in 2008. Animal health and sanitation issues (excluding A-H1N1) did not affect the meat market in 2009 as much as in recent years and should this continue, prospects for the meat sector in 2010 appear brighter as major economies return to a positive growth path, especially in the Southern Hemisphere.

Projection highlights

- Major meat prices expressed in nominal terms are anticipated to be firm the first years of the projection. Nominal prices for beef and pork increase by 22% by 2019 relative to the base 2007-09, whereas poultry prices are expected to be on average 34% higher while lamb prices will rise by 68% (Figure 8.1).
- World meat consumption continues to experience one of the highest rates of growth among the major agricultural commodities, and mostly in the faster growing non-OECD countries. Virtually all of the world growth in meat consumption is projected to occur in developing countries, notably for poultry that grows at 2.8% p.a., and pig meat at 2.3% p.a. Beef consumption may grow the least at 2.0% p.a. Per capita annual consumption growth in the developed area remains low throughout the projection period (Figure 8.2).
- While demand pressure builds, the meat sector is increasingly constrained by higher feed prices and the availability of natural resources. World meat production growth is projected at 1.9% p.a. This growth is expected to originate predominantly outside the OECD area, which will account for 89% of additional sectoral output (Figure 8.2). Driven mostly by an expansion of poultry and beef shipments, world meat exports by 2019 are projected to increase by 22% relative to 2007-09 base period. OECD country exports increase by 7%, while those of non-OECD countries increase by 40% by 2019 relative to the base period (Figure 8.3).

Figure 8.1. **World meat prices in nominal terms remain above historical levels**

Nominal *versus* real meat prices

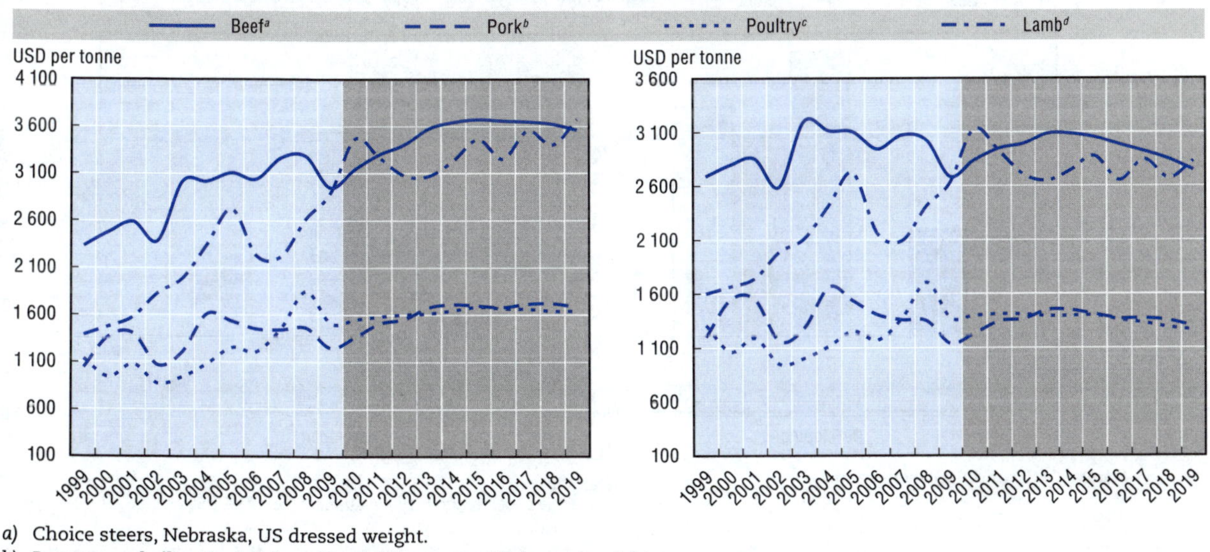

a) Choice steers, Nebraska, US dressed weight.
b) Barrows and gilts, No. 1-3, Iowa/South Minnesota, US dressed weight.
c) Meat of poultry export price, HS0207, Brazil product weight.
d) Lamb schedule price, all grade average, New Zealand dressed weight.

Source: OECD and FAO Secretariats.

StatLink http://dx.doi.org/10.1787/888932286912

Figure 8.2. **Meat production and consumption growth dominated by non-OECD region**

Share of meat production and consumption in OECD and non-OECD countries

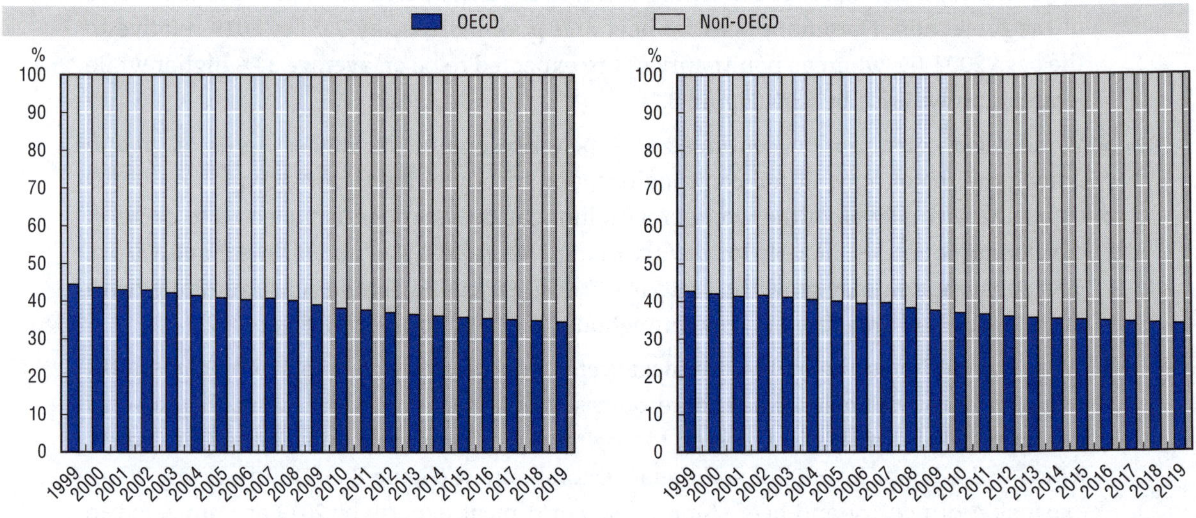

Source: OECD and FAO Secretariats.

StatLink http://dx.doi.org/10.1787/888932286931

Figure 8.3. **World consumer preference for poultry meat**

Overall increase of per capita meat consumption between the 2007-09 base period and 2019

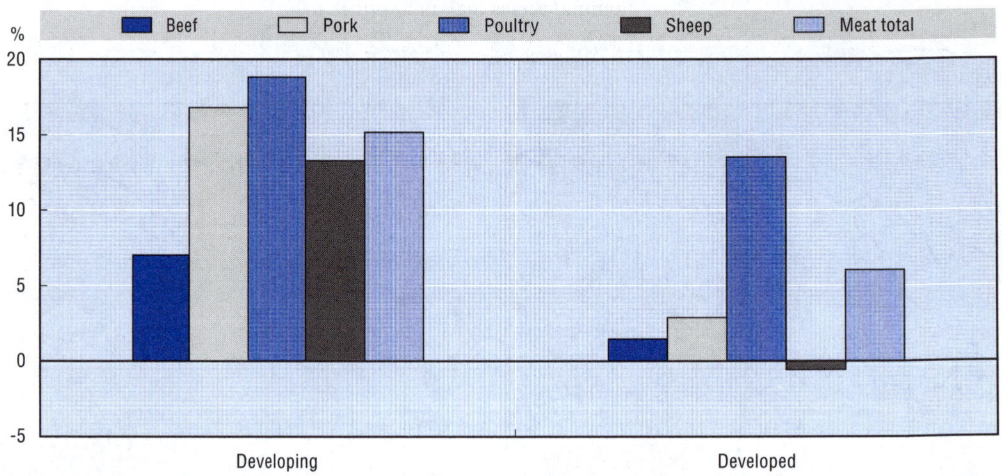

Source: OECD and FAO Secretariats.

StatLink http://dx.doi.org/10.1787/888932286950

Market trends and prospects

Prices

Most prices to increase at the beginning of the Outlook

Real prices for meat are projected to be higher than those observed during the previous decade, as higher costs constrain the expansion of output in the face of increasing demand. Beef prices are anticipated to be firm for the first half of the projections, mainly due to a tight supply following reduced cattle herds. However, an expansion of output in the second half, coupled with a reduction of import demand by the Russian Federation, are anticipated to exert a downward pressure on prices. Pig meat prices in both the Atlantic and Pacific area are not anticipated to increase much beyond 2015 due to an increase in supply from Brazil, USA and China, which are experiencing high productivity gains. Sheep meat prices are anticipated to be weak during the early years, but a reduction of sheep flocks in New Zealand may exert upward pressure on world sheep and lamb prices later in the projection period. Poultry prices expressed in nominal terms are to remain relatively firm throughout, largely reflecting higher feed costs and increased demand.

Higher producer prices are anticipated to boost global meat output. In the short term pig and poultry sectors with shorter production cycle, may respond more rapidly to renewed demand after the recession, while reduced cattle inventories will constrain beef production in the short term. As import demand in all the different meats recover, global meat trade is also set to rebound over the projection period. Accordingly, per capita meat consumption will rise following its marginal decrease of 2009, consistent with improved income prospects.

Production

Growth in production originates mostly from the non-OECD area

On average, annual world meat production growth is projected to slow down to 1.9% during the Outlook period from 2.1% in the previous decade, as the sector is increasingly constrained by higher feed costs and the availability of natural resources. This growth is expected to originate predominantly outside the OECD area, which will be responsible for about 89% of the additional output, while a more stable path of development is observed in OECD countries. Meat production growth in non-OECD countries is anticipated to be dominated by Brazil and China, although the Russian Federation is also anticipated to play an increasing role. The Russian Federation, a major meat importer, is implementing strategy to support public and private investments in the meat sector to bring meat self-sufficiency more than 80% by the end of the projection period (Box 8.1).

Following an initial recovery phase, the meat outlook is characterised by moderate increases in production. In the emerging economies, renewed investment, improved infrastructures and the introduction of modern, intensive and integrated production, processing and transport technologies, are the main factors that may spur higher productivity growth. This is especially true for Brazil, China, India and the Russian Federation, and to some extent in the CIS group of countries. Meat production in other non-OECD countries and regions of the world are also anticipated to grow, including sub-Saharan Africa where the inflow of foreign private investments is on the increase from the so called "land grab" phenomenon, *i.e.* increased animal feed production. However, the global economic crisis might have had somehow impinged on the investment capacity of developing countries. Meat production in the OECD area is anticipated to expand less than 1% p.a., as most farmers already benefit from existing technological advances, and face increasingly stringent animal welfare and food safety regulations.

Except for sub-Saharan Africa, productivity gains of non-ruminants animals are anticipated to be high in all countries and regions, while for ruminants, OECD countries are expected to have higher productivity gains. The contribution of technical change and efficiency to total productivity growth is expected to vary according to the labour/land ratio of the prevailing production systems, with higher land productivity gains in China, and higher labour productivity gains in Brazil.

Box 8.1. **Evolution of Russia Tariffs Quotas (TRQ's) for meat**

The Russian meat production declined rapidly following the collapse of the Soviet Union. As a result, meat imports increase rapidly to sustain domestic demand. In 2002, the Russian Federation implemented meat TRQ's for beef, pig meat and poultry. The life span of the initial TRQs expired by the end of 2009. The quota and related tariffs can be found in Table 8.1. Starting in 2006 the Russian government amended the initial TRQ's as shown in Table 8.2. However, in order to promote domestic production, by restricting over quota imports, the Russian government introduce prohibitive out of quota tariffs for pig meat and poultry.

Table 8.1. **Russian Federation: Meat TRQs for 2002-09, kt, *ad valorem***

	2002-09
Beef and veal import tariff quota, fresh, cooled and frozen (0201 + 0202)	458
Pig meat import tariff quota, fresh, cooled and frozen (0203)	476
Poultry import tariff quota, fresh, cooled and frozen (0105 + 0207)	1 050
Beef and veal	
In quota tariff	15
Out of quota tariff	60
Pig meat	
In quota tariff	15
Out of quota tariff	80
Poultry	
In quota tariff	25
Out of quota tariff	No out of quota possible

Source: Resolutions of the government of the Russian Federation.

StatLink http://dx.doi.org/10.1787/888932287938

Table 8.2. **Russian Federation: Meat TRQs for 2006-09, kt, *ad valorem***

	2006	2007	2008	2009
Beef and veal import tariff quota, fresh, cooled and frozen (0201 + 0202)	461	468	474	480
Pig meat import tariff quota, fresh, cooled and frozen (0203)	476	485	494	532
Poultry import tariff quota, fresh, cooled and frozen (0105 + 0207)	1 131	1 171	1 211	952
Beef and veal				
In quota tariff	15	15	15	15
Out of quota tariff	55	53	50	40
Pig meat				
In quota tariff	15	15	15	15
Out of quota tariff	60	60	60	75
Poultry				
In quota tariff	25	25	25	25
Out of quota tariff	60	60	60	95

Source: Resolutions of the government of the Russian Federation.

StatLink http://dx.doi.org/10.1787/888932287938

In December 2009, the Russian government approved a decree that will regulate meat imports to the Russian Federation in 2010-12. According to the law, the quota on poultry imports will almost be halved – from 952 kt in 2009, the quota for pork will gradually decline, while for beef it will rise (Table 8.3). These TRQs have been developed taking into consideration the different trend of the various meat sectors. For poultry the TRQ has

Box 8.1. **Evolution of Russia Tariffs Quotas (TRQ's) for meat** *(cont.)*

Table 8.3. **Russian Federation: Meat TRQs for the period 2010-12**

	2010	2011	2012
Beef and veal import tariff quota, fresh, cooled and frozen (0201 + 0202)	560	560	560
Pig meat import tariff quota, fresh, cooled and frozen (0203)	490	490	450
Poultry import tariff quota, fresh, cooled and frozen (0105 + 0207)	780	600	550
Beef and veal			
In quota tariff	15	15	15
Out of quota tariff	50	50	50
Pig meat			
In quota tariff	15	15	15
Out of quota tariff	75	75	75
Poultry			
In quota tariff	25	25	25
Out of quota tariff	80	80	80

Source: Resolutions of the government of the Russian Federation.

StatLink http://dx.doi.org/10.1787/888932287938

significantly been reduced to reflect the desire of the Russian government to promote the development of the industry which has grown, in the recent past, by 10% to 15% per year. It is expected with the current level of protection that Russian producer will be able to mostly meet domestic demand for poultry products after 2015. In the case of pig meat, the reduction in the quota reflect the slower growth in demand combine with an inadequate supply level from the domestic market. Nevertheless, it is not expected that Russian producers will catch up on the current gap between supply and demand. Finally, for beef an increase in imports is expected to occur as beef production in the Russian Federation is closely linked to the dairy herd. It is likely that the dairy herd will continue its long term decline leading to a reduction of domestic slaughtering.

In accordance with the new trade regulation for 2010 to 2012, new tariffs have been adopted for all meat products (Table 8.3). The allocation of quotas among countries has been worked out by the Russian authorities (Decree No. 1021 of 16 December 2009).

Environmental constraints as well as higher standards slow production growth

In many OECD countries, animal production systems are already highly intensive, leaving little scope for growth through additional input supply. Changes in feed costs have a significant effect on farm performance, but in addition to these, increasingly stringent public and private food safety and animal health regulation and standards, including animal welfare, and environmental constraints are anticipated to contribute to higher production costs. Nutrition and its impact on health has become a major concern of the meat industries in OECD economies. Consumer conscious on food safety has left governments with an enormous objective of regulating food quality in every step of the food chain. For example major changes are being applied to guidelines for the prevention, inspection and control of outbreaks of pests and diseases, including traceability standards which are already a priority in major meat producing and trading countries. In the US, the recent FCE Farm Act of 2008, the Country of Origin Labeling (COOL) initiative is expected to increase costs on different segments of the supply chain (Box 8.2). A similar labelling policy is also envisaged by the EU.

Box 8.2. US Country of Origin Labelling

The 2002 and 2008 US Farm Bills amended the Agricultural Marketing Act of 1946 to require certain retailers[a] to notify their customers of the country of origin of covered agricultural commodities ("COOL").[b] The COOL legislation went into effect in April 2005 for fish and shellfish and on 30 September 2008 for all other covered commodities with the interim final rule. USDA issued the final rule for mandatory COOL for all covered commodities on 15 January 2009.

COOL is a retail labelling law that provides additional information to consumers at the time of purchase for certain covered commodities, including but not limited to, covered meat products. COOL excludes processed foods, that is, covered commodities that have a change in character resulting from cooking, curing, smoking, or restructuring, or when combined with another food component.

COOL declarations must be specific and accurate. It is ultimately the retailer's responsibility to provide country of origin information to consumers; however, in order for retailers to accurately label covered commodities for COOL, direct and indirect suppliers to retailers must convey country of origin information. Packers and processors rely upon producers to provide country of origin information to initiate country of origin designations. The legislation and regulations specify the criteria that must be met in order to meet the four different types of labels – product of the United States, product from animals of multiple countries of origin, product from animals imported directly for slaughter, product imported directly from a foreign country. For ground meat, all countries from where the meat could reasonably have been derived must be listed. Accurate labelling for imported animals and meat products could likely increase costs in the animal-meat supply chain. The USDA Agricultural Marketing Service has estimated that total first-year implementation costs for all affected US industries would be USD 2.6 billion (USDA/AMS, 2009).

Canada and Mexico have challenged mandatory COOL at the WTO. At their request, a Dispute Settlement Panel was established at the World Trade Organization in November 2009 to determine if the COOL measure is consistent with the United States' international trade obligations.

The precise extent to which COOL is impacting US, Canadian and Mexican livestock and meat markets is unknown, but in this Outlook it is one of the factors contributing to lower projections of Canadian and Mexican live cattle and Canadian live hog exports to the US.

a) Retailers for which the invoice costs of all purchases of perishable agricultural commodities exceed USD 230 000 during a calendar year.

b) For further information on COOL and covered commodities, please see *www.ams.usda.gov/AMSv1.0/Cool.*

Output in the European Union will also be affected by the decoupling of production premiums, following the path that was already established last decade as intervention prices were lowered. Moreover, increased market access granted by the application of new TRQ's following the enlargement of the European Union as well as the Memorandum on beef hormones dispute signed with the United States, should also intensify competition in the domestic market, with an impact on domestic production.

Consumption

Non-OECD consumers renew their appetite for meats

A renewed expansion of the meat sector is expected for non-OECD countries during the projection period, where most of the growth in world consumption originates. Meat

Table 8.4. **Mature developed countries meat market record little per capita consumption growth**

Per annum increase of per capita meat consumption between the 2007-09 base period and 2019

	Global (%)	Developped (%)	Developping (%)
Beef	1.5	0.9	2.0
Pork	1.8	0.8	2.3
Poultry	2.4	1.6	2.7
Sheep	2.1	0.6	2.5
Total meat	**1.9**	**1.1**	**2.4**

Source: OECD and FAO Secretariats.

StatLink http://dx.doi.org/10.1787/888932287919

products are among the most sensitive to economic and demographic change growth. Combined with greater urbanisation, they represent the main drivers of per capita consumption growth. Conversely, the meat sector of OECD countries is expected to grow at a much slower pace due to escalating production costs and a combination of dietary changes and population demographics leading to a slowing down of growth of animal protein consumption (Table 8.4).

Worldwide, the economic recovery will strengthen the intake of animal proteins at the expense of foods of vegetal origin. Non-OECD countries are expected to consume around 86% of the projected global growth in meat production. Much of this expansion will take place in Asia and the Pacific region, especially in China, although some growth is also seen in Latin America, which has traditionally been consuming meat, led by Brazil. Such growth will reflect in particular the rise in consumption of cheaper sources of animal protein, mainly poultry and pig meat. Import dependency in meat products is likewise expected to grow in many dynamic developing countries, as demand surpasses the domestic capacity for meat production throughout the duration of the Outlook.

It is already two decades since the volume of total meat consumed in developing countries overtook that of developed countries, and the gap between them has not ceased to expand. Today, 60% of the meat consumed worldwide is eaten in the developing world, and the projection for this decade is that its increasing share shows no signs of relenting. Meat maintains one of the fastest rates of growth in consumption among all agricultural commodities.

Virtually all of the world growth in meat consumption projected takes place in developing countries, notably in Asia. World poultry consumption increased by an additional 26 Mt r.t.c. by 2019 compared to the base period, increasing per capita consumption by nearly 2 kg per capita r.t.c. by 2019. Developing countries will consume nearly 83% of this increase, *i.e.* of which more than 70% originate from Asian countries. Interestingly, poultry consumption will surpass pig meat consumption in the second half of the projection in the OECD countries.

World pig meat consumption grows from 103 Mt c.w.e. in the base period to 126 Mt c.w.e. in 2019. Again, developing countries will consume the bulk of the increase, or 22 Mt c.w.e., virtually all from Asian countries. Conversely, beef consumption grows at a more moderate rate, 1.5% p.a., and the vast majority of these additional 9.3 Mt c.w.e. will be more evenly distributed among the various developing regions of the world. Although the patterns of consumption are anticipated to be slightly different across countries depending on habit, there seems to be a universal preference for poultry over the other meats (Figure 8.3). Even in OECD countries, where the growth of per capita meat consumption is marginal, the share of poultry increases through the years at the expense of other meats. This trend is expected to continue this decade across the world, although in countries where beef meat production is important and international prices have little impact on domestic production, such as East Africa and Asia (Uzbekistan, Tajikistan, Kyrgyzstan, etc.), beef and sheep meat will remain the preferred types of meat consumed.

Trade

Brazil to dominate non-OECD meat trade

Driven mostly by an expansion of poultry shipments, world meat exports by 2019 are projected to increase by 22% relative to the 2007-09 base period. OECD country exports increase by 7%, while those of non-OECD countries increase by 40%. The bulk of growth in meat trade is expected to originate largely from outside the OECD area, in particular from Brazil which will single-handedly account for over 63% of all the meat exported from non-OECD countries in 2019, and for one-third of world meat exports (Figure 8.4). Alongside Brazil, a handful of major exporters will continue to govern the supply of world meat markets, including the United States, Canada, Argentina and Australia.

Figure 8.4. **Meat exports growth is predominately coming from non-OECD region**

Development of meat exports between 1999 and 2019

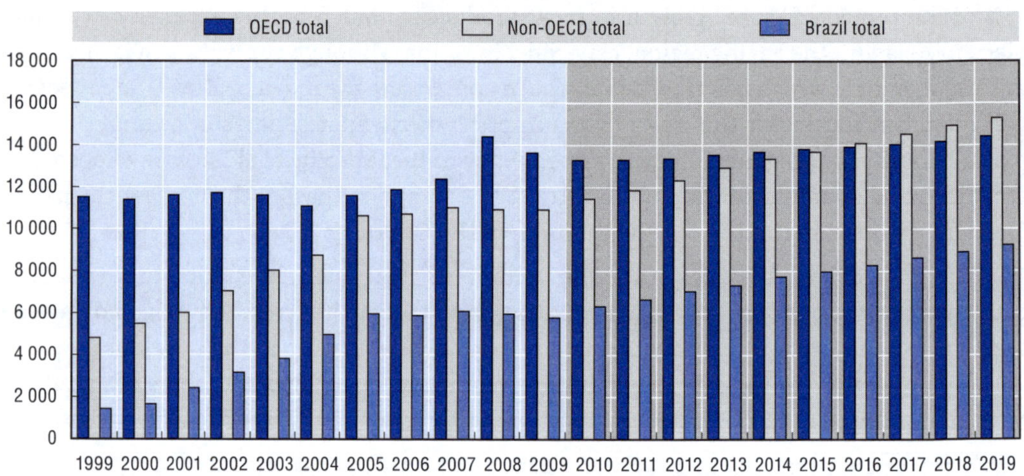

Source: OECD and FAO Secretariats.

StatLink http://dx.doi.org/10.1787/888932286969

Trade prospects are favourable in Southeast Asia, as the creation of a free trade area between China and the ASEAN in January 2010 is anticipated to result in lower mutual trade barriers between countries in the region. This event should stimulate production and trade, as intra-trade duties are below those applied to countries outside the Free Trade Area. Nevertheless, the structure of trade in the region (small scale operations), and its limited infrastructure (limited cold-storage facilities), are unlikely to challenge in the medium term the leading role of traditional suppliers of meat and their by-products, such as the United States, the European Union and Brazil.

In the OECD area, beef exports from North America are expected to increase, as import restrictions imposed on the grounds of bovine spongiform encephalopathy (BSE) are progressively eased. Meat exports from the EU are anticipated to decline over the decade

Figure 8.5. **Biggest net meat imports growth remains in developed countries**

Development of net meat imports from selected countries

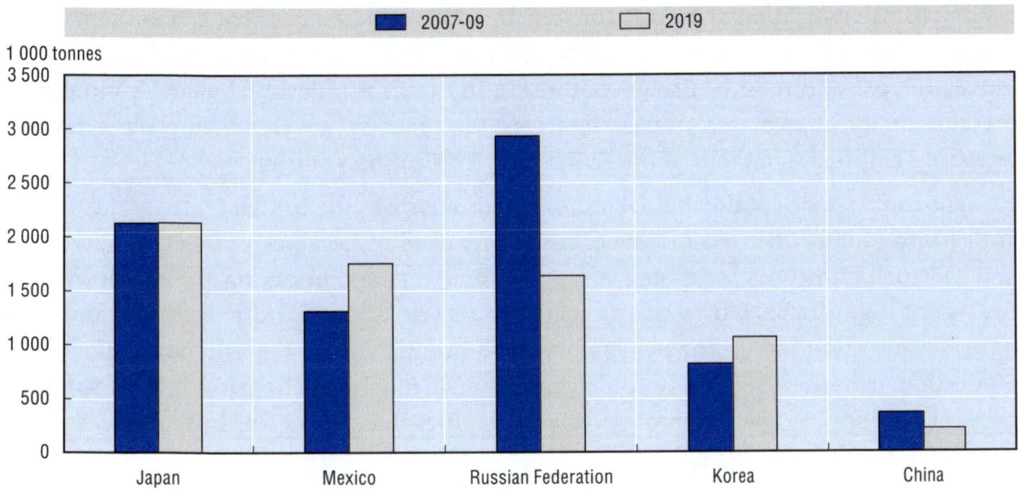

Source: OECD and FAO Secretariats.

StatLink http://dx.doi.org/10.1787/888932286988

due to reduced domestic output following policy reforms, coupled with a growing domestic consumption. The US continue to lead as the largest meat importing country by 2019 with 2.2 Mt, closely followed by the EU import with nearly 2.2 Mt, and by Japan with 2.1 Mt. On a net import basis, Japan, Mexico, Korea remains the major importing countries. The Russian Federation, until recently the largest net meat importer, is expected to decrease its imports for the Outlook period by more than 1 Mt c.w.e., subject to its strategy to expand domestic production to achieve higher self sufficiency by the end of the decade (Figure 8.5).

Agricultural policies, in particular product and process standards, are to become an important driver shaping not only production but also trade. Climate change and the environmental impact of livestock production are rapidly moving up the political agenda. Environmental and animal welfare legislation, and standards at large, add to production costs and constrain the expansion of the sector despite efforts made by the industry to respond to these concerns via research and development. The impact of standards, including those related to production processes, will be increasingly felt in international trade, as OECD countries progressively request these very same standards from imported products. Their impact will likely take the form of increased market segmentation, which will add complexity to the future analysis of global meat markets.

Key issues and uncertainties

The Outlook presented in these chapters is conditional on a number of assumptions. Among these are the continuation of agricultural policies and declared policy reforms, and the macroeconomic environment. Changes in any of these assumptions would lead to a different set of projections for the various meat markets.

An important uncertainty always looming in meat trade relates to sanitary and food safety concerns stemming from outbreaks of animal diseases. In the past decade, such outbreaks have considerably impacted the growth in trade for all the meat. The possibility of future animal disease outbreaks, their duration, intensity, potential consumer reactions and new trade restrictions, is an important uncertainty into the present Outlook. However, as trade has grown, and with greater experience in handling disease outbreaks, the impact of certain disease outbreaks has been considerably reduced. For example, the World Organization for Animal Health (OIE) has developed a system of protocols to allow zoning or compartmentalisation of production which, if followed by a country, enables it to assure sanitary safety of exports from an unaffected zone. This practice has helped minimise the impact of foot and mouth disease outbreaks in Brazil. Similarly, Thailand continues to pressure major importing countries to accept the practice, which would allow exports of uncooked chicken regardless of the Avian Influenza country status.

The rapid growth of the biofuel sectors, particularly in the United States and Europe, is anticipated to affect the availability of feedstuffs used in livestock production, as ethanol and livestock industries compete for grains. Though cereal prices have come down from the record highs in 2008, they are projected to remain above historic levels. At the same time, however, the increasing availability of co-products from the ethanol and biodiesel production, in particular of DDGs, pulps and oilseeds meals, will help to contain the impact on meat production. The net impact varies by biofuel chain as well as by type of meat, and requires detailed monitoring.

Domestic and trade policies remain an important determinant of meat markets. Some countries still have high tariffs on meat products, often including tariff rate quotas (TRQ's) or export taxes. At the same time, domestic policies that directly benefit animal husbandry remain in place in some OECD countries. Outcomes of the WTO negotiations that may commit member countries to lower tariffs on meat, or to decrease their support to animal production, are likely to alter the Outlook. The same holds true for agricultural products that are input to the livestock sector like grains, where changes in prices triggered by policy changes would affect animal production. The Outlook has assumed that the implementation of Country of Origin and Labeling Legislation by the US will give an incentive to US processors to switch from importing live animals from Canada to imports of meat, leading to increased trade of meat and a decreased trade of live animals between these two countries. Due to the recent implementation of the regulation, the degree to which this may take place remains uncertain. In the Russian Federation, the rapid growth of domestic production following the government decision to achieve higher self sufficiency (Box 8.1) is expected to have a strong impact on meat markets in the Outlook. Finally, China's policy on Strategic Meat Reserves from Frozen Meat (which requested provincial frozen meat level reserves high enough to supply China's urban residents for seven days) as well as the new market intervention scheme called the National Swine Price Alert System (which is a price monitoring scheme geared to ensuring sufficient farmers returns) may have an impact in stabilising the price volatility domestically and abroad.

Production of livestock and feedstuffs is responsible for some 80% of all agricultural greenhouse gas (GHG) emissions from sources that include enteric fermentation, manure management, land use change (deforestation and land clearing and burning), soil and the burning of carbon-based fuel. Emissions are expected to increase substantially in the coming decades, as population and income growth increase global demand for meat, dairy and other high value products. It is projected that much of the increase in agriculture-related emissions will take place in Latin America, Asia and Africa. It remains uncertain, but possible that in the medium to long term, livestock production may be subject to carbon mitigation constraints in some countries. There are large differences between regions and meat type in GHG emissions per unit of output. Pricing emissions from livestock production could potentially result insubstantial shifts in production, consumption and trade. At the global level, shifts in production would change the relative price of meat and promote the consumption of meat with lower associated GHG emissions such as poultry.

Chapter 9

Dairy

Market situation

International dairy markets have experienced a dramatic rise and fall in recent years. After a strong run-up in prices, the market situation reversed in the course of 2008. Demand retreated while supply increased in reaction to strong price incentives. This coincided with the commencement of the global economic crises and by early 2009 prices fell by half from mid 2008 levels. As a result, dairy product stocks started to accumulate, most importantly in the United States and the European Union. The dramatic fall in prices from the peaks of 2008 has shaken the global dairy sector. In many countries the relatively stronger market situation witnessed in previous years led to a rethinking and mobilisation of the sector's longer term strategy (for example, the European Commission has established a High Level Expert's Group on milk to discuss medium and longer term arrangements for the dairy sector, given the abolition of the current quota system in 2015).

The depressed situation of early 2009, linked to the global economic crisis, should nevertheless be viewed as a short term phenomena rather than longer term equilibrium. In the course of 2009, international prices started to strengthen, rebounding rapidly at the end of the year. The strong recovery in prices was triggered by increased demand mainly from China and oil exporting countries but also by lower supplies that contracted in some regions in response to low profitability in the previous year and weather impacts on production in the Southern hemisphere producing countries. The European Union also has been restrained in the export of dairy products from higher intervention stocks. In 2010, fundamentals indicate a recovery in demand as markets stabilise with improved economic prospects and market confidence.

Projection highlights

- Over the Outlook to 2019, international dairy prices are expected to grow in nominal terms while remaining relatively flat in real terms (Figures 9.1 and 9.2). On average, world market prices in real terms are expected to stay 15-40% higher compared to the decade preceding the 2007/08 peak. Real butter prices are expected to register the highest gains supported by high vegetable oil prices.
- The dairy sector remains one of the fastest growing sectors covered in the Outlook. By 2019 world milk production is expected to increase by 170 Mt from the 2007-09 base period. The annual growth rate for the projection period is estimated at 2.1%.
- The overwhelming majority (more than 80%) of additional milk is projected to be produced outside the OECD area. World production of WMP, butter, cheese and SMP is expected to grow from the base period by 31%, 28%, 20% and 9 % respectively (Figure 9.3).
- OECD continues to dominate global cheese consumption (nearly three-quarters of the total consumption) which is expected to increase by 20% over the Outlook period. In non-OECD countries demand growth is expected for all dairy products with WMP and butter consumption growing the strongest (38%), followed by cheese (33%) (Figure 9.4).

Figure 9.1. **World dairy prices rising in nominal terms**

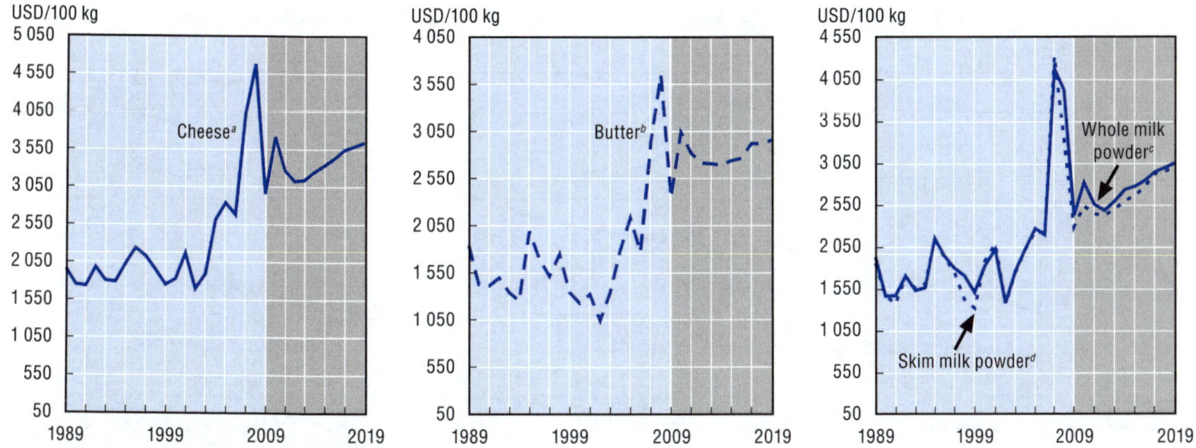

a) F.o.b. export price, cheddar cheese, 39% moisture, Oceania.
b) F.o.b. export price, butter 82% butterfat, Oceania.
c) F.o.b. export price, WMP 26%, Oceania.
d) F.o.b. export price, non fat dry milk, 1.25% butterfat, Oceania.
Source: OECD and FAO Secretariats.

StatLink http://dx.doi.org/10.1787/888932287007

Figure 9.2. **Declining trend in prices in real terms is expected to abate**

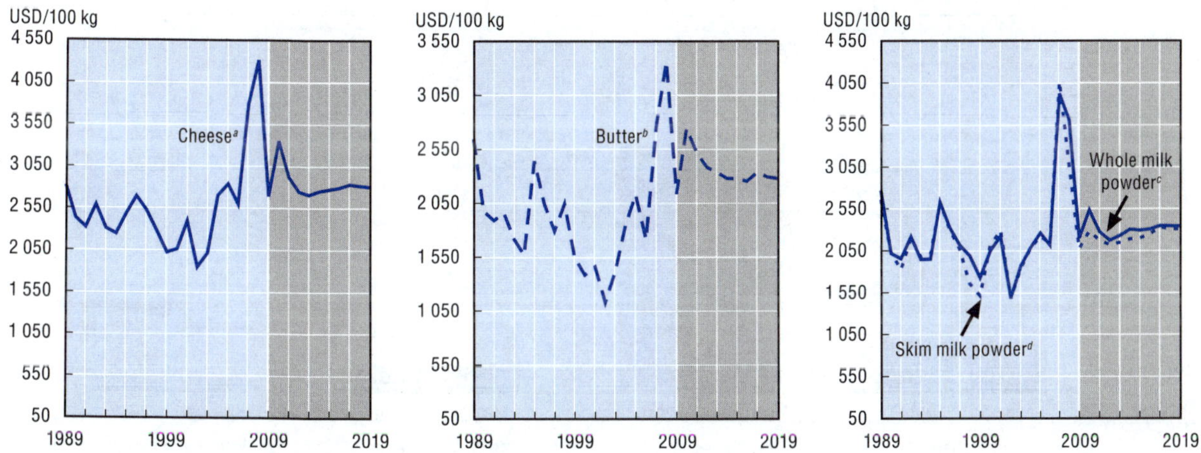

a) F.o.b. export price, cheddar cheese, 39% moisture, Oceania, deflated by US GDP 2005 = 1.
b) F.o.b. export price, butter 82% butterfat, Oceania, deflated by US GDP 2005 = 1.
c) F.o.b. export price, WMP 26%, Oceania, deflated by US GDP 2005 = 1.
d) F.o.b. export price, non fat dry milk,1.25% butterfat, Oceania, deflated by US GDP 2005 = 1.
Source: OECD and FAO Secretariats.

StatLink http://dx.doi.org/10.1787/888932287026

- Trade in dairy products, as a ratio to milk production, is expected to continue to decline towards 6%. This is due to the developing countries' growth of production and consumption of fresh dairy products, which are not traded. Compared to the base period world exports are expected to grow for cheese and WMP (both 14%), SMP (6%) but to decline for butter (–4%) (Figure 9.5). The six largest importers of dairy products continue to cover less than 50% of the world market (Figure 9.6).

Figure 9.3. **Global production of WMP growing rapidly**

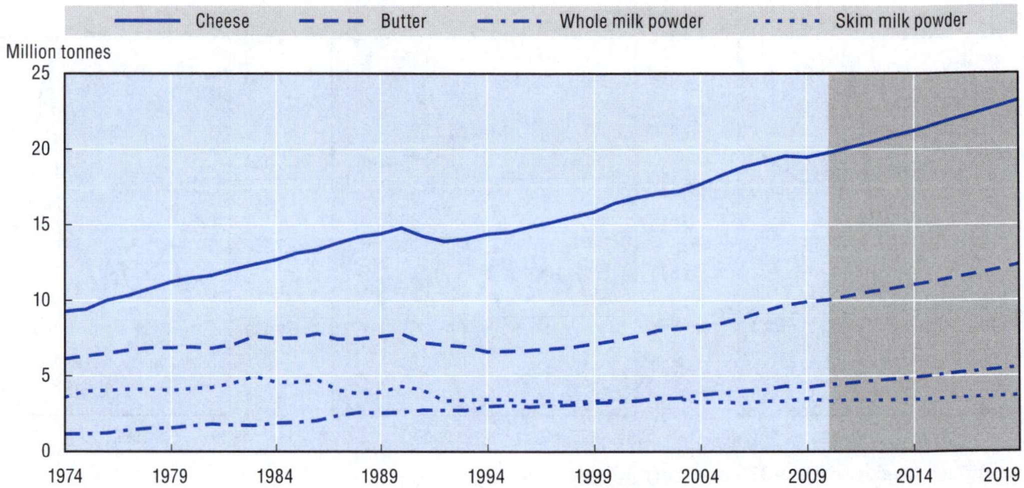

Source: OECD and FAO Secretariats.

StatLink http://dx.doi.org/10.1787/888932287045

Figure 9.4. **Consumption affected by income and product attributes**

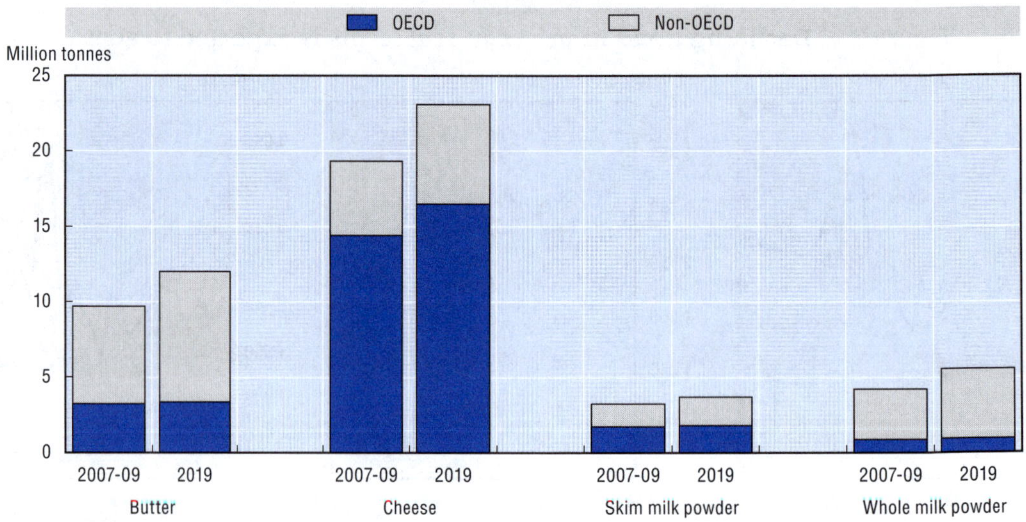

Source: OECD and FAO Secretariats.

StatLink http://dx.doi.org/10.1787/888932287064

Market trends and prospects

Prices

The anticipated return to global economic growth and rising population are expected to underpin the international dairy markets and prices over the Outlook. In the short run, dairy product prices are expected to be tempered by the downsizing of stocks mainly in the US and the EU. Prices are expected to rise steadily from 2012 in nominal terms, by more than 2% per year on average driven by rising demand but also increasing production costs. In real terms, the declining trend in prices is expected to abate, with world prices remaining relatively flat over the projection period (Figures 9.1 and 9.2).

Figure 9.5. **Trade increases mainly for cheese and WMP**

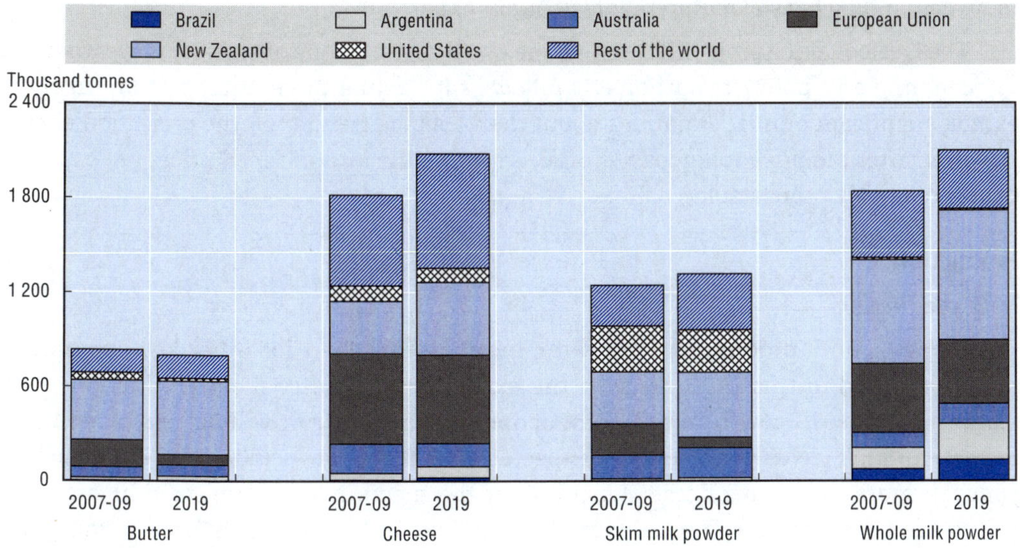

Source: OECD and FAO Secretariats.

StatLink http://dx.doi.org/10.1787/888932287083

Figure 9.6. **Russian Federation a major importer of butter and cheese**

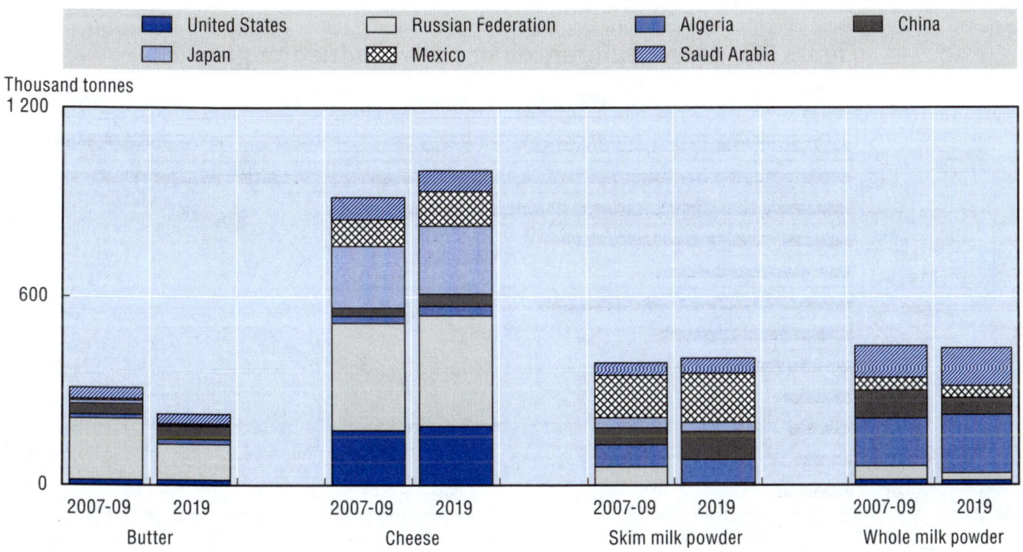

Source: OECD and FAO Secretariats.

StatLink http://dx.doi.org/10.1787/888932287102

Averaging over the Outlook period, world market prices in nominal but also in real terms are expected to stay firm and well above the levels seen in the decade preceding the 2007/08 peak. Milk powder and cheese prices in real terms are anticipated to reach levels about 15% and 23% higher on average respectively. The prices of milk fat components are expected to strengthen the most; average real butter prices may attain levels 40% higher compared to the decade before the price peak. The relative strength of butter prices is partly linked to sustained high energy and vegetable oil prices and to the fact that considerably less butter will be exported from countries such as the European Union or the United States. Emerging exporters concentrate more on milk powder exports which may also be explained by the more demanding logistic requirements of butter exports.

The Outlook price projections reflect the usual assumptions of stability in weather and in economic and policy conditions. It follows that actual price outcomes are likely to exhibit significant annual variations about their average trend than the projection shows. However, after the price rollercoaster of recent years the international prices are expected to stay relatively more stable.

Production

Milk production

Globally, milk production is growing more rapidly than for other key commodity sectors covered in this Outlook, with the exception of poultry meat production. After stagnating in 2009, growth in milk production in the short run will be supported by an increase in milk yields which may be expected after recent cow culling and emergence of younger, more productive herds globally. Over the medium term, projected higher dairy prices are expected to stimulate supply response and steady milk production gains (Figure 9.7). Most of the additional milk is projected to be produced outside the OECD area. China, India, Pakistan, Argentina and Brazil account for more than half of the global milk production gains. Of the 170 Mt of additional milk produced in 2019 as compared to the average level of 2007-09, more than 80%, will come from developing countries.

Figure 9.7. **Large differences in milk production growth**

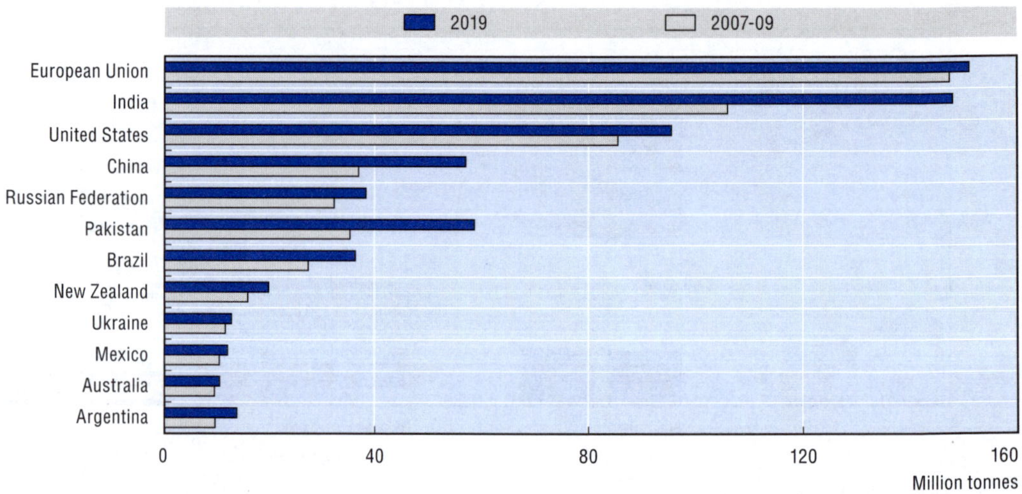

Source: OECD and FAO Secretariats.

StatLink http://dx.doi.org/10.1787/888932287121

Large regional differences in milk production growth are expected to remain a feature of the Outlook. Higher international milk product prices in recent years have created the incentive for expansion in many developing countries. The location of future growth in production depends on the market and policy context in various producing countries, the milk-feed price ratio and competition for land. The milk production decision would also be increasingly influenced by water and environmental concerns, and domestic and climate change policies. The shift in milk production is expected mainly from North to South, and more specifically from Europe and North America to Asia, Latin America and Oceania.

The dairy sector in New Zealand is expected to remain one of the country's key export industries. Milk production is expected to reach nearly 20 Mt increasing from the base period level of 16 Mt. Although the conversion of sheep and beef farms to dairy farms is expected to continue, production gains are expected to come progressively from yield gains as opposed to increased cow inventories. Growing environmental and water constraints, and climate change policy, currently agreed to include agriculture from 2015, will be factors increasingly influencing farm production decisions. After the 5% fall in 2009/10, milk production in Australia is expected to recover gradually mainly in response to improved milk to feed price ratios. Due to drought and reduced milk production margins, the dairy herd has contracted by 25% since 2002. A return to strong growth in the dairy herd, anticipated in previous years, now seems unlikely and the projected modest growth remains sensitive to water availability.

Following the drop in domestic prices, European Union milk deliveries fell in 2009. Deliveries are expected to stabilise in the EU15 in 2010 but fall by a further 1.6% in the EU12. Over the Outlook period, milk supply will not follow the annual 1% quota increases, but grow gradually at about 0.3% per year. Quota abolition in 2015 is expected to have only a limited impact on milk supply. While EU12 milk deliveries are projected to increase by nearly 5%, milk production is projected to fall by more than 5% due to steady decreases in subsistence production and on farm milk consumption over the Outlook period.

In the United States, cow numbers in 2009 fell after several years of expansion. This expansion had reversed a long term trend of declining cow numbers. This long term decline is expected to resume over the medium term but stabilising at the end of the projection period as a result of improved price to feed ratio. Production is expected to grow by more than 1% annually as yield gains more than offset the cow number reduction. Canadian milk deliveries are assumed to follow demand driven milk quota levels growing modestly over the Outlook.

India and Pakistan, two of the largest "traditional" dairy producing countries, are anticipated to continue to expand milk production at trend rates of 3% and 4% respectively over the next decade, driven by expansion in both cow numbers and yields. Such growth has persisted for several decades and an important question is how long it will persist. Dairy products are very much part of the culture of these countries and as incomes grow, particularly in India, milk product demand continues to keep pace. These sectors remain dominated by very small producers, serving local and usually informal markets. In other developing countries of Asia, milk production is less prevalent, but it is also growing at a fast pace, particularly in Thailand, Indonesia, Malaysia and Viet Nam. Milk production should advance in the 3-4% range annually over the decade in these countries. Growth reflects increasing demand with growing urbanisation, but from a small production base. Markets are open to imported powdered milk that, when blended with local supply, satisfies growing demand.

In Latin America, countries have responded to higher incentives with increased production, either decreasing milk deficits or emerging as exporters. Milk production in Argentina is expected to grow strongly over the Outlook at more than 3% annually on average, pushed by increased efficiency on farms and improved profitability of milk *vis-à-vis* arable crops. Brazil's milk production is projected to grow by 2.3% annually, primarily due to increased productivity but also government measures that target continued growth and focus on investment and marketing. Production in countries such as

Box 9.1. China dairy industry growth expected to continue but subject to uncertainty

After years of rapid growth, the increase in milk production in China tumbled in 2009, hit by the aftermath of the melamine crisis. Production is expected to rebound strongly in 2010, by 10%, but then growth is projected to average around 4-5% per year over the period. The Outlook milk and dairy product projections for China are subject to considerable uncertainties and one of the key issues is the speed of the dairy industry restructuring.

The milk farming base in China is very fragmented and the rapid growth of the dairy industry allowed profit seeking intermediaries (often without any background in dairy) to enter the supply chain. The Chinese Government has unveiled a plan to address these structural problems, emphasizing modernisation and increased regulations of the dairy industry. The "China Dairy Industry Policy" announced on 26 June 2009 obliges new dairy processing facilities to source at least 40% of its raw milk from its own herd. The guidelines also require that companies expanding existing facilities must control at least 75% of its raw milk source sent to the improved plants. The government is also revising China's dairy product standards to ensure safety and quality, and regain consumer confidence.

It is expected that small-holders would exit milk farming or combine into larger more efficient operations. The Outlook assumes that the modernisation will also be effective in rectifying the milk quality problem. This is an important assumption which has repercussions on global dairy markets. China is by far the largest producer and consumer of WMP accounting for around 30% of global volumes. Although China was gradually reducing dependence on WMP imports, the melamine crises sparked large WMP imports in 2009. In early 2010, the reports of further sales of contaminated product (from old stocks) aggravated the situation. The Outlook expects that as quality improves, the increase in imports of WMP would ease over the short run and imports would return to historical levels. But, the actual level of the Chinese imports of dairy products is also subject to the uncertainty surrounding exchange rate developments. A hypothetical scenario, run with the OECD-FAO Aglink-Cosimo model, illustrates that a 10% appreciation of the CNY (Yuan) *vis-à-vis* the US dollar (assumed over the entire projection period) would result in additional imports of dairy products and especially of WMP (Figure 9.8). As a result global dairy prices would be lifted by 1-2% on average. A further source of uncertainty for China's dairy imports is the Chinese government recent decision to impose import license requirements on several types of dairy products (fresh milk, milk powders and whey).

Figure 9.8. **Impact of the 10% stronger CNY (Yuan) on the WMP imports to China**

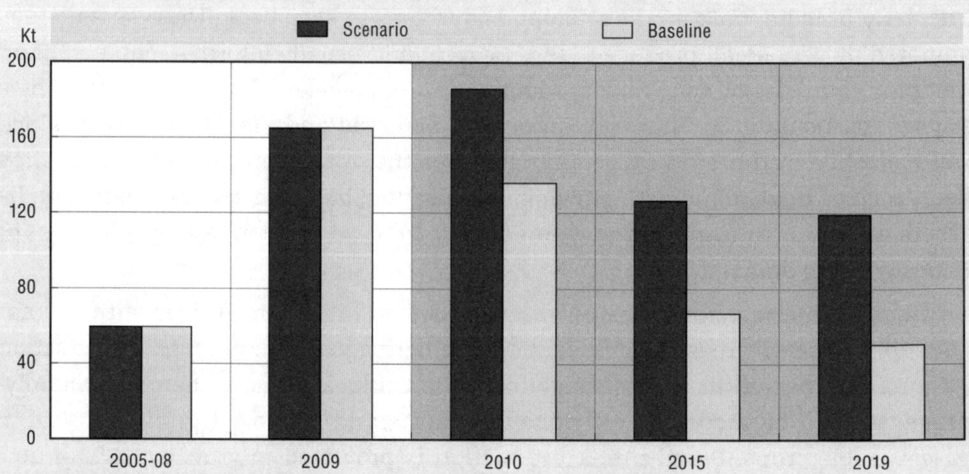

Source: OECD and FAO Secretariats.

StatLink http://dx.doi.org/10.1787/888932287140

Columbia, Peru and Chile may see production advance above the world average growth rate, but below the growth rate for developing countries as a group. Uruguay, which has increased prominence as an emerging exporter of milk products, may see production growth at 1.7% per year.

The dairy sector in Africa continues to grow above 2% per year, but in per capita terms growth is anticipated to be weak. Countries in North Africa, including Algeria and Egypt are projected to sustain relatively high rates of growth, encouraged by programmes to stimulate production. In sub-Saharan Africa, where the milk market is by far dominated by the informal sector, growth is projected to advance about 1% annually in per capita terms.

Dairy product production

After the slowdown in production growth in 2009, global production of dairy products is to continue increasing, following longer term trends, with world production of WMP growing most rapidly. WMP production is expected to record 31% growth, albeit from a lower base level as compared to other products (Figure 9.3). China is expected to remain the most important WMP producer. Despite a significant production drop (16.5%) in 2009, China is expected to contribute more than 40% of additional global supplies over the projection. Brazil and New Zealand are expected to add 25% and 12% to global WMP gains respectively. The main driving force for whole milk powder production is its use for reconstitution in milk production deficit areas and in low-production seasons.

Almost three-quarters of all additional global butter supplies are expected to come from India, with Pakistan and New Zealand adding 10 and 5%, respectively. Nearly 50% of additional cheese output over the projection period is expected to come from the EU and the US who continue to dominate global production accounting for two-thirds of the total. Most of the additional global production of SMP is expected from India (33%), New Zealand (29%), but also Ukraine (21%) and Brazil (20%).

About two-thirds of additional global milk production is to be left in fluid form or to be processed into fresh dairy products since the majority of milk production gains is in the developing world where milk processing remains limited. However, even in developed countries the fresh dairy product market is to remain dynamic with rapid growth of probiotic products, most notably probiotic cultured drinks and yoghurts.

Consumption

Growing demand, especially in developing countries, has underpinned firm international prices. The steady growth in demand was thought to be a permanent feature of the markets. The 2008 international dairy product price spike and global economic crisis were important reminders of the high price and income elasticities of dairy product demand. However, in a symmetric way, lured back by lower prices in 2009 and improved economic conditions, the demand for dairy products started to gradually strengthen.

Increasing population and income, together with the growing popularity of dairy products, particularly among developing country consumers – and boosted by the expansion and penetration of western retail chains – suggests a return to steady growth over the Outlook. The growth in demand will be strongest in China and India. A large increase in consumption is also expected in countries of Latin America – mainly Brazil and Mexico – partly in response to various government food assistance programmes. OECD countries continue to dominate cheese consumption, accounting for three-quarters of the world total. Developing countries now consume about 80% of global WMP consumption and this share is expected to grow to 83% by 2019 (Figure 9.4).

The westernisation of diets and changes in lifestyle are important drivers, particularly for cheese, which has seen rapid growth in demand as an ingredient in fast food products, mainly in pizzas, hamburgers and sandwiches. However, a consequence of the increasing link of cheese demand to the food service industry is that in times when consumer purchasing power declines the negative impact on cheese consumption may intensify. This, to some extent, also holds for butter, especially in developed countries, as demand from restaurants and hotels becomes increasingly important.

Despite substantial per-capita growth in non-OECD economies, consumption per person remains at relatively low levels in absolute terms, with significant potential for growth. Focusing on cheese, one of the fastest growing segments of dairy consumption, the highest per capita consumption is found in the EU (16.7 kg per person) and the US (14.8 kg). In Japan, an important cheese importer, the per capita consumption of only 2 kg is comparatively low but still considerably higher than per capita consumption in many developing counties (Figure 9.9). The developing country region per capita cheese consumption reaches only 0.8 kg per person on average.

Figure 9.9. **Significant growth of cheese consumption in developing regions, albeit from low base**

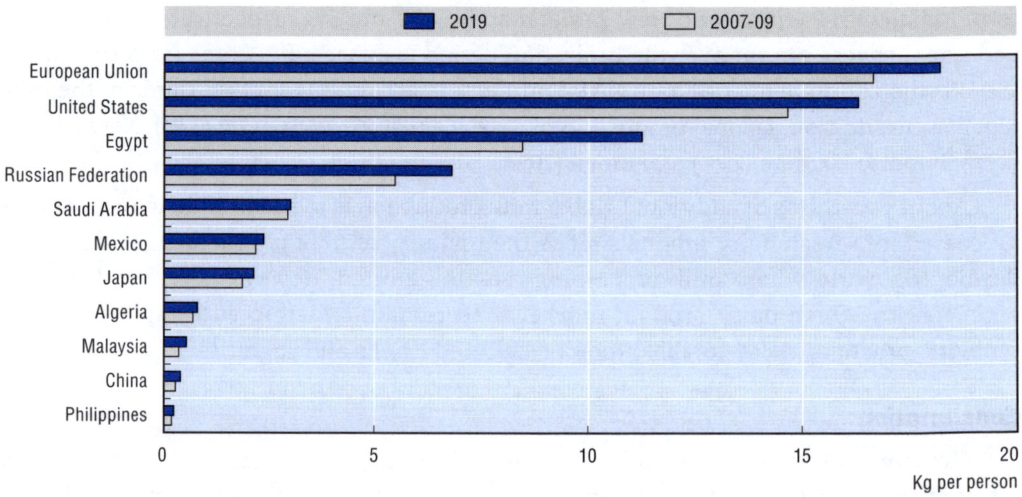

Source: OECD and FAO Secretariats.

StatLink http://dx.doi.org/10.1787/888932287159

Trade

After contracting during the global recession, trade is expected to increase mainly for cheese and WMP fuelled by re-emerging demand in developing countries (Figure 9.5). New Zealand will remain the most significant exporter, with around 35% of global export market share. It is also expected to increase its exports by nearly 30% as compared to the base period. Nevertheless, despite its robust production and export growth New Zealand's long term potential to satisfy growing import demand is limited as the country share in global milk output remains only 2.3%.

Brazil, traditionally regarded as a dairy importing country, is expected to further increase its presence on export markets especially for WMP. Exports from Brazil but also from Argentina are expected to grow by 70-80% over the projection period. On the other hand, EU exports of butter and SMP are expected to fall significantly throughout the Outlook. Although EU export of cheese and WMP would remain relatively stable, compared to base period, the export market share is expected to decline over the Outlook, from 32% to 27% for cheese and from 24% to 19% in the case of WMP.

The Russian Federation and Egypt remain the most important importing countries of butter, although imports to the Russian Federation are anticipated to decline over the Outlook as a result of increased domestic production. The increase in domestic production in the Russian Federation however rest on the assumption of a continuation of government measures aimed at milk production growth. The Russian Federation remains the most important importer of cheese and imports are expected to grow in the next decade by 2% annually. The Russian Federation together with Japan, Mexico and the United States continue to account for 40% of global cheese imports. Mexico, Indonesia and Malaysia are projected to increase their SMP imports, and the Middle East and North Africa regions are to remain key importers of WMP. Algeria, together with Saudi Arabia, is projected to account for around 15% of global WMP imports by 2019 (Figure 9.6).

The share of world milk production traded (intra-EU trade excluded) has been at around 6-7% and international dairy markets are expected to remain thin. In contrast, international trade in dairy products over the base period accounted for about 43% of the production of WMP, 34% of SMP, 9% of cheese and 8.5% of butter. Over the projection period the share of trade to global production is expected to remain relatively stable for cheese (due to trade growth) and SMP (due to production stagnation) but decline for butter to 6.5% (mainly due to the increase in butter production and consumption in India) and for WMP to 37% (due to the increase in production in many developing countries).

Key issues and uncertainties

The key issue in the Outlook for international dairy markets centres around the sustainability of the trend toward higher prices, and their variability around that trend. For a sector which has long been characterised as highly protected, and for which international markets have long been exposed to surplus disposal policies, markets are slowly adjusting to a new market environment. The strong rise in international market prices in 2006-08 sparked interest in dairy market development in many developing countries. The roller-coaster experience in international markets since the beginning of the century has witnessed more ups than downs, and higher prices for imports of milk powders for re-constitution are providing a stimulus to growth of domestic production. In this context, a key uncertainty for the Outlook will be the developments in domestic dairy policies, and above all the outcome of the Doha Development Agenda negotiations. Tariffs and tariff quotas remain highly restrictive, particularly in industrialised countries. Reduced border protection would likely raise dairy prices considerably, stimulating further production growth in developing countries, and also higher trade volumes.

Prospects for the variability of dairy product prices are unclear. The decline/elimination of surplus disposals on international markets may contribute to greater price stability. On the other hand, if an increasing proportion of exports are supplied by regions where pasture conditions vary substantially, increased price variability may result. As noted earlier, the growth of milk production globally is dominated by the growth in developing countries, and international dairy product markets are shrinking relative to global production. Weather conditions remain among the key uncertainties for milk markets, but they may become even more important in the future.

An increasingly important factor also likely to bear on the Outlook is the deterioration of natural grasslands, limited water resources and water pollution. This can be important in many developing countries, particularly in some Asian countries where milk product consumption growth has also been high. Increasing constraints on production growth in this region may stimulate higher trade. Many countries are increasingly concerned about methane-gas emissions and are establishing policies to deal with them. The latter development applies more to developed countries, where milk-producing livestock numbers are also expected to fall 0.3% annually over the Outlook. However, the growth of milk production in developing countries is driven to an important degree by growth in the dairy herd, which is projected to grow 1.6%. In these countries, small-holder dairy farms dominate production in the context of informal markets in which producers meet subsistence requirements and local market needs, providing important support to household and rural income and food security. Increasing resource pressures may limit their growth in future, and require large changes in market structures and technologies, implying that adaptation will be slow and require large investments.

On the demand side, macroeconomic developments represent an important uncertainty for the Outlook. Dairy products are among the most sensitive to income fluctuations, and this is particularly so in many rapidly growing developing countries. A slowdown in global economic growth would moderate import demand and reduce prices. Furthermore, the processes of economic growth, urbanisation, and technology transfer will continue to further change the nature of the dairy industry worldwide. Continued, but slow, convergence in consumption patterns will have a considerable impact on both domestic and international markets. Consumers, not only in developed countries, are increasingly concerned about health and nutrition. As a result, the dairy sector will have to continuously adapt to growing consumers concerns over factors such as products nutrient definition, organic status, health concerns, animal welfare, environmental consideration and management practices. The impact of the melamine crisis in China was a good example of how sensitive dairy product demand can be to food safety concerns. The ability to maintain an image of dairy products as being healthy and safe is a key challenge for the dairy industry in the future.

The tendency to shift manufacturing towards value-added products is expected to rapidly intensify. The evolution of milk pasteurisation, microfiltration and de and re-hydration of milk could have wide reaching impacts on dairy trade and consumption. Different perceptions among countries regarding what qualifies as "fresh" milk remain, but with improved technology, consumers may become increasingly willing, within the limits of remaining quality differences, to replace fresh fluid milk with reconstituted milk and/or UHT. A battle could intensify among products *vis-à-vis* nutrition and health claims and related settings of nutrient profiles and corresponding threshold levels (such as the level of saturated fatty acid). For example, increase dairy product fortification with minerals and vitamins, adding probiotic cultures and development of milk based probiotic dietary supplements are among some alternative avenues.

OECD-FAO Agricultural Outlook 2010-2019

Methodology

This section provides information on the methodological aspects of the generation of the present *Agricultural Outlook*. It discusses the main aspects in the following order: First, a general description of the agricultural baseline projections and the Outlook report is given. Second, the compilation of a consistent set of the assumptions on macroeconomic projections is discussed in more detail. A third part presents an important model element that has been improved for this Outlook, *i.e.* the representation of production costs in the model's supply equations.

The generation of the OECD-FAO *Agricultural Outlook*

The projections presented and analysed in this document are the result of a process that brings together information from a large number of sources. The use of a model jointly developed by the OECD and FAO Secretariats, based on the OECD's Aglink model and extended by FAO's Cosimo model, facilitates consistency in this process. A large amount of expert judgement, however, is applied at various stages of the Outlook process. The *Agricultural Outlook* presents a single, unified assessment, judged by the OECD and FAO Secretariats to be plausible given the underlying assumptions, the procedure of information exchange outlined below and the information to which they had access.

The starting point of the Outlook process is the reply by OECD countries (and some non-member countries) to an annual questionnaire circulated at mid-year. Through these questionnaires, the OECD Secretariat obtains information from these countries on future commodity market developments and on the evolution of their agricultural policies. This information is supplemented by the FAO Secretariat for its members which are not part of the OECD. External sources, such as the World Bank and the UN, are also used to complete the view of the main economic forces determining market developments. This part of the process is aimed at creating a first insight into possible market developments and at establishing the key assumptions which condition the Outlook. The main economic and policy assumptions are summarised in the Macroeconomic and policy assumptions chapter and in specific commodity tables of the present report. The main macroeconomic variables assumed for the Outlook period are based on the December 2009 medium term projections of the OECD's Economics Department for OECD countries, and on the Global Economic Prospects of November 2009 of the World Bank for other countries. While sometimes different from the macroeconomic assumptions provided through the questionnaire replies, it has been judged preferable to use just two consistent sources for these variables. The sources and assumptions for the macroeconomic projections are discussed in more detail further below.

As a next step, the modelling framework jointly developed by the OECD and FAO Secretariats is used to facilitate a consistent integration of this information and to derive an initial set of global market projections (baseline). In addition to quantities produced, consumed and traded, the baseline also includes projections for nominal prices (in local currency units) for the commodities concerned. Unless otherwise stated, prices referred to in the text are also in nominal terms. The data series for the projections is drawn from OECD and FAO databases. For the most part information in these databases has been taken from national statistical sources. For further details on particular series, enquiries should be directed to the OECD and FAO Secretariats.

The model provides a comprehensive dynamic economic and policy specific representation of major world producing and trading countries for the main temperate-zone commodities as well as rice and vegetable oils. The World Sugar Model, formerly a standalone model separate from Aglink was revised during the year and fully integrated into the Aglink-Cosimo modelling system. From the integrated model, a set of long term baseline projections for world and OECD sugar markets, covering raw and white (or refined) sugar was developed. In addition, comprehensive and fully integrated biofuel models have been developed for several OECD member countries as well as for a range of developing countries.* The Aglink and Cosimo country and regional modules are all developed by the OECD and FAO Secretariats in conjunction with country experts and, in some cases, with assistance from other national administrations. The initial baseline results are compared with those obtained from the questionnaire replies and issues arising are discussed in bilateral exchanges with country experts. On the basis of these discussions and of updated information, a second baseline is produced. The information generated is used to prepare market assessments for biofuels, cereals, oilseeds, meats, dairy products and sugar over the course of the Outlook period, which is discussed at the annual meetings of the Group on Commodity Markets of the OECD *Committee for Agriculture*. Following the receipt of comments and final data revisions, a last revision is made to the baseline projections. The revised projections form the basis of a draft of the present *Agricultural Outlook* publication, which is discussed by the *Working Party on Agricultural Policies and Markets of the Committee for Agriculture*, in May 2010, prior to publication. In addition, the Outlook will be used as a basis for analysis presented to the FAO's *Committee on Commodity Problems* and its various *Intergovernmental Commodity Groups*.

The Outlook process implies that the baseline projections presented in this report are conditioned by those developed by OECD countries and other participating economies. It also reconciles inconsistencies between individual country projections through the use of a formal modelling framework. The review process ensures that judgement of country experts is brought to bear on the projections and related analyses. However, the final responsibility for the projections and their interpretation rests with the OECD and FAO Secretariats.

Sources and assumptions for the macroeconomic projections

Population estimates from the 2008 Revision of the *United Nations Population Prospects Database* provide the population data used for all countries and regional aggregates in the Outlook. For the projection period, the medium variant set of estimates was selected for use from the four alternative projection variants (low, medium, high and constant fertility). The *UN Population Prospects Database* was chosen because it represents a comprehensive

* For details on the modeling of biofuels in Aglink-Cosimo and a detailed analysis of the market impacts of biofuel policies, see OECD/IEA, Economic Assessment of Biofuel Support Policies (Paris, OECD).

source of reliable estimates which includes data for non-OECD developing countries. For consistency reasons, the same source is used for both the historical population estimates and the projection data.

The other macroeconomic series used in the Aglink-Cosimo model are real GDP, the GDP deflator, the private consumption expenditure (PCE) deflator, the Brent crude oil price (in US dollars per barrel) and exchange rates expressed as the local currency value of USD 1. Historical data for these series in OECD countries are consistent with those published in the *OECD Economic Outlook*, No. 86, December 2009 and in the OECD Main Economic Indicators. Assumptions made about the future paths of all these variables apart from exchange rates, are based on the recent (December 2009) medium-term macroeconomic projections of the OECD Economics Department and extended from 2016 by holding the 2015 to 2016 annual growth rate constant for the remaining years to 2019. Exchange rates for OECD countries were extended to 2019 from the 2010 projections using the simple assumption of constant rates in real terms.

For non-member economies, historical and projection data for these macroeconomic series were obtained from the World Bank 2010 Global Economic Prospects of November 2009.

The model uses indices for real GDP, consumer prices (PCE deflator) and producer prices (GDP deflator) which are constructed with the base year 2005 value being equal to 1. The assumption of constant real exchange rates implies that a country with higher (lower)inflation relative to the United States (as measured by the US GDP deflator) will have a depreciating (appreciating) currency and therefore an increasing (decreasing) exchange rate over the projection period, since the exchange rate is measured as the local currency value of USD 1. The calculation of the nominal exchange rate uses the percentage growth of the ratio "country-GDP deflator/US GDP deflator".

The oil price used to generate the Outlook is based on information from the *OECD Economic Outlook*, No. 86.

The representation of production costs in Aglink-Cosimo

Changes in production costs are an important variable for farmers' decisions on crop and livestock production quantities, in addition to output returns and, if applicable, policy measures.

While supply in Aglink-Cosimo is largely determined by gross returns, production costs are represented in the model in the form of a cost index used to deflate gross production revenues. In other words, supply equations in the model in most cases depend on gross returns per unit of activity (such as returns per hectare or the meat price) relative to the overall production cost level as expressed by the index. Consequently, equations for harvested areas in crop production and for livestock production quantities take the following general forms:

$$AH = f\left(\frac{RH}{CPCI}\right);\ QP = f\left(\frac{PP}{CPCI}\right)$$

with:

AH	area harvested (crop production);
RH	returns per hectare (crop production);
CPCI	commodity production cost index;
QP	production quantity (livestock production);
PP	producer price (livestock production).

Among others, energy prices, increased by rising crude oil prices, have fostered attention to agricultural production costs in agricultural commodity models. Energy prices can significantly impact on international markets for agricultural products as production costs for both crops and livestock products are highly dependent on energy costs. Fuels for tractors and other machinery, as well as heating and other forms of energy are directly used in the production process. In addition, other inputs such as fertilisers and pesticides, have a high energy content, and costs for these inputs are driven to a significant extent by energy prices. It is therefore important to explicitly consider energy prices in the representation of production costs.

The production cost indices employed in Aglink-Cosimo – one each for crops and for livestock products, respectively, to account for the different shares of input groups in total production costs – is constructed from three sub-indices representing non-tradable inputs, energy inputs, and other tradable inputs, respectively. While the non-tradable sub-index is approximated by the domestic GDP deflator, the energy sub-index is affected by changes in the world crude oil price and the country's exchange rate. Finally, the tradable sub-index is linked to global inflation (approximated by the US GDP deflator) and the country's exchange rate. This relationship is shown in the following equation:

$$
\begin{aligned}
CPCI^{I}_{r,t} = {} & CPCS^{NT,I}_{r,t} * GDPD_{r,t} \\
& + CPCS^{EN,I}_{r,t} * \left(XP^{OIL}_{t} * XR_{r,t}\right) / \left(XP^{OIL}_{bas} * XR_{r,bas}\right) \\
& + \left(1 - CPCS^{NT,I}_{r,t} - CPCS^{EN,I}_{r,t}\right) * XR_{r,t} / XR_{r,bas} * GDPD_{USA,t}
\end{aligned}
$$

with:

$CPCI^{I}$	commodity production cost index for commodity group I;
$CPCS^{NT,I}$	share of non-tradable input in total base commodity production costs for commodity group I;
$CPCS^{EN,I}$	share of energy in total base commodity production costs for commodity group I;
GDPD	deflator for the gross domestic product;
XP^{OIL}	world crude oil price;
XR	nominal exchange rate with respect to the US dollar;
I	commodity group (crops, livestock products);
r,t	region and time index, respectively
bas	base year (2000) value.

Detailed data on the composition of production costs are available to the OECD Secretariat for Argentina, New Zealand and the United States. These data, available from the Secretariat on request, suggest non-tradable and energy shares in crop production costs as shown in the table below. Given that detailed data on other countries are not available, the respective crop production cost shares for Argentina are applied for all non-OECD countries, those for New Zealand are applied also for Australia, and the shares found for the US are applied for all other OECD countries/regions. As no data on livestock production shares are available for Argentina, the shares found for the US are applied to all countries/regions with the exception of New Zealand and Australia.

Production cost shares for:		Argentina (%)	New Zealand (%)	United States (%)
Crop production	Non-tradable	47	66	67
	Energy	43	27	25
	Other tradable	10	7	8
	Applied for:	*All non-OECD countries/regions*	*New Zealand, Australia*	*All other OECD countries/regions*
Livestock production	Non-tradable	n.a.	77	97
	Energy	n.a.	23	3
	Other tradable[a]	n.a.	1	0
	Applied for:	–	*New Zealand, Australia*	*All other countries/regions*

a) Excludes tradable feed.

ANNEX A

Statistical Tables

Table A.1. **Economic assumptions**

Calendar year[a]		Average 2007-09 est.	2010	2011	2012	2013	2014	2015	2016	2017	2018	2019
REAL GDP[b]												
Australia	%	2.4	2.4	3.5	4.0	4.1	4.0	4.0	3.9	3.2	3.2	3.2
Canada	%	0.1	2.0	3.0	2.8	2.3	2.2	2.1	2.1	1.6	1.6	1.6
EU15	%	–0.3	1.0	1.8	2.3	2.2	2.2	2.2	2.2	1.7	1.7	1.7
Japan	%	–1.4	1.5	2.0	1.6	1.1	1.1	1.1	1.2	1.2	1.2	1.2
Korea	%	2.5	4.4	4.2	4.8	4.7	4.6	4.5	4.3	3.8	3.8	3.8
Mexico	%	–0.6	2.7	3.9	4.0	4.0	4.0	4.1	3.9	3.0	3.0	3.0
New Zealand	%	0.4	1.5	2.7	2.7	2.3	2.3	2.2	2.3	2.0	2.0	2.0
Norway	%	1.3	1.3	3.2	3.8	3.6	3.5	3.3	3.2	3.3	3.3	3.3
Switzerland	%	1.2	0.9	1.9	2.6	2.9	3.0	3.0	2.9	2.5	2.5	2.5
Turkey	%	–0.1	3.3	4.2	4.6	4.6	4.6	4.6	4.6	4.6	4.6	4.6
United States	%	0.0	2.5	2.8	2.8	2.6	2.7	2.7	2.7	2.3	2.3	2.3
Argentina	%	4.4	2.3	2.4	4.7	4.7	4.6	4.6	4.6	4.6	4.6	4.6
Brazil	%	3.6	3.6	3.9	3.2	3.2	3.1	3.1	3.1	3.1	3.1	3.1
China	%	10.1	8.7	9.0	8.5	8.4	8.2	8.1	8.2	8.2	8.2	8.2
India	%	7.1	7.5	8.0	6.9	6.9	6.8	6.8	6.8	6.8	6.8	6.8
Russian Federation	%	1.7	3.2	3.4	5.6	5.5	5.3	5.3	5.4	5.4	5.4	5.4
South Africa	%	2.1	2.1	3.5	4.2	4.3	4.3	4.3	4.3	4.3	4.3	4.3
OECD[c, d]	%	–0.2	1.9	2.5	2.7	2.5	2.5	2.5	2.5	2.1	2.1	2.1
PCE DEFLATOR[b]												
Australia	%	3.3	2.4	1.8	2.2	2.5	2.5	2.5	2.5	2.5	2.5	2.5
Canada	%	1.3	1.1	0.9	1.6	2.1	2.1	2.1	2.1	2.1	2.1	2.1
EU15	%	1.9	0.5	0.6	1.5	2.0	2.0	2.0	2.0	2.0	2.0	2.0
Japan	%	–0.5	–1.1	–0.8	0.5	1.1	1.1	1.1	1.1	1.1	1.1	1.1
Korea	%	2.9	2.8	3.0	3.0	3.0	3.0	3.0	3.0	3.0	3.0	3.0
Mexico	%	6.3	4.4	4.8	3.9	3.2	3.2	3.2	3.2	3.2	3.2	3.2
New Zealand	%	2.7	1.4	1.4	1.9	2.1	2.1	2.1	2.1	2.1	2.1	2.1
Norway	%	2.5	1.7	2.2	2.2	2.1	2.1	2.1	2.1	2.1	2.1	2.1
Switzerland	%	1.2	0.7	0.4	0.8	1.1	1.1	1.1	1.1	1.1	1.1	1.1
Turkey	%	8.4	6.8	5.5	6.0	6.0	6.0	6.0	6.0	6.0	6.0	6.0
United States	%	2.1	1.4	1.2	1.7	2.0	2.0	2.0	2.0	2.0	2.0	2.0
Argentina	%	8.6	11.6	14.6	5.0	5.0	5.0	5.6	5.6	5.6	5.6	5.6
Brazil	%	0.2	4.7	2.8	4.5	4.5	4.5	4.7	4.7	4.7	4.7	4.7
China	%	3.2	2.4	3.2	1.9	1.9	1.9	2.0	2.0	2.0	2.0	2.0
India	%	7.9	8.0	6.0	6.0	6.0	6.0	6.0	6.0	6.0	6.0	6.0
Russian Federation	%	11.0	8.8	7.5	7.7	7.2	7.5	7.7	7.7	7.7	7.7	7.7
South Africa	%	7.9	6.6	7.2	6.0	6.0	6.0	6.0	6.0	6.0	6.0	6.0
OECD[c, d]	%	2.1	1.1	1.1	1.8	2.1	2.1	2.1	2.1	2.2	2.2	2.2

For notes, see end of the table.

Table A.1. **Economic assumptions** (*cont.*)

Calendar year[a]		2009 est.	2010	2011	2012	2013	2014	2015	2016	2017	2018	2019
POPULATION												
Australia	%	1.0	1.0	1.0	1.0	1.0	1.0	1.0	1.0	0.9	0.9	0.9
Canada	%	0.9	0.9	0.9	0.9	0.9	0.9	0.9	0.9	0.9	0.9	0.9
EU27	%	0.3	0.3	0.2	0.2	0.2	0.2	0.2	0.1	0.1	0.1	0.1
Japan	%	–0.1	–0.1	–0.1	–0.2	–0.2	–0.2	–0.2	–0.3	–0.3	–0.3	–0.4
Korea	%	0.4	0.3	0.3	0.3	0.3	0.2	0.2	0.2	0.2	0.1	0.1
Mexico	%	1.0	0.9	0.9	0.9	0.9	0.8	0.8	0.8	0.7	0.7	0.7
New Zealand	%	0.9	0.9	0.9	0.9	0.9	0.9	0.8	0.8	0.8	0.8	0.8
Norway	%	0.9	0.9	0.8	0.8	0.7	0.7	0.7	0.7	0.7	0.6	0.6
Switzerland	%	0.4	0.4	0.4	0.4	0.4	0.4	0.4	0.4	0.4	0.4	0.4
Turkey	%	1.2	1.2	1.2	1.1	1.1	1.1	1.0	1.0	1.0	1.0	0.9
United States	%	1.0	0.9	0.9	0.9	0.9	0.9	0.9	0.9	0.8	0.8	0.8
Argentina	%	1.0	1.0	0.9	0.9	0.9	0.9	0.9	0.9	0.8	0.8	0.8
Brazil	%	0.9	0.9	0.8	0.8	0.7	0.7	0.7	0.7	0.6	0.6	0.6
China	%	0.6	0.6	0.6	0.6	0.6	0.6	0.6	0.6	0.5	0.5	0.5
India	%	1.4	1.4	1.3	1.3	1.3	1.2	1.2	1.2	1.1	1.1	1.1
Russia	%	–0.4	–0.4	–0.4	–0.3	–0.3	–0.3	–0.3	–0.4	–0.4	–0.4	–0.4
South Africa	%	0.9	0.8	0.6	0.5	0.4	0.4	0.4	0.4	0.4	0.4	0.4
OECD[c]	%	0.5	0.5	0.5	0.5	0.5	0.4	0.4	0.4	0.4	0.4	0.4
World	%	1.2	1.2	1.1	1.1	1.1	1.1	1.1	1.0	1.0	1.0	1.0

Calendar year[a]		Average 2007-09 est.	2010	2011	2012	2013	2014	2015	2016	2017	2018	2019
GDP DEFLATOR												
Australia	%	3.5	1.8	2.3	2.6	2.6	2.6	2.6	2.6	2.6	2.6	2.6
Canada	%	1.7	2.3	1.4	2.0	2.4	2.4	2.5	2.6	2.7	2.7	2.7
European Union	%	1.6	0.5	0.6	1.5	2.0	2.0	2.0	2.0	2.0	2.0	2.0
Japan	%	–0.4	–1.6	–0.7	0.1	0.9	0.8	0.8	0.8	0.8	0.8	0.8
Korea	%	2.6	0.4	2.0	2.2	1.6	1.5	1.4	1.2	0.9	0.9	0.9
Mexico	%	5.9	4.1	4.7	3.8	3.1	3.1	3.1	3.2	3.2	3.2	3.2
New Zealand	%	3.5	2.5	2.2	2.4	2.0	2.1	2.2	2.2	2.2	2.2	2.2
Norway	%	2.8	3.7	2.9	2.7	2.3	2.2	2.2	2.2	2.2	2.2	2.2
Switzerland	%	1.7	0.5	0.3	0.8	1.2	1.2	1.3	1.3	1.3	1.3	1.3
Turkey	%	8.4	6.8	5.5	6.0	6.0	6.0	6.0	6.0	6.0	6.0	6.0
United States	%	2.1	0.9	1.0	1.6	1.9	1.9	1.9	1.9	1.9	1.9	1.9
Argentina	%	1.3	1.5	4.8	4.0	4.0	4.0	4.0	4.0	4.0	4.0	4.0
Brazil	%	9.4	5.5	6.2	4.7	4.7	4.7	4.7	4.7	4.7	4.7	4.7
China	%	8.6	3.0	4.9	1.7	1.7	1.7	1.7	1.7	1.7	1.7	1.7
India	%	7.9	8.0	6.0	6.0	6.0	6.0	6.0	6.0	6.0	6.0	6.0
Russian Federation	%	6.5	10.7	6.3	6.1	6.1	6.1	6.1	6.1	6.1	6.1	6.1
South Africa	%	7.9	6.6	7.2	6.0	6.0	6.0	6.0	6.0	6.0	6.0	6.0
OECD[c, d]	%	2.0	0.8	1.0	1.7	2.0	2.1	2.1	2.1	2.1	2.1	2.1
WORLD OIL PRICE												
Brent crude oil price	*USD/barrel*	77.0	77.0	77.0	78.5	80.8	83.3	85.8	88.4	91.1	93.9	96.7

est.: Estimate.

a) For OECD member countries, historical data for population, real GDP, private consumption expenditure deflator and GDP deflator were obtained from the OECD *Economic Outlook*, No. 86 (December 2009). For non-member economies, historical macroeconomic data were obtained from the World Bank, November 2009. Assumptions for the projection period draw on the recent medium term macroeconomic projections of the OECD Economics Department, projections of the World Bank, and for population, projections from the *United Nations World Population Prospects Database*, 2008 Revision (medium variant). Data for the European Union are for the euro area aggregates.

b) Annual per cent change. The price index used is the private consumption expenditure deflator.

c) Excludes Iceland.

d) Annual weighted average real GDP and CPI growth rates in OECD countries are based on weights using purchasing power parities (PPPs).

Source: OECD and FAO Secretariats.

StatLink http://dx.doi.org/10.1787/888932287178

Table A.2. **World prices**[a]

		Average 07/08-09/10 est.	10/11	11/12	12/13	13/14	14/15	15/16	16/17	17/18	18/19	19/20
WHEAT												
Price[b]	USD/t	275.9	218.5	200.9	206.2	210.5	214.4	227.4	229.0	226.6	227.7	225.4
COARSE GRAINS												
Price[c]	USD/t	193.6	185.5	182.0	185.8	187.0	191.8	200.7	197.8	193.3	190.0	187.2
RICE												
Price[d]	USD/t	539.0	452.9	436.0	425.9	426.6	429.7	426.2	425.1	425.8	422.5	422.5
OILSEEDS												
Price[e]	USD/t	468.7	409.9	412.0	405.3	401.4	409.0	416.6	416.0	416.5	415.5	418.8
PROTEIN MEALS												
Price[f]	USD/t	374.1	308.0	298.2	283.7	280.0	283.3	287.2	284.6	283.2	283.6	287.9
VEGETABLE OILS												
Price[g]	USD/t	946.7	875.2	897.8	930.2	932.4	954.0	992.1	1 006.4	1 021.7	1 030.1	1 042.7
SUGAR												
Price, raw sugar[h]	USD/t	401.4	397.8	331.1	290.7	275.8	310.6	296.3	281.0	306.8	326.8	371.7
Price, refined sugar[i]	USD/t	462.2	448.4	402.6	368.5	338.3	371.9	360.0	355.0	377.2	395.0	439.2
BEEF AND VEAL												
Price, European Union[j]	EUR/t dw	2 882.5	2 600.3	2 441.1	2 454.1	2 520.8	2 542.0	2 583.1	2 574.2	2 561.5	2 524.7	2 508.4
Price, United States[k]	USD/t dw	3 165.0	3 143.1	3 298.8	3 403.5	3 576.8	3 645.7	3 674.6	3 659.0	3 649.5	3 624.4	3 561.9
Price, Brazil[l]	USD/t pw	3 279.3	3 350.9	3 137.9	3 159.8	3 245.3	3 266.2	3 312.7	3 294.3	3 273.0	3 217.4	3 196.2
PIG MEAT												
Price, European Union[m]	EUR/t dw	1 467.4	1 355.1	1 227.2	1 417.9	1 493.4	1 485.2	1 464.5	1 483.5	1 497.3	1 503.3	1 521.1
Price, United States[n]	USD/t dw	1 384.5	1 364.0	1 506.4	1 554.0	1 677.7	1 708.6	1 694.9	1 679.4	1 714.0	1 717.7	1 681.0
Price, Brazil[o]	USD/t dw	2 269.9	2 233.5	2 399.4	2 396.8	2 493.3	2 397.3	2 401.3	2 401.8	2 318.2	2 311.0	2 259.3
POULTRY MEAT												
Price, European Union[p]	EUR/t rtc	1 192.7	1 132.6	1 125.6	1 118.9	1 130.5	1 127.0	1 121.9	1 109.4	1 098.0	1 086.0	1 090.1
Price, United States[q]	USD/t rtc	1 066.0	1 118.6	1 137.6	1 160.5	1 172.0	1 200.4	1 226.9	1 245.3	1 265.5	1 284.5	1 320.3
Price, Brazil[r]	USD/t pw	1 606.6	1 548.7	1 579.9	1 601.9	1 613.4	1 646.7	1 682.4	1 665.4	1 659.3	1 638.2	1 638.4
SHEEP MEAT												
Price, New Zealand[s]	NZD/t dw	2 575.8	3 467.5	3 268.0	3 076.0	3 070.4	3 236.4	3 460.7	3 251.8	3 555.2	3 400.0	3 673.7
BUTTER												
Price[t]	USD/t	2 978.5	3 042.6	2 820.7	2 716.4	2 709.1	2 693.5	2 741.6	2 765.5	2 921.6	2 919.2	2 958.3
CHEESE												
Price[u]	USD/t	3 886.9	3 716.4	3 272.0	3 126.0	3 139.7	3 250.7	3 337.8	3 427.9	3 541.7	3 589.7	3 640.7
SKIM MILK POWDER												
Price[v]	USD/t	3 308.8	2 530.4	2 434.3	2 417.2	2 493.1	2 590.1	2 653.2	2 759.1	2 917.6	2 942.9	3 000.1
WHOLE MILK POWDER												
Price[w]	USD/t	3 499.7	2 808.0	2 549.8	2 475.5	2 588.4	2 725.3	2 763.1	2 832.2	2 937.8	2 989.3	3 042.4
WHEY POWDER												
Wholesale price, United States[x]	USD/t	853.0	788.1	709.5	705.4	727.2	751.0	767.9	791.2	827.0	835.4	852.0
CASEIN												
Price[y]	USD/t	6 299.8	5 774.5	5 529.4	5 450.6	5 474.2	5 688.8	5 861.8	6 060.6	6 267.8	6 342.7	6 439.8

For notes, see end of the table.

Table A.2. **World prices**[a] *(cont.)*

		Average 07/08-09/10 est.	10/11	11/12	12/13	13/14	14/15	15/16	16/17	17/18	18/19	19/20
ETHANOL												
Price[z]	USD/hl	43.8	47.4	51.4	50.0	50.2	50.7	51.1	52.0	53.3	53.9	54.4
BIODIESEL												
Price[aa]	USD/hl	118.0	117.7	118.5	124.4	126.4	131.8	140.2	142.6	145.0	143.2	144.3

est.: Estimate.

a) This table is a compilation of price information presented in the detailed commodity tables further in this annex. Prices for crops are on marketing year basis and those for meat and dairy products on calendar year basis (*e.g.* 07/08 is calendar year 2007).
b) No. 2 hard red winter wheat, ordinary protein, USA f.o.b. Gulf Ports (June/May), less EEP payments where applicable.
c) No. 2 yellow corn, US f.o.b. Gulf Ports (September/August).
d) Milled, 100%, grade b, Nominal Price Quote (NPQ), f.o.b. Bangkok (January/December).
e) Weighted average oilseed price, European port.
f) Weighted average meal price, European port.
g) Weighted average price of oilseed oils and palm oil, European port.
h) Raw sugar world price, ICE Inc. No. 11 f.o.b., bulk price, October/September.
i) Refined sugar price, White Sugar Futures Contract No. 407, Euronext market, Liffe, London, Europe, October/September.
j) Producer price.
k) Choice steers, 1 100-1 300 lb lw, Nebraska – lw to dw conversion factor 0.63.
l) Brazil meat of bovine export price (HS 0201 and HS 0202).
m) Pig producer price.
n) Barrows and gilts, No. 1-3, 230-250 lb lw, Iowa/South Minnesota – lw to dw conversion factor 0.74.
o) Producer price.
p) Weighted average farm gate live chickens, first choice, lw to rtc conversion of 0.75.
q) Wholesale weighted average broiler price 12 cities.
r) Weighted average wholesale price of differents cuts.
s) Lamb schedule price, all grade average.
t) F.o.b. export price, butter, 82% butterfat, Oceania.
u) F.o.b. export price, cheddar cheese, 39% moisture, Oceania.
v) F.o.b. export price, non-fat dry milk, 1.25% butterfat, Oceania.
w) F.o.b. export price, WMP 26% butterfat, Oceania.
x) Edible dry whey, Wisconsin, plant.
y) Export price, New Zealand.
z) Brazil, Sao Paulo (ex-distillery).
aa) Producer price Germany net of biodiesel tariff.
Source: OECD and FAO Secretariats.

StatLink http://dx.doi.org/10.1787/888932287197

Table A.3. **World trade projections**

IMPORTS			Average 2007-09 est.	2010	2011	2012	2013	2014	2015	2016	2017	2018	2019
Wheat	**World trade**	*kt*	**121 483**	**120 014**	**122 727**	**126 917**	**128 222**	**130 878**	**132 135**	**135 012**	**137 284**	**140 050**	**142 271**
	OECD	*kt*	25 874	24 159	24 184	25 073	24 656	24 841	24 883	24 760	24 832	25 011	25 022
	Developing	*kt*	97 233	97 685	100 453	104 851	106 366	109 135	110 638	113 527	115 886	118 693	121 042
	Least developed countries	*kt*	12 224	13 006	13 412	13 605	14 186	14 692	15 052	15 622	16 122	16 424	16 872
Coarse grains	**World trade**	*kt*	**118 329**	**112 435**	**112 957**	**114 089**	**116 422**	**116 845**	**119 460**	**123 626**	**125 983**	**128 881**	**132 301**
	OECD	*kt*	56 908	50 815	49 976	50 104	51 146	49 878	51 428	53 660	52 623	52 874	53 134
	Developing	*kt*	81 184	82 634	84 214	85 374	86 687	88 647	89 723	91 803	95 807	98 753	102 454
	Least developed countries	*kt*	2 430	2 196	2 075	2 183	2 108	2 161	2 127	2 109	2 322	2 407	2 540
Rice	**World trade**	*kt*	**31 325**	**32 057**	**32 622**	**33 184**	**33 823**	**34 635**	**35 411**	**36 333**	**37 153**	**37 861**	**38 446**
	OECD	*kt*	5 150	5 169	5 238	5 325	5 429	5 570	5 695	5 821	5 921	6 034	6 148
	Developing	*kt*	26 130	26 826	27 329	27 822	28 369	29 021	29 657	30 435	31 132	31 695	32 142
	Least developed countries	*kt*	6 557	6 247	6 582	6 669	6 812	6 945	7 185	7 312	7 394	7 479	7 596
Oilseeds	**World trade**	*kt*	**92 647**	**93 131**	**94 479**	**96 430**	**98 733**	**99 900**	**102 336**	**104 098**	**105 592**	**107 352**	**108 686**
	OECD	*kt*	34 625	33 749	33 843	33 845	34 001	33 785	33 852	33 876	33 910	33 863	33 896
	Developing	*kt*	65 840	67 019	68 463	70 714	72 790	74 249	76 703	78 475	80 023	81 994	83 405
	Least developed countries	*kt*	290	313	335	344	366	377	387	399	414	428	442
Protein meals	**World trade**	*kt*	**66 297**	**69 617**	**71 994**	**73 687**	**75 399**	**77 435**	**78 870**	**80 522**	**82 211**	**83 910**	**85 871**
	OECD	*kt*	41 252	42 563	43 531	43 575	43 867	44 346	44 600	44 971	45 140	45 376	45 730
	Developing	*kt*	27 517	29 907	31 492	33 124	34 528	36 136	37 390	38 713	40 302	41 860	43 590
	Least developed countries	*kt*	409	462	485	509	534	556	576	593	607	620	629
Vegetable oils	**World trade**	*kt*	**56 447**	**58 868**	**61 510**	**62 854**	**64 888**	**67 070**	**68 906**	**71 353**	**73 867**	**76 288**	**78 859**
	OECD	*kt*	16 787	17 699	18 526	19 158	19 796	20 640	21 504	22 346	23 163	23 664	24 231
	Developing	*kt*	39 583	41 468	43 297	44 097	45 504	46 849	47 835	49 457	51 170	53 112	55 108
	Least developed countries	*kt*	3 928	4 145	4 342	4 512	4 709	4 902	5 088	5 289	5 502	5 720	5 942
Sugar	**World trade**	*kt*	**48 712**	**52 966**	**54 815**	**55 066**	**55 589**	**56 954**	**59 022**	**59 778**	**61 306**	**63 171**	**63 234**
	OECD	*kt*	12 431	12 894	13 620	13 560	13 671	14 070	14 335	14 642	15 028	15 278	15 578
	Developing	*kt*	33 053	35 906	37 766	37 892	38 372	39 653	41 686	42 234	43 668	45 630	45 872
	Least developed countries	*kt*	4 834	5 293	5 608	6 001	6 248	6 096	6 197	6 287	6 292	6 280	6 144
Beef[a]	**World trade**	*kt*	**6 623**	**6 739**	**6 727**	**6 850**	**7 308**	**7 412**	**7 619**	**7 828**	**7 951**	**8 076**	**8 168**
	OECD	*kt*	3 342	3 422	3 435	3 481	3 641	3 728	3 848	3 922	3 950	3 991	4 015
	Developing	*kt*	3 073	3 246	3 289	3 370	3 674	3 733	3 840	3 970	4 084	4 172	4 243
	Least developed countries	*kt*	142	174	174	139	197	173	165	169	123	153	81
Pigmeat[a]	**World trade**	*kt*	**5 505**	**5 437**	**5 588**	**5 667**	**5 788**	**5 891**	**5 971**	**6 023**	**6 097**	**6 213**	**6 319**
	OECD	*kt*	2 961	3 018	3 171	3 223	3 258	3 333	3 377	3 395	3 421	3 455	3 485
	Developing	*kt*	2 405	2 563	2 796	2 747	2 833	2 873	2 941	2 968	2 996	3 080	3 121
	Least developed countries	*kt*	108	117	129	110	117	113	111	116	113	125	122
Poultry	**World trade**	*kt*	**9 635**	**9 331**	**9 541**	**9 798**	**9 941**	**10 207**	**10 361**	**10 513**	**10 833**	**11 003**	**11 375**
	OECD	*kt*	2 357	2 518	2 565	2 649	2 733	2 784	2 848	2 910	2 949	3 022	3 078
	Developing	*kt*	6 139	6 190	6 408	6 615	6 720	6 933	7 028	7 097	7 349	7 452	7 723
	Least developed countries	*kt*	566	537	569	577	544	599	602	609	640	644	685
Butter	**World trade**	*kt*	**822**	**755**	**760**	**765**	**749**	**762**	**772**	**779**	**770**	**778**	**781**
	OECD	*kt*	140	111	112	112	92	93	93	93	91	91	91
	Developing	*kt*	445	463	470	476	481	497	510	520	516	525	532
	Least developed countries	*kt*	12	9	12	13	13	14	15	15	15	16	16
Cheese	**World trade**	*kt*	**1 707**	**1 730**	**1 736**	**1 744**	**1 762**	**1 788**	**1 838**	**1 865**	**1 885**	**1 914**	**1 937**
	OECD	*kt*	737	769	758	770	782	788	798	805	808	813	818
	Developing	*kt*	706	741	772	783	798	812	837	855	867	883	903
	Least developed countries	*kt*	14	14	14	14	14	14	14	14	14	14	14
Whole milk powder	**World trade**	*kt*	**1 777**	**1 874**	**1 867**	**1 886**	**1 895**	**1 912**	**1 964**	**1 990**	**1 999**	**2 025**	**2 055**
	OECD	*kt*	101	99	101	102	100	99	101	101	98	96	94
	Developing	*kt*	1 659	1 775	1 772	1 794	1 807	1 823	1 871	1 896	1 906	1 930	1 960
	Least developed countries	*kt*	173	200	207	214	218	224	231	237	243	249	258
Skim milk powder	**World trade**	*kt*	**1 167**	**1 207**	**1 199**	**1 209**	**1 226**	**1 243**	**1 266**	**1 281**	**1 284**	**1 307**	**1 321**
	OECD	*kt*	196	206	202	206	208	207	215	216	209	212	214
	Developing	*kt*	1 038	1 128	1 130	1 140	1 156	1 173	1 194	1 208	1 212	1 233	1 247
	Least developed countries	*kt*	54	61	62	64	66	68	70	71	73	75	77

For notes, see end of the table.

Table A.3. **World trade projections** *(cont.)*

EXPORTS			Average 2007-09 est.	2010	2011	2012	2013	2014	2015	2016	2017	2018	2019
Wheat	OECD	*kt*	78 881	80 927	76 856	77 057	74 531	74 223	73 800	75 148	76 309	77 801	79 127
	Developing	*kt*	15 778	16 677	17 851	17 807	18 958	19 355	19 771	20 339	20 308	20 681	21 031
	Least developed countries	*kt*	117	160	158	153	144	137	133	127	122	119	114
Coarse grains	OECD	*kt*	77 994	71 830	76 199	76 845	77 063	75 272	75 585	77 577	79 368	80 954	83 118
	Developing	*kt*	28 652	31 904	28 553	29 578	31 008	32 964	34 663	36 255	36 998	38 222	39 037
	Least developed countries	*kt*	2 750	3 109	3 017	2 807	2 936	2 878	2 908	2 913	2 677	2 592	2 496
Rice	OECD	*kt*	3 589	3 788	3 842	3 920	3 981	4 033	4 068	4 050	4 073	4 139	4 157
	Developing	*kt*	26 603	28 203	28 716	29 201	29 779	30 540	31 282	32 223	33 020	33 664	34 232
	Least developed countries	*kt*	2 419	2 739	3 037	3 336	3 800	4 243	4 604	5 007	5 488	5 970	6 470
Oilseeds	OECD	*kt*	45 909	48 085	47 772	48 459	49 429	49 031	49 142	48 812	48 509	48 684	48 842
	Developing	*kt*	36 828	45 743	47 082	47 764	48 899	50 068	52 038	53 817	55 253	56 413	57 275
	Least developed countries	*kt*	179	186	166	151	143	138	138	137	132	129	124
Protein meals	OECD	*kt*	11 526	13 203	13 122	13 532	14 061	14 963	15 350	15 693	16 154	16 605	17 138
	Developing	*kt*	57 382	59 065	61 224	62 346	63 424	64 426	65 465	66 699	67 865	69 094	70 446
	Least developed countries	*kt*	200	223	206	215	222	225	257	274	294	308	334
Vegetable oils	OECD	*kt*	4 635	4 959	4 598	4 552	4 653	4 821	5 141	5 411	5 738	6 056	6 378
	Developing	*kt*	48 238	49 272	52 104	53 213	54 948	56 777	58 108	60 105	62 097	64 020	66 101
	Least developed countries	*kt*	191	187	181	178	175	171	169	166	164	162	160
Sugar	OECD	*kt*	6 933	5 754	6 388	6 387	6 536	6 664	6 616	6 332	6 463	6 776	6 846
	Developing	*kt*	45 423	47 648	49 532	49 740	50 208	51 509	53 621	54 621	56 031	57 578	57 541
	Least developed countries	*kt*	1 844	2 172	2 333	2 566	2 772	2 734	2 789	2 842	2 826	2 824	2 776
Beef[a]	OECD	*kt*	3 409	3 409	3 468	3 466	3 566	3 593	3 635	3 659	3 669	3 676	3 691
	Developing	*kt*	4 069	4 212	4 138	4 256	4 541	4 615	4 702	4 884	4 963	5 096	5 133
	Least developed countries	*kt*	3	2	3	3	2	2	2	2	3	2	3
Pigmeat[a]	OECD	*kt*	4 673	4 730	4 832	4 850	4 895	4 968	5 028	5 060	5 116	5 210	5 292
	Developing	*kt*	1 207	1 174	1 243	1 292	1 374	1 392	1 401	1 419	1 415	1 442	1 443
	Least developed countries	*kt*	0	0	0	0	0	0	0	0	0	0	0
Poultry	OECD	*kt*	4 501	4 277	4 122	4 162	4 190	4 220	4 239	4 280	4 330	4 369	4 491
	Developing	*kt*	5 744	6 147	6 618	6 951	7 165	7 512	7 749	7 962	8 346	8 581	8 921
	Least developed countries	*kt*	7	9	9	9	7	6	5	5	4	3	3
Butter	OECD	*kt*	675	602	611	619	604	618	628	634	621	627	629
	Developing	*kt*	92	89	93	95	97	96	97	98	102	103	105
	Least developed countries	*kt*	0	0	0	0	0	0	0	0	0	0	0
Cheese	OECD	*kt*	1 293	1 313	1 310	1 286	1 286	1 293	1 336	1 348	1 334	1 343	1 346
	Developing	*kt*	337	358	351	364	368	374	373	378	393	399	407
	Least developed countries	*kt*	0	0	0	0	0	0	0	0	0	0	1
Whole milk powder	OECD	*kt*	1 236	1 341	1 317	1 330	1 322	1 317	1 364	1 375	1 359	1 368	1 381
	Developing	*kt*	579	552	566	570	587	607	612	626	650	666	684
	Least developed countries	*kt*	7	8	8	8	8	8	8	8	8	8	8
Skim milk powder	OECD	*kt*	995	928	918	924	935	942	958	962	945	958	960
	Developing	*kt*	132	122	126	127	127	129	129	132	138	138	140
	Least developed countries	*kt*	1	1	1	1	1	1	1	1	1	1	1
Biofuel[b]	Ethanol world trade	*mnl*	4 465	5 697	7 176	7 426	7 744	8 756	10 095	10 627	12 365	13 709	14 591
	Biodiesel world trade	*mnl*	2 250	1 888	2 000	2 226	2 452	2 735	3 016	2 980	2 931	2 944	2 957

est.: Estimate.

a) Excludes trade of live animals.

b) Sum of all positive net trade positions.

Source: OECD and FAO Secretariats.

StatLink http://dx.doi.org/10.1787/888932287216

Table A.4. **Main policy assumptions for cereal markets**

Crop year[a]		Average 2007-09 est.	2010	2011	2012	2013	2014	2015	2016	2017	2018	2019
ARGENTINA												
Crops export tax	%	24.7	27.1	27.1	27.1	27.1	27.1	27.1	27.1	27.1	27.1	27.1
Rice export tax	%	1.0	1.0	1.0	1.0	1.0	1.0	1.0	1.0	1.0	1.0	1.0
CANADA												
Tariff-quotas[b]												
Wheat	*kt*	350.0	350.0	350.0	350.0	350.0	350.0	350.0	350.0	350.0	350.0	350.0
In-quota tariff	%	1.1	1.1	1.1	1.1	1.1	1.1	1.1	1.1	1.1	1.1	1.1
Out-of-quota tariff	%	61.7	61.7	61.7	61.7	61.7	61.7	61.7	61.7	61.7	61.7	61.7
Barley	*kt*	399.0	399.0	399.0	399.0	399.0	399.0	399.0	399.0	399.0	399.0	399.0
In-quota tariff	%	0.7	0.7	0.7	0.7	0.7	0.7	0.7	0.7	0.7	0.7	0.7
Out-of-quota tariff	%	58.0	58.0	58.0	58.0	58.0	58.0	58.0	58.0	58.0	58.0	58.0
EUROPEAN UNION[c]												
Cereal support price[d]	*EUR/t*	101.3	101.3	101.3	101.3	101.3	101.3	101.3	101.3	101.3	101.3	101.3
Single farm payment[e]	*EUR/ha*	187.9	187.9	187.9	187.9	187.9	187.9	187.9	187.9	187.9	187.9	187.9
Rice support price[g]	*EUR/t*	150.0	150.0	150.0	150.0	150.0	150.0	150.0	150.0	150.0	150.0	150.0
Compulsory set-aside rate	%	3.3	0.0	0.0	0.0	0.0	0.0	0.0	0.0	0.0	0.0	0.0
Direct payment for rice[f]	*EUR/ha*	458.9	458.9	458.9	187.9	187.9	187.9	187.9	187.9	187.9	187.9	187.9
Wheat tariff-quota[b]	*kt*	3 779.9	3 779.9	3 779.9	3 779.9	3 779.9	3 779.9	3 779.9	3 779.9	3 779.9	3 779.9	3 779.9
Coarse grain tariff-quota[b]	*kt*	3 469.5	3 469.5	3 469.5	3 469.5	3 469.5	3 469.5	3 469.5	3 469.5	3 469.5	3 469.5	3 469.5
Subsidised export limits[b]												
Wheat	*mt*	15.4	15.4	15.4	15.4	15.4	15.4	15.4	15.4	15.4	15.4	15.4
Coarse grains[h]	*mt*	10.8	10.8	10.8	10.8	10.8	10.8	10.8	10.8	10.8	10.8	10.8
JAPAN												
Wheat tariff-quota	*kt*	5 740.0	5 740.0	5 740.0	5 740.0	5 740.0	5 740.0	5 740.0	5 740.0	5 740.0	5 740.0	5 740.0
In-quota tariff	*'000 JPY/t*	0.0	0.0	0.0	0.0	0.0	0.0	0.0	0.0	0.0	0.0	0.0
Out-of-quota tariff	*'000 JPY/t*	55.0	55.0	55.0	55.0	55.0	55.0	55.0	55.0	55.0	55.0	55.0
Barley tariff-quota	*kt*	1 369.0	1 369.0	1 369.0	1 369.0	1 369.0	1 369.0	1 369.0	1 369.0	1 369.0	1 369.0	1 369.0
In-quota tariff	*'000 JPY/t*	0.0	0.0	0.0	0.0	0.0	0.0	0.0	0.0	0.0	0.0	0.0
Out-of-quota tariff	*'000 JPY/t*	39.0	39.0	39.0	39.0	39.0	39.0	39.0	39.0	39.0	39.0	39.0
Rice tariff-quota	*kt*	682.2	682.2	682.2	682.2	682.2	682.2	682.2	682.2	682.2	682.2	682.2
In-quota tariff	*'000 JPY/t*	0.0	0.0	0.0	0.0	0.0	0.0	0.0	0.0	0.0	0.0	0.0
Out-of-quota tariff	*'000 JPY/t*	341.0	341.0	341.0	341.0	341.0	341.0	341.0	341.0	341.0	341.0	341.0
KOREA												
Wheat tariff	%	5.4	5.4	5.4	5.4	5.4	5.4	5.4	5.4	5.4	5.4	5.4
Maize tariff-quota	*kt*	6 102.0	6 102.0	6 102.0	6 102.0	6 102.0	6 102.0	6 102.0	6 102.0	6 102.0	6 102.0	6 102.0
In-quota tariff	%	1.7	1.7	1.7	1.7	1.7	1.7	1.7	1.7	1.7	1.7	1.7
Out-of-quota tariff	%	403.5	403.5	403.5	403.5	403.5	403.5	403.5	403.5	403.5	403.5	403.5
Barley tariff-quota	*kt*	53.6	53.6	53.6	53.6	53.6	53.6	53.6	53.6	53.6	53.6	53.6
In-quota tariff	%	22.5	22.5	22.5	22.5	22.5	22.5	22.5	22.5	22.5	22.5	22.5
Out-of-quota tariff	%	359.3	359.3	359.3	359.3	359.3	359.3	359.3	359.3	359.3	359.3	359.3
Rice quota[i]	*kt*	295.7	340.9	363.5	386.1	408.7	408.7	408.7	408.7	408.7	408.7	408.7
In-quota tariff	%	5	5	5	5	5	5	5	5	5	5	5
MERCOSUR												
Wheat tariff	%	10.0	10.0	10.0	10.0	10.0	10.0	10.0	10.0	10.0	10.0	10.0
Coarse grain tariff	%	8.0	8.0	8.0	8.0	8.0	8.0	8.0	8.0	8.0	8.0	8.0
Rice tariff	%	10.5	10.0	10.0	10.0	10.0	10.0	10.0	10.0	10.0	10.0	10.0
MEXICO												
Wheat NAFTA tariff	%	0.0	0.0	0.0	0.0	0.0	0.0	0.0	0.0	0.0	0.0	0.0
Maize tariff-quota	*kt*	2 501	2 501	2 501	2 501	2 501	2 501	2 501	2 501	2 501	2 501	2 501
In-quota tariff	%	50.0	50.0	50.0	50.0	50.0	50.0	50.0	50.0	50.0	50.0	50.0
Out-of-quota tariff	%	194.0	194.0	194.0	194.0	194.0	194.0	194.0	194.0	194.0	194.0	194.0
Barley tariff-quota	*kt*	4.7	4.7	4.7	4.7	4.7	4.7	4.7	4.7	4.7	4.7	4.7
In-quota tariff	%	50.0	50.0	50.0	50.0	50.0	50.0	50.0	50.0	50.0	50.0	50.0
Out-of-quota tariff	%	115.2	115.2	115.2	115.2	115.2	115.2	115.2	115.2	115.2	115.2	115.2

For notes, see end of the table.

Table A.4. **Main policy assumptions for cereal markets** *(cont.)*

Crop year[a]		Average 2007-09 est.	2010	2011	2012	2013	2014	2015	2016	2017	2018	2019
UNITED STATES												
ACRE participation rate												
Wheat	%	0.0	10.0	10.0	10.0	10.0	10.0	10.0	10.0	10.0	10.0	10.0
Coarse grains	%	0.0	10.0	10.0	10.0	10.0	10.0	10.0	10.0	10.0	10.0	10.0
Rice	%	0.0	0.0	0.0	0.0	0.0	0.0	0.0	0.0	0.0	0.0	0.0
Wheat loan rate	USD/t	101.0	108.0	108.0	108.0	108.0	108.0	108.0	108.0	108.0	108.0	108.0
Maize loan rate	USD/t	76.8	76.8	76.8	76.8	76.8	76.8	76.8	76.8	76.8	76.8	76.8
Prod. flex. contract payment												
Wheat	USD/t	16.9	16.9	16.9	16.9	16.9	16.9	16.9	16.9	16.9	16.9	16.9
Maize	USD/t	10.3	10.3	10.3	10.3	10.3	10.3	10.3	10.3	10.3	10.3	10.3
CRP areas[j]	mha	6.9	6.4	6.1	6.0	6.0	6.1	6.3	6.4	6.4	6.4	3.1
Wheat	mha	3.6	3.3	3.2	3.1	3.1	3.2	3.3	3.3	3.3	3.3	0.0
Coarse grains	mha	3.4	3.0	2.9	2.9	2.9	2.9	3.0	3.1	3.1	3.1	3.1
Subsidised export limits[b]												
Wheat	mt	14.5	14.5	14.5	14.5	14.5	14.5	14.5	14.5	14.5	14.5	14.5
Coarse grains	mt	1.6	1.6	1.6	1.6	1.6	1.6	1.6	1.6	1.6	1.6	1.6
Wheat EEP payment[k]	USD/t	0	0	0	0	0	0	0	0	0	0	0
CHINA												
Wheat tariff-quota	kt	9 636	9 636	9 636	9 636	9 636	9 636	9 636	9 636	9 636	9 636	9 636
In-quota tariff	%	2.3	2.3	2.3	2.3	2.3	2.3	2.3	2.3	2.3	2.3	2.3
Out-of-quota tariff	%	65.0	65.0	65.0	65.0	65.0	65.0	65.0	65.0	65.0	65.0	65.0
Coarse grains tariff	%	2.0	2.0	2.0	2.0	2.0	2.0	2.0	2.0	2.0	2.0	2.0
Maize tariff-quota	kt	7 200	7 200	7 200	7 200	7 200	7 200	7 200	7 200	7 200	7 200	7 200
In-quota tariff	%	3.7	3.7	3.7	3.7	3.7	3.7	3.7	3.7	3.7	3.7	3.7
Out-of-quota tariff	%	41.7	41.7	41.7	41.7	41.7	41.7	41.7	41.7	41.7	41.7	41.7
Rice tariff-quota	kt	5 320	5 320	5 320	5 320	5 320	5 320	5 320	5 320	5 320	5 320	5 320
In-quota tariff	%	2.3	2.3	2.3	2.3	2.3	2.3	2.3	2.3	2.3	2.3	2.3
Out-of-quota tariff	%	51.7	51.7	51.7	51.7	51.7	51.7	51.7	51.7	51.7	51.7	51.7
INDIA												
Input subsidy rate coarse grains[l]	INR/t	1 144	1 249	1 249	1 249	1 249	1 249	1 249	1 249	1 249	1 249	1 249
Input subsidy rate rice[l]	INR/t	649	671	671	671	671	671	671	671	671	671	671
Input subsidy rate wheat[l]	INR/t	1 707	1 621	1 621	1 621	1 621	1 621	1 621	1 621	1 621	1 621	1 621
Minimum support price												
Maize	INR/t	3 600	0	0	0	0	0	0	0	0	0	0
Rice	INR/t	9 133	9 800	9 800	9 800	9 800	9 800	9 800	9 800	9 800	9 800	9 800
Wheat	INR/t	10 533	10 800	10 800	10 800	10 800	10 800	10 800	10 800	10 800	10 800	10 800
Wheat export subsidy	INR/t	1 941	1 941	1 941	1 941	1 941	1 941	1 941	1 941	1 941	1 941	1 941
Wheat tariff	%	88	88	88	88	88	88	88	88	88	88	88
Maize tariff	%	50	50	50	50	50	50	50	50	50	50	50
Rice tariff	%	30	30	30	30	30	30	30	30	30	30	30
Barley tariff	%	100	100	100	100	100	100	100	100	100	100	100

Note: The source for tariffs and Tariff Rate Quotas is AMAD (*Agricultural Market Access Database*). The tariff and TRQ data are based on Most Favoured Nation rates scheduled with the WTO and exclude those under preferential or regional agreements, which may be substantially different. Tariffs are simple averages of several product lines. Specific rates are converted to *ad valorem* rates using world prices in the Outlook. Import quotas are based on global commitments scheduled in the WTO rather than those allocated to preferential partners under regional or other agreements. For Mexico, the NAFTA in-quota tariff on maize and barley is zero, while the tariff-rate quota becomes unlimited in 2003 for barley and 2008 for maize.
est.: Estimate.

a) Beginning crop marketing year – see "Glossary of Terms" for definitions.
b) Year beginning 1 July.
c) EU farmers also benefit from the Single Farm Payment (SFP) Scheme, which provides flat-rate payments independent from current production decisions and market developments. For the new member states, payments are phased in with the assumption of maximum top-ups from national budgets up to 2013 through the Single Area Payment (SAP), and through the (SFP) from 2014. Due to modulation, an increasing share of the total SFP will go to rural development spending rather than directly to farmers.
d) Common intervention price for soft wheat, barley, maize and sorghum.
e) Actual payments are made per eligible hectare (all agricultural land) based on historical reference or regional average.
f) Actual payments made per hectare based on programme yields; will be integrated in the single farm payment as of 2012.
g) Subject to a purchase limit of 75 000 tonnes per year.
h) The export volume excludes 0.4 mt of exported potato starch. The original limit on subsidised exports is 10.8 mt.
i) Husked rice basis.
j) Includes wheat, barley, maize, oats and sorghum.
k) Average per tonne of total exports.
l) Indian input subsidies consist of those for electricty, fertiliser and irrigation.

Source: OECD and FAO Secretariats.

StatLink http://dx.doi.org/10.1787/888932287235

Table A.5. **World cereal projections**

Crop year[a]		Average 07/08-09/10 est.	10/11	11/12	12/13	13/14	14/15	15/16	16/17	17/18	18/19	19/20
WHEAT												
OECD[b]												
Production	*mt*	267.9	273.7	277.5	273.5	274.1	278.1	281.1	286.0	288.3	290.7	293.6
Consumption	*mt*	210.3	217.7	223.6	225.0	226.7	230.2	233.2	234.8	236.2	237.8	239.2
Closing stocks	*mt*	52.6	63.5	64.7	61.2	58.8	57.3	56.3	57.1	57.7	57.9	58.3
Non-OECD												
Production	*mt*	387.9	403.1	412.1	412.5	418.3	425.2	430.6	437.5	442.5	447.3	452.5
Consumption	*mt*	428.8	442.4	451.9	460.0	465.0	471.3	476.4	482.0	488.0	494.2	500.9
Closing stocks	*mt*	140.3	166.8	176.0	176.7	176.2	175.7	175.0	177.2	179.5	181.7	183.7
WORLD[c]												
Production	*mt*	655.9	676.7	689.6	686.0	692.4	703.3	711.7	723.6	730.8	738.1	746.2
Consumption	*mt*	639.1	660.0	675.5	684.9	691.7	701.5	709.6	716.8	724.1	731.9	740.1
Closing stocks	*mt*	192.9	230.3	240.7	238.0	234.9	233.0	231.3	234.3	237.2	239.6	241.9
Price[d]	*USD/t*	275.9	218.5	200.9	206.2	210.5	214.4	227.4	229.0	226.6	227.7	225.4
COARSE GRAINS												
OECD[b]												
Production	*mt*	579.7	591.9	609.1	608.7	616.8	624.7	631.0	638.8	644.4	648.8	651.2
Consumption	*mt*	551.0	574.4	579.8	585.2	592.5	600.2	608.5	613.9	615.5	618.4	620.9
Closing stocks	*mt*	99.3	100.9	104.0	100.8	99.3	98.4	96.7	97.7	99.8	102.1	102.3
Non-OECD												
Production	*mt*	524.0	550.8	556.6	569.0	582.2	595.8	608.5	623.4	634.8	648.0	659.9
Consumption	*mt*	537.2	559.7	576.1	587.4	600.8	613.2	625.3	638.2	652.5	667.1	681.6
Closing stocks	*mt*	98.4	105.6	105.9	107.7	108.5	109.9	110.8	113.4	116.0	118.4	120.2
WORLD[c]												
Production	*mt*	1 103.7	1 142.7	1 165.7	1 177.7	1 199.0	1 220.4	1 239.5	1 262.2	1 279.2	1 296.7	1 311.1
Consumption	*mt*	1 088.1	1 134.2	1 155.9	1 172.5	1 193.3	1 213.4	1 233.8	1 252.1	1 268.0	1 285.6	1 302.5
Closing stocks	*mt*	197.7	206.5	209.9	208.6	207.8	208.4	207.5	211.1	215.8	220.5	222.6
Price[e]	*USD/t*	193.6	185.5	182.0	185.8	187.0	191.8	200.7	197.8	193.3	190.0	187.2
RICE												
OECD[b]												
Production	*mt*	22.6	23.0	22.6	22.8	23.1	23.2	23.4	23.3	23.3	23.4	23.5
Consumption	*mt*	24.4	24.2	24.3	24.5	24.6	24.7	24.9	25.0	25.1	25.3	25.4
Closing stocks	*mt*	5.8	6.4	6.1	5.9	5.8	5.9	6.0	6.0	6.0	6.0	6.0
Non-OECD												
Production	*mt*	432.0	454.3	458.4	461.8	465.7	473.0	478.2	483.2	488.3	493.4	498.2
Consumption	*mt*	427.4	447.2	455.1	458.1	463.0	469.7	475.3	480.7	485.8	491.1	496.1
Closing stocks	*mt*	93.6	99.6	101.5	104.0	105.3	107.2	108.5	109.5	110.3	111.0	111.2
WORLD[c]												
Production	*mt*	454.6	477.3	481.0	484.6	488.8	496.2	501.5	506.5	511.6	516.8	521.7
Consumption	*mt*	451.8	471.4	479.4	482.6	487.6	494.5	500.2	505.7	510.9	516.3	521.5
Closing stocks	*mt*	99.4	106.0	107.6	109.8	111.1	113.0	114.5	115.5	116.3	117.0	117.3
Price[f]	*USD/t*	539.0	452.9	436.0	425.9	426.6	429.7	426.2	425.1	425.8	422.5	422.5

est.: Estimate.

a) Beginning crop marketing year – see "Glossary of Terms" for definitions.

b) Excludes Iceland but includes the EU8 members that are not members of the OECD.

c) Source of historic data is USDA.

d) No. 2 hard red winter wheat, ordinary protein, USA f.o.b. Gulf Ports (June/May), less EEP payments where applicable.

e) No. 2 yellow corn, US f.o.b. Gulf Ports (September/August).

f) Milled, 100%, grade b, Nominal Price Quote (NPQ), f.o.b. Bangkok (January/December).

Source: OECD and FAO Secretariats.

StatLink http://dx.doi.org/10.1787/888932287254

Table A.6. **Wheat projections**

Crop year[a]	PRODUCTION (kt)		Growth[b] (%)		IMPORTS (kt)		Growth[b] (%)		EXPORTS (kt)		Growth[b] (%)	
	Average 2007-09 est.	2019	2000-09	2010-19	Average 2007-09 est.	2019	2000-09	2010-19	Average 2007-09 est.	2019	2000-09	2010-19
WORLD	**655 855**	**746 183**	**1.71**	**1.10**	**121 483**	**142 271**	**2.11**	**1.83**	**123 478**	**146 024**	**2.41**	**1.77**
DEVELOPED	**353 931**	**405 750**	**1.65**	**1.32**	**24 250**	**21 229**	**0.68**	**–0.59**	**107 700**	**124 994**	**3.45**	**1.67**
NORTH AMERICA	**86 441**	**86 628**	**1.58**	**0.35**	**3 212**	**3 444**	**3.89**	**1.64**	**45 601**	**43 617**	**0.85**	**0.02**
Canada	25 060	28 350	2.26	0.05	37	42	–10.62	0.00	17 442	18 980	2.90	–0.23
United States	61 381	58 278	1.28	0.50	3 175	3 402	4.25	1.66	28 159	24 637	–0.43	0.29
EUROPE	**217 567**	**247 605**	**1.83**	**1.28**	**9 289**	**8 073**	**–2.39**	**–0.99**	**41 969**	**50 267**	**9.71**	**2.89**
EU27	136 576	150 094	0.94	0.86	6 755	5 651	0.37	–0.72	18 115	13 995	3.49	–2.62
Russian Federation	54 174	66 543	3.72	2.43	350	587	–8.25	–1.91	15 705	25 638	28.33	8.15
Ukraine	19 713	22 796	5.51	0.79	124	11	–35.72	–0.09	7 446	9 723	29.81	1.17
OCEANIA DEVELOPED	**19 414**	**26 637**	**–1.11**	**2.04**	**314**	**363**	**2.80**	**0.82**	**12 472**	**18 228**	**–2.60**	**1.83**
Australia	19 052	26 225	–1.15	2.05	0	0	–	–	12 472	18 228	–2.60	1.83
New Zealand	362	412	1.09	1.52	314	363	2.80	0.82	0	0	47.58	–
OTHER DEVELOPED[c]	**30 508**	**44 880**	**3.77**	**3.29**	**11 435**	**9 349**	**2.96**	**–1.02**	**7 658**	**12 882**	**7.98**	**3.23**
Japan	878	957	2.34	1.28	5 272	4 953	–0.59	–0.60	0	0	–	–
South Africa	2 008	2 020	–1.92	1.34	1 263	1 752	10.06	2.08	218	137	6.49	–2.08
DEVELOPING	**301 924**	**340 433**	**1.74**	**0.84**	**97 233**	**121 042**	**2.50**	**2.31**	**15 778**	**21 031**	**–3.86**	**2.43**
AFRICA	**20 436**	**30 072**	**5.50**	**2.51**	**32 572**	**39 876**	**2.61**	**2.58**	**711**	**542**	**4.76**	**–1.91**
NORTH AFRICA	**19 019**	**27 402**	**5.70**	**2.25**	**21 420**	**23 498**	**1.89**	**2.10**	**350**	**295**	**4.06**	**–0.66**
Algeria	2 943	3 945	10.25	2.87	5 633	5 865	1.14	0.62	0	0	0.00	–0.02
Egypt	8 053	9 636	3.38	0.84	8 427	11 288	4.10	3.97	0	0	–119.57	–0.10
SUB-SAHARAN AFRICA	**1 417**	**2 669**	**2.99**	**5.66**	**11 152**	**16 379**	**3.96**	**3.30**	**361**	**247**	**5.48**	**–3.23**
LATIN AMERICA and CARIBBEAN	**23 236**	**30 495**	**–0.87**	**1.96**	**19 341**	**23 929**	**0.88**	**1.73**	**7 308**	**12 229**	**–8.00**	**3.16**
Argentina	10 908	15 221	–5.60	2.30	0	0	–	–	5 249	8 934	–13.41	3.53
Brazil	5 129	6 703	7.60	2.31	6 170	8 389	–0.34	2.69	488	765	79.46	2.94
Chile	1 178	1 302	–6.66	1.43	1 007	1 202	19.31	0.47	0	0	0.00	–0.47
Mexico	3 943	4 282	2.75	0.25	3 219	3 708	0.70	1.07	714	1 072	4.19	0.94
Uruguay	1 022	1 659	20.52	2.61	17	0	–120.52	–3.67	513	1 134	84.75	3.67
ASIA and PACIFIC	**258 252**	**279 867**	**1.72**	**0.56**	**45 320**	**57 236**	**3.13**	**2.37**	**7 759**	**8 260**	**–0.58**	**1.77**
Bangladesh	844	1 328	–10.33	4.44	2 243	2 415	4.91	0.36	0	0	–148.90	–0.01
China[d]	112 035	112 176	2.55	–0.30	315	590	–9.66	3.93	1 680	4 053	3.18	6.14
India	78 317	86 894	0.99	0.75	871	2 112	33.22	24.49	108	177	–64.38	–1.38
Indonesia	0	0	0.00	1.61	5 267	6 333	4.01	2.09	30	30	5.90	–0.02
Iran, Islamic Republic of	12 600	16 027	3.43	2.21	4 233	2 281	–5.19	–4.98	867	275	103.67	2.81
Korea	8	9	11.11	0.19	2 829	3 241	–2.61	1.62	11	11	–11.09	0.00
Malaysia	0	0	0.00	1.95	1 300	1 440	–0.25	1.01	90	90	–6.51	–0.01
Pakistan	22 942	26 504	2.23	1.71	1 907	3 275	30.13	3.81	1 667	570	10.97	–3.67
Saudi Arabia	1 819	511	–2.81	–8.47	1 009	3 183	31.02	5.59	0	0	–149.69	–0.14
Turkey	18 508	22 802	–0.68	1.16	2 532	1 760	14.98	0.32	1 900	2 159	11.65	–0.32
LEAST DEVELOPED COUNTRIES (LDC)	**7 638**	**11 237**	**3.92**	**3.43**	**12 224**	**16 872**	**3.84**	**2.98**	**117**	**114**	**–11.29**	**–3.99**
OECD	**267 906**	**293 642**	**0.79**	**0.83**	**25 874**	**25 022**	**1.34**	**0.31**	**78 881**	**79 127**	**1.05**	**–0.09**
NON-OECD	**387 949**	**452 541**	**2.39**	**1.28**	**95 609**	**117 248**	**2.37**	**2.18**	**44 596**	**66 898**	**6.06**	**4.41**

For notes, see end of the table.

Table A.6. **Wheat projections** *(cont.)*

Crop year[a]	CONSUMPTION (kt)		Growth[b] (%)		FOOD USE (kt)		Growth[b] (%)		PER CAPITA (kg)		Growth[b] (%)	
	Average 2007-09 est.	2019	2000-09	2010-19	Average 2007-09 est.	2019	2000-09	2010-19	Average 2007-09 est.	2019	2000-09	2010-19
WORLD	**639 120**	**740 115**	**0.97**	**1.21**	**454 604**	**513 980**	**1.10**	**1.07**	**67.5**	**67.8**	**–0.12**	**0.00**
DEVELOPED	**264 376**	**300 784**	**0.82**	**1.00**	**130 741**	**138 712**	**0.15**	**0.58**	**97.3**	**100.1**	**–0.25**	**0.31**
NORTH AMERICA	**39 510**	**46 276**	**–0.36**	**0.72**	**28 334**	**31 359**	**0.04**	**0.91**	**82.2**	**82.5**	**–0.95**	**0.04**
Canada	7 610	9 383	0.58	0.84	2 796	3 125	–0.50	0.66	84.1	85.0	–1.51	–0.25
United States	31 900	36 894	–0.58	0.70	25 538	28 234	0.10	0.94	81.9	82.2	–0.89	0.07
EUROPE	**184 159**	**204 871**	**0.60**	**0.83**	**81 148**	**83 593**	**–0.03**	**0.34**	**112.6**	**116.6**	**–0.09**	**0.40**
EU27	124 836	141 603	0.63	1.03	55 183	58 548	0.36	0.53	111.8	117.5	0.07	0.46
Russian Federation	39 268	41 234	1.05	0.16	14 807	14 157	–1.06	–0.05	104.7	104.1	–0.60	0.30
Ukraine	11 657	12 972	–0.91	0.82	5 673	5 396	–1.00	–0.38	123.3	124.9	–0.25	0.17
OCEANIA DEVELOPED	**7 358**	**8 773**	**3.54**	**2.44**	**2 109**	**2 185**	**0.68**	**0.98**	**83.4**	**77.8**	**–0.48**	**0.04**
Australia	6 696	7 997	3.76	2.57	1 753	1 767	0.70	0.93	83.2	75.3	–0.48	–0.04
New Zealand	662	776	1.47	1.19	356	418	0.55	1.23	84.1	90.2	–0.56	0.41
OTHER DEVELOPED[c]	**33 349**	**40 864**	**3.24**	**1.98**	**19 150**	**21 575**	**1.03**	**1.00**	**75.7**	**82.6**	**0.52**	**0.74**
Japan	6 169	5 889	–0.27	–0.32	5 131	4 899	–0.42	–0.32	40.3	39.5	–0.46	–0.07
South Africa	2 944	3 615	0.79	1.40	2 854	3 508	1.87	1.38	57.5	66.9	0.63	0.97
DEVELOPING	**374 744**	**439 331**	**1.08**	**1.35**	**323 863**	**375 268**	**1.50**	**1.25**	**60.0**	**60.5**	**0.07**	**0.00**
AFRICA	**51 902**	**69 008**	**3.36**	**2.42**	**46 540**	**62 208**	**3.00**	**2.50**	**49.2**	**51.5**	**0.65**	**0.29**
NORTH AFRICA	**39 573**	**50 455**	**3.11**	**2.05**	**34 724**	**44 276**	**2.66**	**2.10**	**141.9**	**148.7**	**0.74**	**0.34**
Algeria	8 160	9 749	2.95	1.46	7 160	8 427	2.01	1.43	208.3	209.9	0.52	0.04
Egypt	16 113	20 921	2.92	2.21	14 946	19 700	2.60	2.35	183.3	202.4	0.73	0.78
SUB-SAHARAN AFRICA	**12 329**	**18 552**	**4.20**	**3.52**	**11 815**	**17 932**	**4.04**	**3.56**	**16.8**	**19.7**	**1.54**	**1.20**
LATIN AMERICA and CARIBBEAN	**35 172**	**41 881**	**1.71**	**1.41**	**30 933**	**37 459**	**2.04**	**1.55**	**53.7**	**58.5**	**0.81**	**0.61**
Argentina	5 671	6 288	2.43	0.76	4 696	5 442	3.68	1.00	117.8	123.8	2.72	0.13
Brazil	10 492	14 052	1.84	2.21	9 828	13 368	1.97	2.31	51.2	64.3	0.78	1.62
Chile	2 221	2 501	0.99	1.04	2 053	2 274	1.39	0.92	122.2	122.9	0.33	0.06
Mexico	6 448	6 919	1.48	0.57	4 855	5 259	1.70	0.80	44.7	44.2	0.65	0.00
Uruguay	493	517	5.15	0.72	370	400	2.77	0.74	110.4	114.9	2.66	0.38
ASIA and PACIFIC	**287 670**	**328 442**	**0.63**	**1.12**	**246 391**	**275 601**	**1.17**	**0.95**	**63.6**	**63.3**	**–0.08**	**–0.09**
Bangladesh	3 086	3 723	–0.45	1.52	2 976	3 576	0.77	1.46	18.6	19.5	–0.81	0.24
China[d]	103 424	108 269	–0.71	0.41	88 416	86 956	–0.07	–0.18	66.1	61.0	–0.73	–0.76
India	77 713	89 465	1.24	1.06	70 013	81 890	1.78	1.20	59.3	60.5	0.23	0.00
Indonesia	5 103	6 197	3.82	1.90	4 343	5 290	2.98	2.04	19.1	21.0	1.71	1.14
Iran, Islamic Republic of	15 817	17 963	1.26	1.02	12 183	13 889	1.28	1.18	166.2	167.5	0.14	0.07
Korea	2 879	3 237	–2.48	1.59	2 069	2 101	–0.23	–0.01	43.0	42.5	–0.66	–0.22
Malaysia	1 204	1 346	0.38	0.96	1 007	1 108	1.06	0.87	37.3	35.0	–0.77	–0.52
Pakistan	23 215	29 169	1.90	2.06	20 632	25 744	1.98	2.00	116.6	115.9	–0.23	–0.05
Saudi Arabia	2 778	3 590	3.74	2.29	2 558	3 326	4.55	2.34	101.5	107.0	2.20	0.46
Turkey	19 273	22 402	–0.34	1.27	14 633	16 562	1.39	1.10	198.0	199.3	0.08	0.06
LEAST DEVELOPED COUNTRIES (LDC)	**19 590**	**27 704**	**3.91**	**2.98**	**17 857**	**25 285**	**3.60**	**3.06**	**23.8**	**26.7**	**1.34**	**0.93**
OECD	**210 319**	**239 185**	**0.40**	**0.99**	**115 412**	**124 114**	**0.44**	**0.68**	**92.3**	**94.5**	**–0.17**	**0.25**
NON-OECD	**428 801**	**500 930**	**1.26**	**1.31**	**339 192**	**389 867**	**1.33**	**1.19**	**61.8**	**62.2**	**–0.03**	**–0.01**

est.: Estimate.

a) Beginning crop marketing year – see "Glossary of Terms" for definitions.

b) Least-squares growth rate (see glossary).

c) Includes other developed Asian countries: Kazakhstan, Kyrgyzstan, Tajikistan, Turkmenistan, Uzbekistan, Armenia, Azerbaijan and Georgia.

d) Refers to mainland only. The economies of Chinese Taipei, Hong Kong (China) and Macau (China) are included in the other Asia Pacific aggregate.

Source: OECD and FAO Secretariats.

StatLink http://dx.doi.org/10.1787/888932287273

Table A.7. **Coarse grain projections**

Crop year[a]	PRODUCTION (kt) Average 2007-09 est.	PRODUCTION (kt) 2019	Growth[b] (%) 2000-09	Growth[b] (%) 2010-19	IMPORTS (kt) Average 2007-09 est.	IMPORTS (kt) 2019	Growth[b] (%) 2000-09	Growth[b] (%) 2010-19	EXPORTS (kt) Average 2007-09 est.	EXPORTS (kt) 2019	Growth[b] (%) 2000-09	Growth[b] (%) 2010-19
WORLD	**1 103 742**	**1 311 082**	**2.88**	**1.55**	**118 329**	**132 301**	**1.14**	**1.87**	**120 459**	**138 791**	**1.20**	**1.77**
DEVELOPED	**621 372**	**706 165**	**2.21**	**1.13**	**37 144**	**29 847**	**–0.29**	**0.51**	**91 807**	**99 754**	**1.67**	**1.22**
NORTH AMERICA	**367 817**	**422 676**	**3.21**	**1.20**	**5 203**	**4 385**	**–2.38**	**0.55**	**64 774**	**65 915**	**1.44**	**0.87**
Canada	25 800	27 512	1.14	0.72	2 411	1 830	–6.14	1.39	5 810	6 468	5.33	1.17
United States	342 017	395 164	3.37	1.24	2 792	2 555	1.86	0.00	58 964	59 446	1.13	0.84
EUROPE	**223 820**	**247 994**	**0.87**	**0.94**	**11 298**	**6 948**	**0.58**	**4.78**	**19 282**	**24 240**	**2.14**	**2.44**
EU27	152 043	160 374	0.29	0.60	9 793	5 394	2.62	5.74	7 358	9 823	–6.80	3.36
Russian Federation	34 201	38 709	0.99	1.12	323	349	–11.68	3.02	2 375	1 719	5.25	10.58
Ukraine	19 862	27 270	3.75	2.68	33	21	–21.75	–1.70	7 969	11 717	15.23	3.07
OCEANIA DEVELOPED	**12 436**	**14 655**	**0.14**	**1.81**	**27**	**32**	**–4.91**	**–0.96**	**5 519**	**6 935**	**0.30**	**2.00**
Australia	11 816	14 014	0.14	1.85	0	0	–	–	5 509	6 935	0.30	2.00
New Zealand	620	641	0.80	0.94	27	32	–2.23	–0.96	10	0	12.92	–
OTHER DEVELOPED[c]	**17 300**	**20 840**	**1.76**	**1.57**	**20 617**	**18 482**	**–0.76**	**–0.83**	**2 232**	**2 664**	**2.70**	**–0.87**
Japan	230	200	–1.02	–0.97	19 827	18 100	–0.74	–0.85	0	0	–	–
South Africa	11 549	12 528	1.21	0.59	538	196	–10.28	–0.28	1 678	2 097	1.58	0.29
DEVELOPING	**482 370**	**604 918**	**3.79**	**2.07**	**81 184**	**102 454**	**1.68**	**2.31**	**28 652**	**39 037**	**–0.10**	**3.33**
AFRICA	**91 390**	**120 892**	**3.97**	**2.29**	**15 886**	**20 820**	**1.93**	**2.93**	**2 805**	**2 542**	**10.85**	**–1.61**
NORTH AFRICA	**23 995**	**31 086**	**5.09**	**2.04**	**11 377**	**15 728**	**1.61**	**3.54**	**200**	**62**	**16.33**	**–1.60**
Algeria	1 059	1 918	14.43	2.17	2 313	2 441	2.28	2.45	0	0	0.00	–0.06
Egypt	8 092	9 071	1.02	1.38	4 851	7 196	–0.51	2.39	0	0	0.00	–0.06
SUB-SAHARAN AFRICA	**67 395**	**89 806**	**3.60**	**2.37**	**4 509**	**5 092**	**2.83**	**1.22**	**2 605**	**2 480**	**10.47**	**–1.61**
LATIN AMERICA and CARIBBEAN	**126 200**	**156 490**	**2.97**	**2.38**	**25 575**	**30 304**	**2.69**	**1.70**	**20 779**	**30 995**	**3.31**	**4.80**
Argentina	20 628	26 404	0.65	3.18	91	0	28.58	–	11 454	15 451	–0.86	3.83
Brazil	55 891	72 699	3.57	3.30	1 357	1 106	3.34	0.35	7 564	13 575	10.40	6.70
Chile	1 849	1 928	6.23	0.95	1 926	3 360	6.08	4.67	33	3	–19.82	–4.67
Mexico	32 175	35 957	2.79	0.54	10 564	11 404	–0.11	0.43	187	226	45.03	0.22
Uruguay	917	1 383	10.39	2.41	87	51	–2.35	–3.77	195	414	10.30	3.77
ASIA and PACIFIC	**264 781**	**327 536**	**4.13**	**1.84**	**39 724**	**51 330**	**0.97**	**2.44**	**5 068**	**5 500**	**–10.83**	**–0.57**
Bangladesh	933	564	34.55	0.28	50	135	–18.06	14.17	0	0	0.00	–0.35
China[d]	165 496	210 355	4.56	1.91	1 336	2 949	–5.96	5.82	656	256	–35.92	–8.61
India	37 747	42 620	2.21	2.03	20	5	–39.30	0.00	1 926	1 509	103.47	–0.03
Indonesia	15 755	18 692	6.80	0.75	145	450	–32.20	16.51	733	1 092	29.44	–3.35
Iran, Islamic Republic of	3 743	4 595	0.36	2.46	4 117	5 194	10.27	2.14	0	0	0.00	–0.05
Korea	367	380	0.15	0.00	8 240	7 779	–1.36	0.00	0	0	–	–
Malaysia	80	98	2.44	1.68	2 567	2 942	0.37	1.14	8	7	–11.09	–1.14
Pakistan	3 850	4 171	7.67	1.75	26	62	–11.02	4.21	0	0	–73.66	–2.40
Saudi Arabia	374	451	–1.55	1.65	9 010	12 437	3.71	2.80	0	0	0.00	–0.07
Turkey	11 398	13 615	1.33	1.55	782	2 177	–7.10	5.02	94	186	–19.71	–0.27
LEAST DEVELOPED COUNTRIES (LDC)	**46 727**	**64 140**	**5.11**	**2.58**	**2 430**	**2 540**	**1.02**	**1.60**	**2 750**	**2 496**	**12.55**	**–1.99**
OECD	**579 737**	**651 156**	**2.24**	**1.03**	**56 908**	**53 134**	**–0.02**	**0.74**	**77 994**	**83 118**	**0.34**	**1.17**
NON-OECD	**524 006**	**659 927**	**3.63**	**2.10**	**61 421**	**79 167**	**2.16**	**2.71**	**42 465**	**55 674**	**2.98**	**2.75**

For notes, see end of the table.

Table A.7. **Coarse grain projections** *(cont.)*

Crop year[a]	CONSUMPTION (kt)		Growth[b] (%)		FEED USE (kt)		Growth[b] (%)		PER CAPITA (kg)		Growth[b] (%)	
	Average 2007-09 est.	2019	2000-09	2010-19	Average 2007-09 est.	2019	2000-09	2010-19	Average 2007-09 est.	2019	2000-09	2010-19
WORLD	**1 088 145**	**1 302 511**	**2.48**	**1.54**	**638 217**	**741 287**	**0.97**	**1.55**	**29.3**	**30.4**	**0.80**	**0.41**
DEVELOPED	**557 153**	**635 786**	**2.06**	**1.01**	**350 681**	**366 137**	**–0.17**	**0.68**	**22.7**	**22.9**	**0.11**	**0.04**
NORTH AMERICA	**303 296**	**361 494**	**3.24**	**1.20**	**166 256**	**175 759**	**–0.98**	**1.05**	**23.9**	**22.1**	**–0.75**	**–0.55**
Canada	21 995	22 809	–0.88	0.69	17 233	17 267	–2.10	0.45	66.6	54.6	–1.93	–1.15
United States	281 302	338 685	3.62	1.24	149 023	158 492	–0.84	1.11	19.3	18.6	–0.29	–0.35
EUROPE	**211 535**	**230 173**	**0.84**	**0.82**	**155 853**	**159 782**	**0.69**	**0.28**	**23.0**	**24.5**	**0.54**	**0.44**
EU27	150 958	155 604	0.91	0.50	115 602	107 943	0.75	–0.42	18.5	19.8	–0.25	0.60
Russian Federation	31 648	37 303	1.34	0.76	19 608	23 468	1.37	0.89	29.1	33.1	0.59	0.43
Ukraine	11 775	15 369	–0.90	2.43	6 832	10 133	–2.49	3.30	43.9	45.1	5.28	0.72
OCEANIA DEVELOPED	**6 974**	**7 752**	**0.49**	**1.63**	**5 702**	**6 640**	**–1.44**	**2.36**	**12.4**	**12.8**	**–0.56**	**0.04**
Australia	6 338	7 079	0.47	1.71	5 164	6 076	–1.69	2.51	10.8	11.3	0.77	0.02
New Zealand	636	674	0.79	0.85	538	565	1.28	0.85	20.2	20.5	–3.79	0.13
OTHER DEVELOPED[c]	**35 348**	**36 366**	**0.40**	**0.31**	**22 869**	**23 955**	**0.51**	**0.34**	**21.2**	**20.9**	**0.09**	**–0.25**
Japan	20 001	18 123	–0.64	–0.96	14 956	13 303	–0.56	–1.28	1.5	1.5	–0.67	0.00
South Africa	10 218	10 595	1.20	0.26	4 290	4 836	1.57	0.71	97.1	91.3	–0.46	–0.70
DEVELOPING	**530 992**	**666 726**	**2.93**	**2.07**	**287 536**	**375 150**	**2.50**	**2.47**	**30.9**	**32.1**	**0.88**	**0.43**
AFRICA	**105 237**	**139 166**	**3.39**	**2.50**	**20 907**	**28 991**	**3.94**	**2.89**	**74.7**	**77.9**	**0.48**	**0.36**
NORTH AFRICA	**35 168**	**46 716**	**3.66**	**2.53**	**16 518**	**22 321**	**3.17**	**2.57**	**65.9**	**72.0**	**1.97**	**0.95**
Algeria	3 389	4 336	4.86	2.29	2 570	3 348	6.09	2.46	20.0	21.0	0.39	0.54
Egypt	13 046	16 249	0.69	1.85	8 668	10 716	0.24	1.87	46.8	49.1	–0.04	0.42
SUB-SAHARAN AFRICA	**70 069**	**92 449**	**3.26**	**2.48**	**4 388**	**6 670**	**7.32**	**4.06**	**77.7**	**79.9**	**0.03**	**0.16**
LATIN AMERICA and CARIBBEAN	**129 755**	**155 663**	**2.89**	**1.79**	**77 389**	**94 891**	**2.11**	**2.26**	**65.5**	**69.4**	**2.98**	**0.21**
Argentina	9 221	10 953	2.59	2.32	6 602	7 888	2.09	2.58	9.6	11.0	2.09	1.01
Brazil	48 496	60 170	2.84	2.46	37 451	47 324	2.98	2.77	33.0	36.9	2.00	0.95
Chile	3 794	5 268	6.29	3.22	3 110	4 516	6.67	3.59	18.9	20.2	1.69	0.85
Mexico	42 522	47 141	1.90	0.54	15 667	16 037	–2.47	0.13	204.5	217.0	4.30	–0.03
Uruguay	773	1 010	6.74	1.89	318	461	3.47	3.08	33.4	36.6	1.22	1.09
ASIA and PACIFIC	**296 000**	**371 897**	**2.78**	**2.03**	**189 241**	**251 268**	**2.51**	**2.50**	**15.1**	**13.9**	**–0.61**	**–0.28**
Bangladesh	983	699	17.69	1.91	257	344	9.70	2.28	4.5	1.9	24.15	0.34
China[d]	161 966	211 755	2.85	2.07	111 257	152 869	2.13	2.53	11.3	9.6	–3.30	–1.76
India	35 811	41 174	1.73	2.16	8 912	13 645	3.27	4.08	21.1	18.8	0.00	0.10
Indonesia	14 830	18 053	4.18	1.45	4 900	6 794	6.44	2.72	28.9	28.2	0.34	–0.24
Iran, Islamic Republic of	7 843	9 774	4.99	2.33	7 444	9 351	5.27	2.40	1.4	1.4	–0.46	0.08
Korea	8 774	8 158	–1.02	–0.01	6 188	5 547	–1.80	–0.01	10.9	10.6	2.86	–0.22
Malaysia	2 665	3 029	0.85	1.17	2 494	2 790	0.92	1.17	1.7	2.0	–1.62	1.26
Pakistan	3 809	4 297	7.04	1.80	1 595	1 606	13.19	1.25	8.7	9.1	–0.13	0.64
Saudi Arabia	9 584	12 803	4.30	2.60	9 372	12 576	4.40	2.64	3.8	3.3	–1.82	–1.34
Turkey	12 753	15 535	1.44	1.99	10 262	12 104	1.69	2.19	16.9	16.5	0.05	–0.29
LEAST DEVELOPED COUNTRIES (LDC)	**46 477**	**64 224**	**4.22**	**2.82**	**4 413**	**6 442**	**10.45**	**3.36**	**46.7**	**52.4**	**0.96**	**0.85**
OECD	**550 977**	**620 919**	**2.16**	**0.92**	**339 344**	**343 436**	**–0.36**	**0.49**	**33.9**	**35.8**	**2.09**	**0.23**
NON-OECD	**537 168**	**681 592**	**2.81**	**2.14**	**298 872**	**397 851**	**2.65**	**2.55**	**28.2**	**29.3**	**0.47**	**0.49**

est.: Estimate.

a) Beginning crop marketing year – see "Glossary of Terms" for definitions.

b) Least-squares growth rate (see glossary).

c) Includes other developed Asian countries: Kazakhstan, Kyrgyzstan, Tajikistan, Turkmenistan, Uzbekistan, Armenia, Azerbaijan and Georgia.

d) Refers to mainland only. The economies of Chinese Taipei, Hong Kong (China) and Macau (China) are included in the other Asia Pacific aggregate.

Source: OECD and FAO Secretariats.

StatLink http://dx.doi.org/10.1787/888932287292

Table A.8. **Rice projections**

Crop year[a]	PRODUCTION (kt) Average 2007-09 est.	PRODUCTION (kt) 2019	Growth[b] (%) 2000-09	Growth[b] (%) 2010-19	IMPORTS (kt) Average 2007-09 est.	IMPORTS (kt) 2019	Growth[b] (%) 2000-09	Growth[b] (%) 2010-19	EXPORTS (kt) Average 2007-09 est.	EXPORTS (kt) 2019	Growth[b] (%) 2000-09	Growth[b] (%) 2010-19
WORLD	**454 558**	**521 701**	**1.85**	**1.03**	**31 325**	**38 446**	**2.35**	**2.11**	**30 114**	**38 304**	**2.09**	**2.12**
DEVELOPED	**18 177**	**18 997**	**–0.47**	**0.40**	**5 196**	**6 304**	**2.67**	**2.17**	**3 511**	**4 072**	**–2.29**	**0.97**
NORTH AMERICA	**6 546**	**7 361**	**0.28**	**1.12**	**1 038**	**1 391**	**6.63**	**2.85**	**3 129**	**3 627**	**0.17**	**1.08**
Canada	0	0	–	–	364	484	5.47	2.65	0	0	–	–
United States	6 546	7 361	0.28	1.12	675	906	7.35	2.96	3 129	3 627	0.17	1.08
EUROPE	**3 270**	**3 473**	**1.21**	**0.66**	**2 025**	**2 661**	**0.78**	**2.11**	**169**	**127**	**–4.97**	**–0.23**
EU27	2 684	2 682	0.41	0.07	1 525	2 142	1.97	2.96	134	96	–7.60	–0.47
Russian Federation	505	698	5.96	3.17	201	137	–7.50	–6.74	25	24	10.04	1.70
Ukraine	68	78	3.56	1.26	100	138	2.85	2.90	7	5	55.24	–2.88
OCEANIA DEVELOPED	**69**	**428**	**–33.23**	**4.21**	**349**	**201**	**13.37**	**5.12**	**65**	**109**	**–23.07**	**1.00**
Australia	69	428	–33.23	4.21	310	160	14.40	6.73	65	109	–23.07	1.00
New Zealand	0	0	–	–	39	42	2.34	1.31	0	0	–	–
OTHER DEVELOPED[c]	**8 292**	**7 735**	**–0.41**	**–0.52**	**1 783**	**2 051**	**1.53**	**1.63**	**147**	**208**	**–12.98**	**–0.02**
Japan	7 853	7 231	–0.60	–0.64	740	682	–0.80	0.00	139	200	–13.52	0.00
South Africa	2	2	0.00	–0.27	850	1 053	4.16	2.28	0	0	0.00	–0.02
DEVELOPING	**436 382**	**502 704**	**1.95**	**1.05**	**26 130**	**32 142**	**2.28**	**2.10**	**26 603**	**34 232**	**2.77**	**2.26**
AFRICA	**15 684**	**27 032**	**4.55**	**5.09**	**8 604**	**10 568**	**2.17**	**1.94**	**478**	**502**	**–2.89**	**18.64**
NORTH AFRICA	**4 576**	**4 097**	**1.64**	**–0.10**	**314**	**1 052**	**4.96**	**11.21**	**467**	**0**	**–2.67**	**–95.09**
Algeria	1	1	0.00	–0.36	74	96	3.04	2.53	0	0	0.00	–0.03
Egypt	4 555	4 078	1.64	–0.10	20	695	26.41	22.78	467	0	–2.67	–158.11
SUB-SAHARAN AFRICA	**11 109**	**22 935**	**5.86**	**6.33**	**8 290**	**9 516**	**2.07**	**1.26**	**11**	**502**	**–11.55**	**20.38**
LATIN AMERICA and CARIBBEAN	**17 635**	**22 111**	**2.10**	**1.59**	**3 606**	**4 679**	**1.74**	**3.05**	**2 238**	**3 119**	**7.35**	**2.32**
Argentina	846	1 319	6.63	3.03	7	0	–	–	501	818	11.28	2.70
Brazil	8 094	9 535	1.76	0.88	590	810	–4.61	3.44	471	764	43.06	1.87
Chile	81	96	–1.57	0.96	121	185	5.38	3.08	0	0	52.59	–3.08
Mexico	280	386	6.27	3.07	786	719	2.20	–0.78	10	15	40.65	10.74
Uruguay	878	1 101	2.65	1.96	0	0	–19.05	–2.37	804	956	1.44	2.37
ASIA and PACIFIC	**403 062**	**453 561**	**1.85**	**0.83**	**13 919**	**16 896**	**2.53**	**1.94**	**23 887**	**30 611**	**2.55**	**2.11**
Bangladesh	30 250	34 641	2.67	1.18	895	890	1.90	10.69	11	10	31.79	–0.48
China[d]	133 836	132 426	0.98	–0.55	326	632	0.68	4.85	1 025	1 453	–7.53	0.38
India	94 347	109 982	1.45	1.17	77	128	32.20	1.76	2 609	1 610	–4.25	–2.46
Indonesia	38 078	45 153	2.28	1.17	780	95	–21.32	–18.83	13	817	46.18	20.10
Iran, Islamic Republic of	2 062	2 435	5.07	1.56	1 167	1 821	4.44	3.94	0	0	0.00	–0.10
Korea	4 644	4 723	–1.49	0.14	307	409	12.87	1.48	100	100	–	0.00
Malaysia	1 551	1 941	1.46	2.05	900	865	6.70	–0.85	0	0	76.69	0.02
Pakistan	6 297	7 959	4.94	2.06	1	1	–29.91	1.27	2 828	3 340	5.29	2.59
Saudi Arabia	0	0	0.00	1.08	1 000	1 408	3.25	2.07	0	0	0.00	–0.02
Turkey	432	562	10.41	2.15	177	283	–6.62	3.12	13	9	15.25	–0.57
LEAST DEVELOPED COUNTRIES (LDC)	**67 668**	**92 820**	**4.11**	**2.91**	**6 557**	**7 596**	**2.90**	**2.07**	**2 419**	**6 470**	**17.06**	**9.62**
OECD	**22 589**	**23 469**	**–0.67**	**0.33**	**5 150**	**6 148**	**3.08**	**2.01**	**3 589**	**4 157**	**–2.65**	**0.98**
NON-OECD	**431 969**	**498 232**	**1.99**	**1.06**	**26 175**	**32 298**	**2.21**	**2.13**	**26 524**	**34 147**	**2.85**	**2.26**

For notes, see end of the table.

Table A.8. **Rice projections** *(cont.)*

Crop year[a]	CONSUMPTION (kt)		Growth[b] (%)		PER CAPITA (kg)		Growth[b] (%)	
	Average 2007-09 est.	2019	2000-09	2010-19	Average 2007-09 est.	2019	2000-09	2010-19
WORLD	**451 800**	**521 526**	**1.27**	**1.11**	**57.6**	**59.2**	**–0.12**	**0.20**
DEVELOPED	**20 085**	**21 190**	**0.71**	**0.74**	**14.5**	**14.9**	**0.03**	**0.48**
NORTH AMERICA	**4 418**	**5 110**	**1.54**	**1.38**	**12.8**	**13.4**	**0.55**	**0.50**
Canada	364	484	5.47	2.65	10.9	13.2	4.46	1.73
United States	4 055	4 626	1.24	1.25	13.0	13.5	0.26	0.38
EUROPE	**5 256**	**6 012**	**1.65**	**1.25**	**7.3**	**8.4**	**1.59**	**1.30**
EU27	4 199	4 734	1.82	1.23	8.5	9.5	1.53	1.16
Russian Federation	684	811	0.15	0.87	4.8	6.0	0.61	1.22
Ukraine	161	211	2.60	2.58	3.4	4.8	3.35	3.14
OCEANIA DEVELOPED	**360**	**505**	**–1.48**	**3.55**	**14.2**	**18.0**	**–2.65**	**2.61**
Australia	321	463	–1.87	3.78	15.3	19.7	–3.05	2.81
New Zealand	39	42	2.34	1.31	9.2	9.0	1.23	0.48
OTHER DEVELOPED[c]	**10 051**	**9 563**	**–0.01**	**–0.01**	**37.3**	**34.4**	**–1.11**	**–0.29**
Japan	8 564	7 699	–0.52	–0.50	63.1	58.2	–1.23	–0.27
South Africa	862	1 053	4.11	2.35	16.4	19.1	3.16	1.95
DEVELOPING	**431 715**	**500 336**	**1.30**	**1.13**	**68.3**	**69.2**	**–0.30**	**0.04**
AFRICA	**23 751**	**36 976**	**3.90**	**4.03**	**21.9**	**27.2**	**1.58**	**1.98**
NORTH AFRICA	**4 320**	**5 221**	**2.22**	**1.72**	**14.9**	**16.0**	**1.50**	**0.41**
Algeria	75	96	2.96	2.54	2.2	2.4	1.46	1.15
Egypt	4 006	4 847	2.09	1.73	40.9	45.0	1.50	0.64
SUB-SAHARAN AFRICA	**19 431**	**31 756**	**4.29**	**4.46**	**24.3**	**30.8**	**1.54**	**2.21**
LATIN AMERICA and CARIBBEAN	**19 391**	**23 650**	**1.81**	**1.80**	**31.9**	**35.0**	**0.70**	**0.90**
Argentina	356	501	0.85	3.60	8.9	11.4	–0.10	2.73
Brazil	8 647	9 583	1.12	1.01	44.9	45.9	–0.07	0.31
Chile	198	280	2.25	2.32	11.6	14.8	1.31	1.48
Mexico	1 016	1 087	2.18	0.36	9.4	9.1	1.12	–0.44
Uruguay	107	142	11.09	2.09	10.8	13.6	3.07	1.65
ASIA and PACIFIC	**388 573**	**439 710**	**1.13**	**0.88**	**85.0**	**85.8**	**–0.31**	**0.00**
Bangladesh	30 634	35 484	2.79	1.35	152.7	165.0	0.95	0.72
China[d]	130 066	131 398	–0.53	–0.35	78.5	71.9	–1.30	–0.87
India	91 381	108 597	1.28	1.31	73.7	77.1	–0.38	0.17
Indonesia	38 045	44 815	0.89	1.23	156.7	153.9	0.20	–0.20
Iran, Islamic Republic of	3 199	4 255	4.16	2.53	37.7	45.4	3.03	1.63
Korea	4 918	5 022	–0.58	0.17	74.9	74.0	–2.24	–0.09
Malaysia	2 447	2 803	2.84	1.15	81.6	79.5	1.19	–0.24
Pakistan	3 453	4 569	3.49	1.73	16.8	17.5	1.00	–0.37
Saudi Arabia	1 023	1 394	3.23	2.78	38.8	43.2	1.20	0.93
Turkey	597	835	2.47	2.65	7.5	9.3	0.82	1.65
LEAST DEVELOPED COUNTRIES (LDC)	**71 855**	**93 665**	**3.97**	**2.50**	**73.5**	**82.0**	**0.76**	**1.08**
OECD	**24 382**	**25 407**	**0.41**	**0.55**	**18.0**	**17.9**	**–0.63**	**0.15**
NON-OECD	**427 417**	**496 119**	**1.32**	**1.14**	**66.6**	**67.9**	**–0.19**	**0.10**

est.: Estimate.

a) Beginning crop marketing year – see "Glossary of Terms" for definitions.

b) Least-squares growth rate (see glossary).

c) Includes other developed Asian countries: Kazakhstan, Kyrgyzstan, Tajikistan, Turkmenistan, Uzbekistan, Armenia, Azerbaijan and Georgia.

d) Refers to mainland only. The economies of Chinese Taipei, Hong Kong (China) and Macau (China) are included in the other Asia Pacific aggregate.

Source: OECD and FAO Secretariats.

StatLink http://dx.doi.org/10.1787/888932287311

Table A.9. **Main policy assumptions for oilseed markets**

Crop year[a]		Average 07/08-09/10 est.	10/11	11/12	12/13	13/14	14/15	15/16	16/17	17/18	18/19	19/20
ARGENTINA												
Oilseed export tax	%	27.3	35.0	35.0	35.0	35.0	35.0	35.0	35.0	35.0	35.0	35.0
Oilseed meal export tax	%	24.0	32.0	32.0	32.0	32.0	32.0	32.0	32.0	32.0	32.0	32.0
Oilseed oil export tax	%	24.0	32.0	32.0	32.0	32.0	32.0	32.0	32.0	32.0	32.0	32.0
AUSTRALIA												
Tariffs												
Soybean oil	%	8	8	8	8	8	8	8	8	8	8	8
Rapeseed oil	%	8	8	8	8	8	8	8	8	8	8	8
CANADA												
Tariffs												
Rapeseed oil	%	6.4	6.4	6.4	6.4	6.4	6.4	6.4	6.4	6.4	6.4	6.4
EUROPEAN UNION												
Single farm payment[b]	*EUR/ha*	188	188	188	188	188	188	188	188	188	188	188
Compulsory set-aside rate	%	3.3	0	0	0	0	0	0	0	0	0	0
Tariffs												
Soybean oil	%	6	6	6	6	6	6	6	6	6	6	6
Rapeseed oil	%	6	6	6	6	6	6	6	6	6	6	6
JAPAN												
New output payments												
Soybeans	*bn JPY*	23.5	23.5	23.5	23.5	23.5	23.5	23.5	23.5	23.5	23.5	23.5
Tariffs												
Soybean oil	*JPY/kg*	10.9	10.9	10.9	10.9	10.9	10.9	10.9	10.9	10.9	10.9	10.9
Rapeseed oil	*JPY/kg*	10.9	10.9	10.9	10.9	10.9	10.9	10.9	10.9	10.9	10.9	10.9
KOREA												
Soybean tariff-quota	*kt*	1 032	1 032	1 032	1 032	1 032	1 032	1 032	1 032	1 032	1 032	1 032
In-quota tariff	%	5	5	5	5	5	5	5	5	5	5	5
Out-of-quota tariff	%	487	487	487	487	487	487	487	487	487	487	487
Soybean (for food) mark up	*'000 KRW/t*	176	161	155	150	146	141	138	134	130	127	123
MEXICO												
Tariffs												
Soybeans	%	33	33	33	33	33	33	33	33	33	33	33
Soybean meal	%	23.8	23.8	23.8	23.8	23.8	23.8	23.8	23.8	23.8	23.8	23.8
Soybean oil	%	45	45	45	45	45	45	45	45	45	45	45
UNITED STATES												
ACRE participation rate												
Soybeans	%	0.0	10.0	10.0	10.0	10.0	10.0	10.0	10.0	10.0	10.0	10.0
Soybeans loan rate	*USD/t*	183.7	183.7	183.7	183.7	183.7	183.7	183.7	183.7	183.7	183.7	183.7
CRP area												
Soybeans	*mha*	2.1	1.8	1.7	1.7	1.7	1.7	1.8	1.8	1.8	1.8	1.8
Tariffs												
Rapeseed	%	3	3	3	3	3	3	3	3	3	3	3
Soybean meal	%	2.2	2.2	2.2	2.2	2.2	2.2	2.2	2.2	2.2	2.2	2.2
Rapeseed meal	%	1.2	1.2	1.2	1.2	1.2	1.2	1.2	1.2	1.2	1.2	1.2
Soybean oil	%	12.7	12.7	12.7	12.7	12.7	12.7	12.7	12.7	12.7	12.7	12.7
Rapeseed oil	%	3.2	3.2	3.2	3.2	3.2	3.2	3.2	3.2	3.2	3.2	3.2
Subsidised export limits												
Oilseed oils	*kt*	141	141	141	141	141	141	141	141	141	141	141

For notes, see end of the table.

Table A.9. **Main policy assumptions for oilseed markets** *(cont.)*

Crop year[a]		Average 07/08-09/10 est.	10/11	11/12	12/13	13/14	14/15	15/16	16/17	17/18	18/19	19/20
CHINA												
Soybeans support price	*CNY/t*	1 453.5	1 504.1	1 553.5	1 602.8	1 654.2	1 710.6	1 779.6	1 850.5	1 922.3	1 995.0	2 067.1
Tariffs												
Soybeans	*%*	2.4	2.4	2.4	2.4	2.4	2.4	2.4	2.4	2.4	2.4	2.4
Soybean meal	*%*	6.3	6.3	6.3	6.3	6.3	6.3	6.3	6.3	6.3	6.3	6.3
Soybean oil in-quota tariff	*%*	9	9	9	9	9	9	9	9	9	9	9
Vegetable oil tariff-quota	*kt*	7 998.1	7 998.1	7 998.1	7 998.1	7 998.1	7 998.1	7 998.1	7 998.1	7 998.1	7 998.1	7 999.1
INDIA												
Input subsidy rate, oilseeds[c]	*INR/t*	4 888	4 888.3	4 888.3	4 888.3	4 888.3	4 888.3	4 888.3	4 888.3	4 888.3	4 888.3	4 888.3
Soybean tariff	*%*	30	30	30	30	30	30	30	30	30	30	30
Rapeseed tariff	*%*	30	30	30	30	30	30	30	30	30	30	30
Sunflower tariff	*%*	30	30	30	30	30	30	30	30	30	30	30
Oilseed tariff	*%*	30	30	30	30	30	30	30	30	30	30	30
Soybean meal tariff	*%*	100	100	100	100	100	100	100	100	100	100	100
Rapeseed meal tariff	*%*	100	100	100	100	100	100	100	100	100	100	100
Sunflower meal tariff	*%*	100	100	100	100	100	100	100	100	100	100	100
Soybean oil tariff	*%*	21	0	0	0	0	0	0	0	0	0	0
Rapeseed oil tariff	*%*	33	8	8	8	8	8	8	8	8	8	8
Sunflower oil tariff	*%*	19	0	0	0	0	0	0	0	0	0	0
Palm oil tariff	*%*	19	0	0	0	0	0	0	0	0	0	0

Note: The source for tariffs and Tariff Rate Quotas is AMAD (Agricultural Market Access Database). The tariff and TRQ data are based on Most Favoured Nation rates scheduled with the WTO and exclude those under preferential or regional agreements, which may be substantially different. Tariffs are simple averages of several product lines. Specific rates are converted to *ad valorem* rates using world prices in the Outlook. Import quotas are based on global commitments scheduled in the WTO rather than those allocated to preferential partners under regional or other agreements. For Mexico, the NAFTA tariffs on soybeans, oil meals and soybean oil are zero after 2003.

est.: Estimate.

a) Beginning crop marketing year – see "Glossary of Terms" for definitions.

b) EU farmers benefit from the Single Farm Payment (SFP) Scheme, which provides flat-rate payments independent from current production decisions and market developments. For the new member states, payments are phased in with the assumption of maximum top-ups from national budgets up to 2013 through the Single Area Payment (SAP), and through the (SFP) from 2014. Due to modulation, an increasing share of the total SFP will go to rural development spending rather than directly to farmers.

c) Indian input subsidies consist of those for electricty, fertiliser and irrigation.

Source: OECD and FAO Secretariats.

StatLink http://dx.doi.org/10.1787/888932287330

Table A.10. **World oilseed projections**

		Average 07/08-09/10 est.	10/11	11/12	12/13	13/14	14/15	15/16	16/17	17/18	18/19	19/20
OILSEED (crop year[a])												
OECD[b]												
Production	*mt*	137.1	147.9	146.1	150.3	152.3	153.6	156.0	158.3	160.0	162.2	164.3
Consumption	*mt*	129.0	132.8	131.6	135.1	137.3	139.2	141.2	143.2	145.3	147.4	149.4
Crush	*mt*	114.4	117.8	116.5	119.7	121.7	123.6	125.7	127.6	129.7	131.6	133.6
Closing stocks	*mt*	16.1	17.9	18.5	19.1	18.6	17.8	17.4	17.5	17.6	17.6	17.5
Non-OECD												
Production	*mt*	250.0	271.4	278.0	284.8	290.6	296.8	304.1	311.4	318.1	324.8	331.0
Consumption	*mt*	267.5	281.2	288.0	295.6	302.5	308.6	315.8	322.5	329.0	335.8	342.4
Crush	*mt*	211.3	223.4	229.2	235.8	241.7	246.9	253.1	258.9	264.4	270.1	275.7
Closing stocks	*mt*	17.1	17.9	18.2	18.4	18.4	18.1	18.1	18.3	18.4	18.6	18.5
WORLD[c]												
Production	*mt*	387.1	419.3	424.1	435.1	442.9	450.3	460.2	469.7	478.2	486.9	495.3
Consumption	*mt*	396.5	414.0	419.6	430.7	439.8	447.8	456.9	465.8	474.3	483.2	491.8
Crush	*mt*	325.7	341.1	345.7	355.5	363.4	370.5	378.8	386.5	394.1	401.8	409.3
Closing stocks	*mt*	33.1	35.9	36.7	37.5	37.0	35.9	35.5	35.8	36.0	36.1	36.0
Price[d]	*USD/t*	468.7	409.9	412.0	405.3	401.4	409.0	416.6	416.0	416.5	415.5	418.8
PROTEIN MEALS (marketing year)												
OECD[b]												
Production	*mt*	79.4	81.2	80.1	82.3	83.7	85.0	86.4	87.7	89.1	90.4	91.7
Consumption	*mt*	109.1	110.6	110.5	112.3	113.5	114.4	115.6	116.9	118.0	119.1	120.3
Closing stocks	*mt*	1.2	1.2	1.2	1.2	1.2	1.2	1.2	1.2	1.2	1.2	1.2
Non-OECD		0.0	0.0	0.0	0.0	0.0	0.0	0.0	0.0	0.0	0.0	0.0
Production	*mt*	151.9	159.8	164.3	169.3	173.8	177.9	182.6	187.0	191.2	195.6	199.9
Consumption	*mt*	116.9	124.8	128.5	133.8	138.6	143.0	147.8	152.2	156.7	161.3	165.8
Closing stocks	*mt*	6.2	6.0	6.0	6.1	6.2	6.2	6.3	6.4	6.5	6.6	6.6
WORLD[c]												
Production	*mt*	231.3	240.9	244.4	251.6	257.6	262.9	268.9	274.7	280.2	286.0	291.6
Consumption	*mt*	226.1	235.4	239.0	246.1	252.1	257.4	263.4	269.1	274.7	280.5	286.1
Closing stocks	*mt*	7.4	7.1	7.2	7.3	7.3	7.4	7.5	7.6	7.7	7.8	7.8
Price[e]	*USD/t*	374.1	308.0	298.2	283.7	280.0	283.3	287.2	284.6	283.2	283.6	287.9
VEGETABLE OILS (marketing year)												
OECD[b]												
Production	*mt*	30.4	31.9	31.8	32.7	33.3	33.8	34.5	35.1	35.7	36.3	36.9
Consumption	*mt*	42.7	44.7	45.8	47.3	48.5	49.7	50.9	52.0	53.1	53.9	54.7
Closing stocks	*mt*	2.2	1.8	1.8	1.8	1.8	1.8	1.8	1.8	1.8	1.8	1.8
Non-OECD												
Production	*mt*	101.5	109.2	113.2	117.6	121.7	125.6	129.8	133.9	137.9	141.9	146.0
Consumption	*mt*	90.3	97.8	101.7	104.5	108.0	111.4	114.9	118.4	121.9	125.7	129.6
Closing stocks	*mt*	8.2	8.4	7.7	8.0	8.3	8.6	8.8	9.1	9.4	9.8	10.1
WORLD[c]												
Production	*mt*	131.9	141.1	145.0	150.3	155.1	159.5	164.3	169.0	173.6	178.2	182.8
of which: Palm oil	*mt*	131.9	141.1	145.0	150.3	155.1	159.5	164.3	169.0	173.6	178.2	182.8
Consumption	*mt*	133.0	142.5	147.5	151.8	156.4	161.0	165.8	170.4	175.0	179.6	184.3
Closing stocks	*mt*	10.4	10.3	9.6	9.8	10.2	10.4	10.5	10.9	11.2	11.6	11.9
Price[f]	*USD/t*	946.7	875.2	897.8	930.2	932.4	954.0	992.1	1 006.4	1 021.7	1 030.1	1 042.7

est.: Estimate.

a) Beginning crop marketing year – see "Glossary of Terms" for definitions.

b) Excludes Iceland but includes the EU8 members that are not members of the OECD.

c) Source of historic data is USDA.

d) Weighted average oilseed price, European port.

e) Weighted average protein meal price, European port.

f) Weighted average price of oilseed oils and palm oil, European port.

Source: OECD and FAO Secretariats.

StatLink http://dx.doi.org/10.1787/888932287349

Table A.11. **Oilseed projections**

Crop year[a]	PRODUCTION (kt)		Growth[b] (%)		IMPORTS (kt)		Growth[b] (%)		EXPORTS (kt)		Growth[b] (%)	
	Average 2007-09 est.	2019	2000-09	2010-19	Average 2007-09 est.	2019	2000-09	2010-19	Average 2007-09 est.	2019	2000-09	2010-19
WORLD	**387 122**	**495 300**	**3.27**	**1.90**	**92 647**	**108 686**	**4.66**	**1.77**	**85 717**	**112 307**	**3.90**	**1.71**
DEVELOPED	**158 194**	**197 579**	**3.27**	**1.69**	**26 807**	**25 281**	**–1.50**	**–0.33**	**48 889**	**55 032**	**4.13**	**0.81**
NORTH AMERICA	**105 052**	**123 442**	**1.99**	**1.09**	**1 872**	**1 733**	**0.79**	**0.27**	**44 260**	**46 620**	**4.25**	**0.11**
Canada	14 623	18 898	8.43	1.59	657	610	–6.87	0.15	8 124	8 938	9.93	0.36
United States	90 429	104 543	1.21	1.00	1 215	1 122	8.21	0.34	36 136	37 682	3.31	0.06
EUROPE	**45 715**	**64 533**	**7.38**	**2.88**	**18 245**	**17 672**	**–1.50**	**–0.26**	**3 565**	**6 692**	**9.46**	**6.91**
EU27	27 019	34 575	5.29	2.05	17 341	16 651	–1.71	–0.38	730	621	1.33	–0.97
Russian Federation	8 018	11 635	10.29	2.65	320	437	29.45	4.22	240	834	1.67	10.38
Ukraine	8 655	15 230	13.65	5.11	12	11	0.47	–0.25	2 287	4 780	21.67	8.47
OCEANIA DEVELOPED	**2 123**	**3 054**	**–3.64**	**1.66**	**36**	**38**	**2.09**	**0.00**	**898**	**1 586**	**–10.19**	**1.67**
Australia	2 119	3 050	–3.65	1.66	26	28	3.40	0.00	897	1 586	–10.20	1.67
New Zealand	4	4	0.00	0.00	10	10	0.73	–0.01	0	0	–1.75	–0.87
OTHER DEVELOPED[c]	**5 304**	**6 550**	**2.87**	**2.36**	**6 654**	**5 838**	**–2.02**	**–0.69**	**166**	**135**	**2.75**	**7.64**
Japan	278	289	0.12	0.12	6 273	5 558	–2.33	–0.43	0	0	–6.43	0.00
South Africa	1 383	1 653	0.87	3.90	179	51	10.20	–20.89	100	96	2.75	20.61
DEVELOPING	**228 928**	**297 721**	**3.26**	**2.04**	**65 840**	**83 405**	**8.28**	**2.49**	**36 828**	**57 275**	**3.57**	**2.63**
AFRICA	**12 021**	**14 322**	**0.38**	**1.74**	**2 091**	**2 996**	**12.12**	**3.42**	**412**	**273**	**–1.95**	**–4.33**
NORTH AFRICA	**833**	**796**	**–1.30**	**0.21**	**1 964**	**2 848**	**13.34**	**3.47**	**44**	**34**	**–0.90**	**–2.58**
Algeria	121	129	2.05	–0.51	100	153	11.72	5.37	0	0	0.00	–0.13
Egypt	490	413	–2.82	–0.22	1 258	1 879	20.37	3.47	10	9	–3.31	–0.50
SUB-SAHARAN AFRICA	**11 188**	**13 526**	**0.51**	**1.83**	**126**	**148**	**0.48**	**2.45**	**368**	**240**	**–2.07**	**–4.55**
LATIN AMERICA and CARIBBEAN	**117 486**	**166 838**	**4.65**	**2.47**	**8 681**	**8 403**	**0.08**	**1.32**	**34 744**	**55 610**	**3.77**	**2.79**
Argentina	45 798	63 741	4.36	1.47	1 928	955	21.38	2.24	7 457	8 699	–2.50	–3.51
Brazil	61 176	89 152	4.46	3.33	70	61	–38.97	2.34	22 001	39 551	4.26	5.03
Chile	82	95	0.84	1.27	234	253	4.79	0.84	8	7	–1.22	–0.84
Mexico	331	362	1.01	0.03	5 024	5 275	–1.29	0.56	5	2	–10.61	–0.63
Uruguay	6 135	8 292	6.44	2.10	33	29	14.48	–1.82	4 059	5 667	7.35	1.82
ASIA and PACIFIC	**99 421**	**116 561**	**2.13**	**1.49**	**55 069**	**72 006**	**9.98**	**2.60**	**1 673**	**1 391**	**1.07**	**–1.09**
Bangladesh	293	333	–0.26	1.11	155	320	0.32	5.85	0	0	0.00	–0.14
China[d]	53 884	59 860	0.70	1.09	41 107	54 240	13.88	2.70	1 128	978	0.41	–1.32
India	32 165	41 079	5.78	2.18	168	227	32.82	0.41	295	183	9.06	–0.81
Indonesia	1 882	2 205	0.26	1.33	1 470	1 980	1.43	1.76	5	5	–0.67	–0.06
Iran, Islamic Republic of	726	875	3.26	1.55	825	1 196	6.55	3.82	6	5	0.04	–0.20
Korea	143	159	1.83	0.00	1 474	1 666	–1.40	0.78	0	0	–0.30	0.00
Malaysia	5	5	0.00	–0.23	605	586	–2.71	0.38	27	24	–6.63	0.00
Pakistan	4 647	5 488	2.37	1.38	907	2 091	5.42	7.12	12	20	114.40	–0.24
Saudi Arabia	4	4	0.00	–0.14	63	78	6.84	3.10	0	0	0.00	–0.03
Turkey	2 009	2 234	–0.66	0.85	1 919	2 250	10.60	2.25	7	3	–1.75	–0.08
LEAST DEVELOPED COUNTRIES (LDC)	**7 831**	**9 224**	**0.47**	**1.49**	**290**	**442**	**0.51**	**3.67**	**179**	**124**	**–6.52**	**–3.73**
OECD	**137 133**	**164 310**	**2.42**	**1.29**	**34 625**	**33 896**	**–1.14**	**0.03**	**45 909**	**48 842**	**3.76**	**0.15**
NON-OECD	**249 989**	**330 990**	**3.77**	**2.22**	**58 022**	**74 790**	**9.90**	**2.65**	**39 808**	**63 466**	**4.06**	**3.07**

For notes, see end of the table.

Table A.11. **Oilseed projections** *(cont.)*

Crop year[a]	CONSUMPTION (kt)		Growth[b] (%)		DOMESTIC CRUSH (kt)		Growth[b] (%)	
	Average 2007-09 est.	2019	2000-09	2010-19	Average 2007-09 est.	2019	2000-09	2010-19
WORLD	**396 515**	**491 781**	**3.44**	**1.95**	**325 685**	**409 291**	**3.78**	**2.07**
DEVELOPED	**139 153**	**167 859**	**1.95**	**1.72**	**123 010**	**149 470**	**2.21**	**1.83**
NORTH AMERICA	**65 718**	**78 549**	**0.51**	**1.81**	**58 227**	**69 642**	**0.92**	**1.96**
Canada	6 943	10 483	3.27	2.31	6 061	9 552	4.67	2.56
United States	58 775	68 066	0.22	1.73	52 166	60 090	0.55	1.87
EUROPE	**60 390**	**75 557**	**4.09**	**1.78**	**53 825**	**68 327**	**4.21**	**1.87**
EU27	43 654	50 666	2.26	1.22	39 895	47 063	2.53	1.31
Russian Federation	8 085	11 238	11.56	2.29	7 528	10 476	11.39	2.29
Ukraine	6 372	10 447	12.30	3.93	4 694	8 228	12.12	4.50
OCEANIA DEVELOPED	**1 287**	**1 506**	**4.13**	**1.60**	**1 163**	**1 363**	**5.17**	**1.78**
Australia	1 273	1 492	4.17	1.61	1 155	1 354	5.20	1.79
New Zealand	13	13	0.59	0.02	9	8	1.42	0.01
OTHER DEVELOPED[c]	**11 758**	**12 247**	**0.01**	**0.80**	**9 794**	**10 139**	**–0.25**	**0.75**
Japan	6 539	5 845	–2.21	–0.42	5 605	5 026	–2.47	–0.40
South Africa	1 437	1 606	2.27	1.08	1 264	1 421	2.43	1.14
DEVELOPING	**257 363**	**323 921**	**4.33**	**2.07**	**202 675**	**259 820**	**4.84**	**2.21**
AFRICA	**13 727**	**17 042**	**1.80**	**2.15**	**7 780**	**9 567**	**2.62**	**2.01**
NORTH AFRICA	**2 776**	**3 609**	**7.60**	**2.74**	**2 206**	**2 879**	**8.66**	**2.78**
Algeria	223	282	5.53	2.29	149	194	4.30	2.43
Egypt	1 755	2 282	9.91	2.74	1 457	1 893	11.48	2.74
SUB-SAHARAN AFRICA	**10 951**	**13 433**	**0.63**	**2.00**	**5 574**	**6 688**	**0.77**	**1.69**
LATIN AMERICA and CARIBBEAN	**92 394**	**119 578**	**4.68**	**2.27**	**83 000**	**108 533**	**4.84**	**2.37**
Argentina	40 628	56 022	6.12	2.51	39 641	54 978	6.31	2.55
Brazil	39 868	49 596	4.41	2.20	33 950	42 536	4.20	2.39
Chile	309	342	3.79	1.00	273	301	3.84	0.94
Mexico	5 362	5 635	–1.15	0.52	4 756	5 145	–0.37	0.53
Uruguay	2 094	2 647	4.84	2.60	1 852	2 356	4.91	2.67
ASIA and PACIFIC	**151 242**	**187 301**	**4.37**	**1.94**	**111 896**	**141 720**	**5.00**	**2.10**
Bangladesh	445	653	–0.25	3.20	378	560	–0.27	3.27
China[d]	91 993	113 277	4.73	1.85	70 093	89 694	5.55	2.15
India	32 226	41 104	5.92	2.23	25 672	32 476	5.68	2.06
Indonesia	3 353	4 179	0.74	1.58	102	102	3.44	0.44
Iran, Islamic Republic of	1 556	2 065	4.91	2.81	1 365	1 817	5.14	2.86
Korea	1 616	1 824	–1.04	0.71	1 107	1 276	–1.87	1.13
Malaysia	585	567	–2.47	0.39	498	460	–2.40	–0.05
Pakistan	5 526	7 554	2.68	2.72	4 104	5 547	2.94	2.58
Saudi Arabia	66	82	6.38	2.93	47	58	6.98	3.01
Turkey	3 975	4 480	3.91	1.54	3 058	3 382	3.05	1.26
LEAST DEVELOPED COUNTRIES (LDC)	**7 939**	**9 542**	**0.70**	**1.67**	**4 681**	**5 840**	**0.84**	**1.90**
OECD	**129 007**	**149 417**	**0.95**	**1.43**	**114 423**	**133 569**	**1.27**	**1.54**
NON-OECD	**267 509**	**342 364**	**4.83**	**2.18**	**211 263**	**275 721**	**5.36**	**2.33**

est.: Estimate.

a) Beginning crop marketing year – see "Glossary of Terms" for definitions.

b) Least-squares growth rate (see glossary).

c) Includes other developed Asian countries: Kazakhstan, Kyrgyzstan, Tajikistan, Turkmenistan, Uzbekistan, Armenia, Azerbaijan and Georgia.

d) Refers to mainland only. The economies of Chinese Taipei, Hong Kong (China) and Macau (China) are included in the other Asia Pacific aggregate.

Source: OECD and FAO Secretariats.

StatLink http://dx.doi.org/10.1787/888932287368

Table A.12. **Protein meal projections**

Marketing year	PRODUCTION (kt)		Growth[a] (%)		IMPORTS (kt)		Growth[a] (%)		EXPORTS (kt)		Growth[a] (%)	
	Average 2007-09 est.	2019	2000-09	2010-19	Average 2007-09 est.	2019	2000-09	2010-19	Average 2007-09 est.	2019	2000-09	2010-19
WORLD	**231 290**	**291 576**	**3.80**	**2.16**	**66 297**	**85 871**	**4.42**	**2.26**	**71 646**	**91 296**	**5.08**	**2.11**
DEVELOPED	**81 708**	**97 113**	**1.53**	**1.73**	**38 780**	**42 281**	**3.54**	**0.64**	**14 264**	**20 849**	**6.49**	**3.09**
NORTH AMERICA	**43 221**	**51 605**	**0.75**	**1.96**	**3 521**	**5 244**	**6.88**	**2.51**	**10 743**	**16 445**	**5.11**	**3.30**
Canada	3 887	6 061	3.24	2.47	1 306	1 189	3.44	0.74	2 158	4 073	9.45	3.55
United States	39 334	45 545	0.53	1.90	2 215	4 055	9.37	3.11	8 585	12 371	4.30	3.23
EUROPE	**31 639**	**38 496**	**3.09**	**1.61**	**30 612**	**30 844**	**2.24**	**0.19**	**3 312**	**4 139**	**12.37**	**2.13**
EU27	25 313	29 044	1.68	1.13	28 532	28 285	1.78	0.04	664	646	6.90	0.59
Russian Federation	3 353	4 663	12.03	2.31	812	1 125	15.12	3.01	893	636	23.34	–2.70
Ukraine	2 066	3 602	12.77	4.50	70	45	–0.33	–4.43	1 597	2 720	11.86	4.43
OCEANIA DEVELOPED	**672**	**785**	**5.51**	**1.78**	**1 682**	**2 039**	**20.00**	**0.51**	**14**	**13**	**–6.00**	**0.00**
Australia	667	780	5.54	1.79	683	830	10.55	1.29	14	13	–6.00	0.00
New Zealand	5	5	2.69	0.00	998	1 209	35.83	0.00	0	0	0.00	–6.57
OTHER DEVELOPED[b]	**6 176**	**6 227**	**–0.71**	**0.60**	**2 965**	**4 154**	**9.26**	**2.08**	**195**	**252**	**9.15**	**7.24**
Japan	3 861	3 456	–2.87	–0.40	1 800	2 052	9.94	0.38	8	0	–2.95	–
South Africa	738	812	4.03	1.14	1 107	2 059	8.64	4.29	27	21	18.55	–1.52
DEVELOPING	**149 582**	**194 463**	**5.22**	**2.39**	**27 517**	**43 590**	**5.76**	**4.09**	**57 382**	**70 446**	**4.77**	**1.83**
AFRICA	**4 799**	**5 817**	**3.38**	**2.01**	**2 517**	**3 790**	**3.69**	**3.53**	**455**	**458**	**–2.25**	**1.06**
NORTH AFRICA	**1 608**	**2 083**	**10.43**	**2.78**	**2 174**	**3 248**	**3.15**	**3.46**	**6**	**6**	**5.06**	**–0.19**
Algeria	93	119	5.35	2.38	624	769	8.65	1.96	0	0	0.00	–0.05
Egypt	1 090	1 417	13.87	2.75	528	773	–8.60	2.16	2	2	0.00	–0.06
SUB-SAHARAN AFRICA	**3 191**	**3 734**	**0.80**	**1.60**	**343**	**542**	**7.69**	**4.00**	**449**	**452**	**–2.33**	**1.08**
LATIN AMERICA and CARIBBEAN	**63 018**	**82 893**	**4.93**	**2.38**	**6 140**	**8 415**	**5.50**	**2.61**	**44 087**	**57 080**	**4.43**	**2.13**
Argentina	30 067	41 929	6.70	2.54	0	0	0.00	0.00	28 897	39 313	6.81	2.33
Brazil	25 929	32 560	4.19	2.39	62	48	–25.66	4.06	13 326	15 854	0.73	1.91
Chile	199	216	4.23	0.94	845	1 430	6.97	5.02	8	5	103.11	–5.02
Mexico	3 304	3 640	–1.39	0.58	473	936	13.08	3.71	12	15	–	0.00
Uruguay	1 460	1 841	4.98	2.70	9	5	6.38	–2.82	737	984	0.95	2.82
ASIA and PACIFIC	**81 765**	**105 752**	**5.57**	**2.41**	**18 861**	**31 386**	**6.15**	**4.59**	**12 839**	**12 908**	**6.43**	**0.62**
Bangladesh	249	360	0.01	3.19	243	436	4.83	4.61	0	0	0.00	–0.11
China[c]	47 968	62 327	6.37	2.46	608	7 121	20.24	23.62	977	0	–1.82	–
India	16 503	20 740	6.06	2.05	38	32	7.40	0.01	5 532	5 653	10.45	–0.16
Indonesia	3 251	4 926	9.76	3.89	2 477	2 210	6.22	–1.68	2 681	3 280	9.44	1.68
Iran, Islamic Republic of	1 004	1 318	5.59	2.86	1 288	2 345	13.63	5.16	73	36	8.98	–5.02
Korea	836	972	–1.91	1.17	3 274	3 900	3.77	0.77	0	0	–	–
Malaysia	3 104	4 486	3.00	3.45	866	672	6.46	–2.45	2 336	3 077	4.03	2.45
Pakistan	2 492	3 312	2.77	2.58	397	726	25.25	4.61	66	63	10.81	–1.17
Saudi Arabia	25	31	7.12	3.00	737	1 110	3.55	3.57	0	0	0.00	–0.26
Turkey	1 740	1 892	3.77	1.26	828	1 524	4.87	4.30	40	14	–1.67	–2.92
LEAST DEVELOPED COUNTRIES (LDC)	**2 462**	**3 000**	**1.00**	**1.80**	**409**	**629**	**3.67**	**3.47**	**200**	**334**	**–5.02**	**5.33**
OECD	**79 364**	**91 701**	**0.81**	**1.51**	**41 252**	**45 730**	**3.33**	**0.72**	**11 526**	**17 138**	**5.16**	**3.17**
NON-OECD	**151 927**	**199 874**	**5.67**	**2.48**	**25 045**	**40 142**	**6.42**	**4.31**	**60 120**	**74 157**	**5.10**	**1.88**

For notes, see end of the table.

Table A.12. **Protein meal projections** *(cont.)*

Marketing year	CONSUMPTION (kt)		Growth[a] (%)	
	Average 2007-09 est.	2019	2000-09	2010-19
WORLD	**226 088**	**286 074**	**3.62**	**2.21**
DEVELOPED	**106 240**	**118 543**	**1.62**	**1.11**
NORTH AMERICA	**36 024**	**40 409**	**0.13**	**1.52**
Canada	3 035	3 179	0.24	0.63
United States	32 989	37 230	0.13	1.60
EUROPE	**58 919**	**65 196**	**2.24**	**0.88**
EU27	53 168	56 686	1.67	0.58
Russian Federation	3 279	5 152	10.56	3.31
Ukraine	521	920	9.99	4.06
OCEANIA DEVELOPED	**2 332**	**2 810**	**14.42**	**0.83**
Australia	1 341	1 596	8.17	1.54
New Zealand	991	1 214	35.03	–0.04
OTHER DEVELOPED[b]	**8 965**	**10 127**	**1.78**	**1.07**
Japan	5 673	5 508	0.12	–0.11
South Africa	1 815	2 848	6.82	3.39
DEVELOPING	**119 848**	**167 531**	**5.66**	**3.06**
AFRICA	**6 858**	**9 144**	**3.96**	**2.67**
NORTH AFRICA	**3 774**	**5 322**	**5.77**	**3.19**
Algeria	718	888	8.15	2.02
Egypt	1 613	2 186	2.32	2.54
SUB-SAHARAN AFRICA	**3 084**	**3 822**	**2.00**	**1.98**
LATIN AMERICA and CARIBBEAN	**25 030**	**34 214**	**6.30**	**2.89**
Argentina	1 193	2 618	7.31	6.18
Brazil	12 643	16 754	8.93	2.90
Chile	1 030	1 638	6.20	4.44
Mexico	3 770	4 561	–0.24	1.16
Uruguay	728	861	10.49	2.51
ASIA and PACIFIC	**87 960**	**124 173**	**5.62**	**3.14**
Bangladesh	492	795	2.12	3.95
China[c]	47 606	69 448	6.70	3.45
India	11 085	15 101	4.53	3.01
Indonesia	3 097	3 848	7.45	2.01
Iran, Islamic Republic of	2 210	3 622	9.52	4.42
Korea	4 113	4 872	2.32	0.84
Malaysia	1 643	2 074	3.27	2.56
Pakistan	2 814	3 973	4.12	3.00
Saudi Arabia	762	1 141	3.63	3.56
Turkey	2 555	3 400	4.17	2.54
LEAST DEVELOPED COUNTRIES (LDC)	**2 668**	**3 294**	**1.99**	**1.80**
OECD	**109 140**	**120 295**	**1.27**	**0.99**
NON-OECD	**116 948**	**165 779**	**6.23**	**3.19**

est.: Estimate.

a) Least-squares growth rate (see glossary).

b) Includes other developed Asian countries: Kazakhstan, Kyrgyzstan, Tajikistan, Turkmenistan, Uzbekistan, Armenia, Azerbaijan and Georgia.

c) Refers to mainland only. The economies of Chinese Taipei, Hong Kong (China) and Macau (China) are included in the other Asia Pacific aggregate.

Source: OECD and FAO Secretariats.

StatLink http://dx.doi.org/10.1787/888932287387

Table A.13. **Vegetable oil projections**

Marketing year	PRODUCTION (kt)		Growth[a] (%)		IMPORTS (kt)		Growth[a] (%)		EXPORTS (kt)		Growth[a] (%)	
	Average 2007-09 est.	2019	2000-09	2010-19	Average 2007-09 est.	2019	2000-09	2010-19	Average 2007-09 est.	2019	2000-09	2010-19
WORLD	**131 915**	**182 833**	**5.19**	**2.90**	**56 447**	**78 859**	**7.54**	**3.19**	**55 373**	**77 124**	**6.36**	**3.27**
DEVELOPED	**34 191**	**43 545**	**3.80**	**2.05**	**16 864**	**23 751**	**7.83**	**3.54**	**7 135**	**11 023**	**7.09**	**4.22**
NORTH AMERICA	**12 380**	**15 358**	**1.53**	**2.09**	**3 339**	**3 651**	**10.14**	**0.64**	**3 069**	**4 901**	**6.58**	**5.18**
Canada	2 206	3 593	5.62	2.81	310	314	5.27	0.07	1 480	2 444	10.27	3.30
United States	10 175	11 766	0.78	1.88	3 030	3 337	10.77	0.70	1 589	2 456	4.22	7.65
EUROPE	**18 862**	**25 097**	**6.00**	**2.20**	**11 593**	**18 244**	**8.04**	**4.52**	**3 651**	**5 805**	**8.31**	**3.99**
EU27	13 425	16 770	4.46	1.69	9 492	16 299	8.75	5.22	873	842	–6.39	0.30
Russian Federation	2 847	3 971	11.15	2.29	1 077	1 096	0.59	0.75	749	1 746	30.92	6.41
Ukraine	2 097	3 615	11.77	4.50	413	225	17.98	–4.17	1 873	3 038	16.41	4.17
OCEANIA DEVELOPED	**402**	**446**	**7.01**	**1.76**	**315**	**98**	**3.32**	**–0.35**	**40**	**36**	**–3.16**	**–1.97**
Australia	395	440	7.04	1.79	222	6	3.90	–4.97	39	36	–3.18	–1.99
New Zealand	7	6	4.99	0.00	93	93	2.21	0.00	0	0	0.46	–0.20
OTHER DEVELOPED[b]	**2 548**	**2 644**	**0.47**	**0.66**	**1 618**	**1 757**	**3.78**	**1.17**	**375**	**281**	**1.93**	**–2.54**
Japan	1 499	1 371	–1.43	–0.40	733	694	4.57	–0.35	278	208	0.74	–2.70
South Africa	387	429	1.83	1.13	620	758	3.93	1.98	76	58	12.92	–1.98
DEVELOPING	**97 724**	**139 288**	**5.72**	**3.18**	**39 583**	**55 108**	**7.43**	**3.03**	**48 238**	**66 101**	**6.26**	**3.12**
AFRICA	**4 475**	**5 361**	**1.59**	**1.74**	**5 412**	**7 761**	**5.24**	**3.44**	**539**	**395**	**3.96**	**–3.43**
NORTH AFRICA	**498**	**645**	**7.67**	**2.76**	**2 581**	**3 447**	**3.16**	**2.74**	**88**	**60**	**19.09**	**–2.55**
Algeria	55	70	3.93	2.36	601	759	5.90	2.01	30	25	43.53	–2.01
Egypt	295	378	10.80	2.76	1 251	1 753	2.57	2.99	55	33	20.35	–2.99
SUB-SAHARAN AFRICA	**3 978**	**4 717**	**0.97**	**1.61**	**2 832**	**4 314**	**7.50**	**4.03**	**451**	**336**	**1.77**	**–3.58**
LATIN AMERICA and CARIBBEAN	**19 449**	**26 761**	**5.08**	**2.68**	**3 359**	**3 832**	**4.85**	**1.01**	**9 296**	**10 329**	**2.52**	**1.92**
Argentina	7 978	10 944	5.46	2.61	13	13	6.05	–0.04	5 851	6 531	2.65	1.74
Brazil	6 804	9 257	4.70	2.37	310	453	8.33	3.69	2 027	2 324	–0.40	2.89
Chile	70	76	2.99	0.94	321	389	8.17	2.03	1	1	–7.19	–2.03
Mexico	1 396	1 524	3.09	0.86	639	869	3.23	1.65	44	41	1.93	0.00
Uruguay	369	470	4.80	2.79	14	11	0.00	–3.26	200	246	5.03	3.26
ASIA and PACIFIC	**73 800**	**107 165**	**6.18**	**3.38**	**30 811**	**43 514**	**8.18**	**3.16**	**38 403**	**55 377**	**7.29**	**3.42**
Bangladesh	179	253	0.41	3.03	1 138	1 933	5.41	5.08	0	0	0.00	–0.12
China[c]	16 014	20 960	4.53	2.33	9 180	15 391	17.99	4.58	126	135	3.58	–0.05
India	6 759	8 371	4.61	2.00	7 554	11 156	5.01	3.58	85	67	3.71	–0.32
Indonesia	23 873	38 101	10.90	4.28	52	50	5.56	–0.25	17 769	27 197	12.59	4.07
Iran, Islamic Republic of	311	410	5.06	2.86	1 212	1 362	0.21	1.40	136	99	–4.26	–1.40
Korea	200	230	–1.71	1.13	642	705	4.99	0.43	3	2	–10.04	0.00
Malaysia	19 827	30 183	4.87	3.82	1 373	941	7.60	–3.46	17 012	24 957	4.60	3.46
Pakistan	807	1 085	4.04	2.58	2 006	2 700	3.46	2.81	98	51	120.46	–2.81
Saudi Arabia	10	12	5.87	2.96	518	647	8.53	2.40	18	12	2.20	–0.02
Turkey	926	1 017	4.27	1.26	1 010	1 207	6.78	1.08	265	268	27.80	–1.08
LEAST DEVELOPED COUNTRIES (LDC)	**2 155**	**2 436**	**0.39**	**1.19**	**3 928**	**5 942**	**6.93**	**3.97**	**191**	**160**	**–1.47**	**–1.71**
OECD	**30 379**	**36 876**	**2.78**	**1.71**	**16 787**	**24 231**	**8.18**	**3.57**	**4 635**	**6 378**	**3.33**	**3.56**
NON-OECD	**101 537**	**145 957**	**6.01**	**3.22**	**39 660**	**54 628**	**7.28**	**3.02**	**50 738**	**70 746**	**6.69**	**3.25**

For notes, see end of the table.

Table A.13. **Vegetable oil projections** *(cont.)*

Marketing year	CONSUMPTION (kt)		Growth[a] (%)		FOOD VEGETABLE USE PER CAPITA (kg)		Growth[a] (%)	
	Average 2007-09 est.	2019	2000-09	2010-19	Average 2007-09 est.	2019	2000-09	2010-19
WORLD	**133 010**	**184 283**	**5.64**	**2.84**	**16.7**	**19.1**	**3.10**	**1.23**
DEVELOPED	**44 106**	**56 258**	**4.74**	**2.25**	**24.8**	**26.0**	**1.69**	**0.46**
NORTH AMERICA	**12 800**	**14 101**	**2.26**	**0.81**	**33.9**	**34.0**	**0.06**	**–0.17**
Canada	1 020	1 463	0.84	1.39	30.4	37.8	–0.27	0.48
United States	11 780	12 638	2.39	0.74	34.3	33.6	0.09	–0.24
EUROPE	**26 811**	**37 530**	**6.54**	**2.99**	**24.3**	**26.1**	**2.94**	**0.70**
EU27	22 061	32 225	6.83	3.39	25.8	26.9	1.95	0.52
Russian Federation	3 156	3 319	4.68	0.12	22.3	24.4	5.14	0.47
Ukraine	647	800	6.06	2.40	13.9	18.3	6.81	2.96
OCEANIA DEVELOPED	**680**	**509**	**5.97**	**1.63**	**14.7**	**14.2**	**–3.66**	**0.59**
Australia	581	410	6.73	2.06	12.9	12.8	–5.15	1.13
New Zealand	99	99	2.37	0.00	23.4	21.3	1.26	–0.83
OTHER DEVELOPED[b]	**3 816**	**4 118**	**1.96**	**1.12**	**14.8**	**15.5**	**1.30**	**0.86**
Japan	1 969	1 857	0.46	–0.10	15.5	15.0	0.42	0.15
South Africa	941	1 128	3.41	1.86	18.0	20.5	1.58	1.49
DEVELOPING	**88 904**	**128 025**	**6.12**	**3.11**	**14.7**	**17.5**	**3.94**	**1.61**
AFRICA	**9 369**	**12 718**	**3.50**	**2.95**	**9.7**	**10.4**	**1.05**	**0.76**
NORTH AFRICA	**3 017**	**4 028**	**3.62**	**2.82**	**12.2**	**13.4**	**1.70**	**1.06**
Algeria	628	803	5.60	2.17	18.1	19.8	4.11	0.78
Egypt	1 504	2 097	3.35	3.05	18.3	21.4	1.48	1.48
SUB-SAHARAN AFRICA	**6 351**	**8 690**	**3.44**	**3.02**	**8.9**	**9.4**	**0.79**	**0.68**
LATIN AMERICA and CARIBBEAN	**13 525**	**20 257**	**6.79**	**2.71**	**19.4**	**20.8**	**3.19**	**1.00**
Argentina	2 143	4 425	14.96	3.93	22.8	22.0	2.78	–0.11
Brazil	5 103	7 386	7.43	2.27	22.6	24.4	3.99	1.73
Chile	391	464	7.13	1.85	23.0	24.8	6.07	1.00
Mexico	1 993	2 352	3.26	1.16	18.3	19.8	2.20	0.36
Uruguay	132	215	20.15	4.05	15.0	18.7	6.52	1.98
ASIA and PACIFIC	**66 010**	**95 050**	**6.39**	**3.22**	**15.3**	**19.0**	**4.65**	**1.96**
Bangladesh	1 335	2 183	4.96	4.80	8.3	11.8	3.38	3.60
China[c]	24 909	36 214	7.97	3.24	18.4	25.2	7.29	2.67
India	14 228	19 420	4.63	2.84	11.8	14.1	2.98	1.66
Indonesia	6 136	10 860	7.24	4.81	18.3	25.0	2.76	2.82
Iran, Islamic Republic of	1 396	1 672	1.88	1.92	18.8	20.0	0.74	0.81
Korea	839	933	3.20	0.60	17.4	18.9	2.76	0.39
Malaysia	4 199	6 086	6.73	3.43	22.0	31.6	3.51	3.19
Pakistan	2 726	3 727	3.08	2.84	15.2	16.6	0.88	0.80
Saudi Arabia	499	645	8.60	2.43	19.6	20.6	6.24	0.56
Turkey	1 699	1 955	4.14	1.49	20.4	23.0	0.96	0.46
LEAST DEVELOPED COUNTRIES (LDC)	**5 911**	**8 215**	**4.60**	**3.19**	**7.7**	**8.5**	**2.17**	**1.09**
OECD	**42 743**	**54 717**	**4.65**	**2.28**	**25.4**	**26.3**	**1.19**	**0.28**
NON-OECD	**90 267**	**129 566**	**6.15**	**3.09**	**14.8**	**17.6**	**4.08**	**1.63**

est.: Estimate.

a) Least-squares growth rate (see glossary).

b) Includes other developed Asian countries: Kazakhstan, Kyrgyzstan, Tajikistan, Turkmenistan, Uzbekistan, Armenia, Azerbaijan and Georgia.

c) Refers to mainland only. The economies of Chinese Taipei, Hong Kong (China) and Macau (China) are included in the other Asia Pacific aggregate.

Source: OECD and FAO Secretariats.

StatLink http://dx.doi.org/10.1787/888932287406

Table A.14. **Main policy assumptions for sugar markets**

Crop year[a]		Average 07/08-09/10 est.	10/11	11/12	12/13	13/14	14/15	15/16	16/17	17/18	18/19	19/20
ARGENTINA												
Tariff, sugar	*ARS/t*	35.0	35.0	35.0	35.0	35.0	35.0	35.0	35.0	35.0	35.0	35.0
BANGLADESH												
Tariff, white sugar	*%*	32.5	32.5	32.5	32.5	32.5	32.5	32.5	32.5	32.5	32.5	32.5
BRAZIL												
Tariff, raw sugar	*%*	35.0	35.0	35.0	35.0	35.0	35.0	35.0	35.0	35.0	35.0	35.0
Tariff, white sugar	*%*	35.0	35.0	35.0	35.0	35.0	35.0	35.0	35.0	35.0	35.0	35.0
Ethanol blending ratio with gazoline	*%*	25.0	25.0	25.0	25.0	25.0	25.0	25.0	25.0	25.0	25.0	25.0
CANADA												
Tariff, raw sugar	*CAD/t*	24.7	24.7	24.7	24.7	24.7	24.7	24.7	24.7	24.7	24.7	24.7
Tariff, white sugar	*CAD/t*	30.9	30.9	30.9	30.9	30.9	30.9	30.9	30.9	30.9	30.9	30.9
CHINA[b]												
TRQ sugar	*kt*	1 954	1 954	1 954	1 954	1 954	1 954	1 954	1 954	1 954	1 954	1 954
Tariff, in-quota, raw sugar	*%*	15.0	15.0	15.0	15.0	15.0	15.0	15.0	15.0	15.0	15.0	15.0
Tariff, in-quota, white sugar	*%*	20.0	20.0	20.0	20.0	20.0	20.0	20.0	20.0	20.0	20.0	20.0
Tariff, over-quota	*%*	50.0	50.0	50.0	50.0	50.0	50.0	50.0	50.0	50.0	50.0	50.0
EUROPEAN UNION[c]												
Reference price, white sugar[d]	*EUR/t*	526	404	404	404	404	404	404	404	404	404	404
Production quota[e]	*mt wse*	16.2	13.4	13.4	13.4	13.4	13.4	13.4	13.4	13.4	13.4	13.4
Subsidised export limits												
Quantity limit	*kt rse*	1 431	1 431	1 431	1 431	1 431	1 431	1 431	1 431	1 431	1 431	1 431
Value limit	*000 EUR*	534 426	534 426	534 426	534 426	534 426	534 426	534 426	534 426	534 426	534 426	534 426
Tariff, raw sugar	*EUR/t*	339	339	339	339	339	339	339	339	339	339	339
Tariff, white sugar	*EUR/t*	419	419	419	419	419	419	419	419	419	419	419
INDIA												
Intervention price, sugarcane	*INR/t*	750	750	750	750	750	750	750	750	750	750	750
Applied tariff, raw sugar	*%*	60.0	60.0	60.0	60.0	60.0	60.0	60.0	60.0	60.0	60.0	60.0
INDONESIA												
Tariff, white sugar	*%*	25.0	25.0	25.0	25.0	25.0	25.0	25.0	25.0	25.0	25.0	25.0
JAPAN												
Minimum stabilisation price, raw sugar	*JPY/kg*	151.8	151.8	151.8	151.8	151.8	151.8	151.8	151.8	151.8	151.8	151.8
Tariff, raw sugar	*JPY/kg*	71.8	71.8	71.8	71.8	71.8	71.8	71.8	71.8	71.8	71.8	71.8
Tariff, white sugar	*JPY/kg*	103.1	103.1	103.1	103.1	103.1	103.1	103.1	103.1	103.1	103.1	103.1
KOREA												
Tariff, raw sugar	*%*	18.0	18.0	18.0	18.0	18.0	18.0	18.0	18.0	18.0	18.0	18.0
MEXICO												
Ad valorem tariff, raw sugar	*%*	156	156	156	156	156	156	156	156	156	156	156
RUSSIAN FEDERATION												
Minimum tariff, raw sugar	*%*	37.2	35.2	42.3	48.2	50.8	45.1	47.3	49.8	45.6	42.8	37.7
Minimum tariff, white sugar	*%*	57.6	55.7	62.1	67.8	73.9	67.2	69.4	70.4	66.3	63.3	56.9
UNITED STATES[c]												
Loan rate, raw sugar	*USD/t*	398.7	407.9	413.4	413.4	413.4	413.4	413.4	413.4	413.4	413.4	413.4
Loan rate, white sugar	*USD/t*	508.9	524.1	531.2	531.2	531.2	531.2	531.2	531.2	531.2	531.2	531.2
TRQ, raw sugar	*kt rse*	1 227	1 184	1 186	1 191	1 193	1 195	1 198	1 202	1 205	1 207	1 209
TRQ, refined sugar	*kt rse*	49.0	49.0	49.0	49.0	49.0	49.0	49.0	49.0	49.0	49.0	49.0
Raw sugar 2nd tier WTO tariff	*USD/t*	338.6	338.6	338.6	338.6	338.6	338.6	338.6	338.6	338.6	338.6	338.6
White sugar 2nd tier WTO tariff	*USD/t*	357.4	357.4	357.4	357.4	357.4	357.4	357.4	357.4	357.4	357.4	357.4
Raw sugar 2nd tier NAFTA tariff	*USD/t*	11.1	0.0	0.0	0.0	0.0	0.0	0.0	0.0	0.0	0.0	0.0

For notes, see end of the table.

Table A.14. **Main policy assumptions for sugar markets** *(cont.)*

Crop year[a]		Average 07/08-09/10 est.	10/11	11/12	12/13	13/14	14/15	15/16	16/17	17/18	18/19	19/20
SOUTH AFRICA												
Tariff, raw sugar	%	105.0	105.0	105.0	105.0	105.0	105.0	105.0	105.0	105.0	105.0	105.0
TANZANIA												
Applied tariff, white sugar	%	25.0	25.0	25.0	25.0	25.0	25.0	25.0	25.0	25.0	25.0	25.0
VIET NAM												
Applied tariff, white sugar	%	40.0	40.0	40.0	40.0	40.0	40.0	40.0	40.0	40.0	40.0	40.0

Note: The source for tariffs (except United States and the Russian Federation) is AMAD. The source for the Russian Federation and United States tariffs is ERS, USDA.

est.: Estimate.

a) Beginning crop marketing year – see the "Glossary of Terms" for definitions.

b) Refers to mainland only.

c) In addition, price based special safeguard actions may apply.

d) Reference price for consumers.

e) Production that receives official support.

Source: OECD and FAO Secretariats.

StatLink http://dx.doi.org/10.1787/888932287425

Table A.15. **World sugar projections (in raw sugar equivalent)**

Crop year[a]		Average 07/08-09/10 est.	10/11	11/12	12/13	13/14	14/15	15/16	16/17	17/18	18/19	19/20
OECD												
Production	kt rse	37 425	37 390	36 491	36 786	37 100	37 316	37 044	36 743	36 754	36 897	37 048
Consumption	kt rse	43 488	43 638	43 945	44 068	44 154	44 447	44 642	44 881	45 157	45 435	45 736
Closing stocks	kt rse	20 326	20 720	20 499	20 390	20 472	20 746	20 866	21 038	21 200	21 163	21 208
Non-OECD												
Production	kt rse	123 688	137 436	140 149	143 581	143 602	144 669	144 096	149 584	155 222	158 840	163 029
Consumption	kt rse	116 822	123 302	127 078	130 546	134 075	136 431	138 646	141 130	144 772	148 253	152 087
Closing stocks	kt rse	49 250	49 036	53 363	57 713	58 593	57 914	54 133	52 765	53 138	53 712	54 409
WORLD												
Production	kt rse	161 113	174 826	176 640	180 367	180 703	181 985	181 139	186 327	191 975	195 737	200 077
Consumption	kt rse	160 310	166 940	171 023	174 614	178 228	180 878	183 288	186 011	189 928	193 689	197 823
Closing stocks	kt rse	69 576	69 757	73 862	78 103	79 065	78 660	74 999	73 803	74 338	74 875	75 617
Price, raw sugar[b]	USD/t	401.4	397.8	331.1	290.7	275.8	310.6	296.3	281.0	306.8	326.8	371.7
Price, white sugar[c]	USD/t	462.2	448.4	402.6	368.5	338.3	371.9	360.0	355.0	377.2	395.0	439.2

est.: Estimate.

a) Beginning crop marketing year – see the "Glossary of Terms" for definitions.

b) Raw sugar world price, ICE Inc. No. 11 f.o.b., bulk price, October/September.

c) Refined sugar price, White Sugar Futures Contract No. 407, Euronext market, Liffe, London, Europe, October/September.

Source: OECD and FAO Secretariats.

StatLink http://dx.doi.org/10.1787/888932287444

Table A.16. **World sugar projections (in raw sugar equivalent)**

Crop year[a]	PRODUCTION (kt)		Growth[b] (%)		IMPORTS (kt)		Growth[b] (%)		EXPORTS (kt)		Growth[b] (%)	
	Average 2007-09 est.	2019	2000-09	2010-19	Average 2007-09 est.	2019	2000-09	2010-19	Average 2007-09 est.	2019	2000-09	2010-19
WORLD	**161 113**	**200 077**	**2.18**	**1.41**	**48 712**	**63 234**	**2.60**	**2.05**	**53 237**	**64 746**	**3.04**	**2.00**
DEVELOPED	**38 691**	**39 523**	**−0.75**	**0.24**	**15 659**	**17 363**	**−0.43**	**0.33**	**7 814**	**7 205**	**−3.59**	**0.40**
NORTH AMERICA	**7 161**	**7 201**	**−1.06**	**−0.08**	**3 841**	**4 803**	**4.72**	**1.60**	**264**	**251**	**10.22**	**−0.03**
Canada	101	137	0.90	6.62	1 355	1 414	0.49	0.27	66	69	20.81	−0.10
United States	7 060	7 064	−1.10	−0.18	2 486	3 390	7.62	2.20	198	181	7.80	0.00
EUROPE	**23 288**	**23 104**	**−0.93**	**0.23**	**7 745**	**8 311**	**−2.70**	**−0.31**	**2 818**	**1 139**	**−7.92**	**−4.09**
EU27	16 235	14 434	−3.42	−0.70	3 847	5 047	9.26	2.48	1 867	212	−11.59	−12.77
Russian Federation	3 533	4 088	9.99	0.88	2 333	1 967	−11.13	−2.32	112	27	−1.96	−0.17
Ukraine	1 814	2 230	1.59	2.70	297	293	−15.17	−8.43	4	9	−52.97	10.24
OCEANIA DEVELOPED	**4 890**	**5 337**	**0.51**	**0.12**	**256**	**285**	**−0.07**	**0.63**	**3 510**	**4 257**	**−0.22**	**1.13**
Australia	4 890	5 337	0.51	0.12	12	8	8.70	0.28	3 489	4 237	−0.22	1.14
New Zealand	0	0	–	–	244	277	−0.37	0.64	20	20	−0.79	0.00
OTHER DEVELOPED[c]	**3 352**	**3 881**	**−0.52**	**1.14**	**3 816**	**3 963**	**0.26**	**0.27**	**1 222**	**1 559**	**−2.67**	**3.00**
Japan	921	962	0.72	0.31	1 427	1 323	−1.02	−0.64	2	2	−2.60	0.76
South Africa	2 299	2 720	−1.20	1.33	152	283	−10.45	7.13	875	1 211	−5.42	4.02
DEVELOPING	**122 422**	**160 554**	**3.27**	**1.72**	**33 053**	**45 872**	**4.26**	**2.77**	**45 423**	**57 541**	**4.59**	**2.21**
AFRICA	**8 225**	**11 380**	**1.88**	**2.70**	**8 987**	**10 892**	**3.38**	**1.25**	**3 898**	**4 615**	**1.99**	**0.99**
NORTH AFRICA	**2 609**	**3 419**	**1.40**	**2.51**	**4 013**	**5 010**	**2.93**	**1.60**	**356**	**300**	**0.05**	**−1.42**
Algeria	0	0	0.00	0.00	1 206	1 680	2.61	2.40	48	33	105.24	−1.20
Egypt	1 808	2 358	2.25	2.46	1 136	1 194	1.03	0.81	153	145	−3.19	−0.98
SUB-SAHARAN AFRICA	**5 617**	**7 961**	**2.11**	**2.78**	**4 974**	**5 882**	**3.75**	**0.97**	**3 541**	**4 315**	**2.23**	**1.18**
LATIN AMERICA and CARIBBEAN	**56 287**	**71 722**	**4.14**	**1.81**	**2 357**	**2 232**	**5.93**	**0.44**	**29 647**	**41 075**	**5.93**	**2.45**
Argentina	2 415	2 750	4.82	1.13	14	0	–	–	513	624	15.18	0.89
Brazil	36 059	47 328	7.23	2.04	0	4	–	0.00	23 121	32 685	9.27	2.53
Chile	344	474	−4.53	2.98	591	526	14.65	−0.55	1	1	−53.25	0.55
Mexico	5 437	5 681	0.38	0.75	364	959	28.97	8.31	911	1 705	19.22	6.75
Uruguay	7	10	−13.70	2.78	101	114	−1.20	0.08	2	2	−3.23	−0.08
ASIA and PACIFIC	**57 909**	**77 452**	**2.55**	**1.50**	**21 710**	**32 747**	**4.46**	**3.52**	**11 878**	**11 851**	**2.16**	**1.87**
China[d]	14 229	18 242	6.52	1.80	1 051	2 202	0.39	3.69	71	49	−8.51	−3.84
India	21 036	31 891	0.96	1.20	3 334	5 961	24.50	7.38	2 152	159	−13.70	6.49
Indonesia	3 055	4 455	6.82	2.43	1 467	2 225	−3.77	3.20	1	1	−6.79	−0.08
Iran, Islamic Republic of	751	692	−4.97	2.97	800	1 246	−1.75	3.66	0	0	30.74	−0.09
Korea	0	0	–	–	1 543	1 960	−0.26	2.12	314	349	−0.07	1.34
Malaysia	32	41	−15.98	1.22	1 442	1 936	1.72	2.64	194	99	−9.30	−0.12
Pakistan	4 278	5 195	1.74	2.45	372	1 280	10.33	5.27	218	25	5.81	0.54
Saudi Arabia	0	0	0.00	−0.26	1 367	1 703	5.73	1.68	539	564	10.34	−0.02
Thailand	7 884	8 995	2.79	0.86	0	0	0.00	−0.03	4 586	5 884	1.02	1.37
Turkey	2 164	2 681	−0.34	1.53	178	236	24.94	9.58	32	15	−36.17	−9.58
LEAST DEVELOPED COUNTRIES (LDC)	**3 208**	**5 276**	**3.89**	**4.23**	**4 834**	**6 144**	**4.64**	**1.46**	**1 844**	**2 776**	**5.96**	**2.50**
OECD	**37 425**	**37 048**	**−1.71**	**−0.03**	**12 431**	**15 578**	**4.90**	**1.97**	**6 933**	**6 846**	**−3.59**	**1.17**
NON-OECD	**123 688**	**163 029**	**3.62**	**1.77**	**36 281**	**47 656**	**1.89**	**2.07**	**46 304**	**57 901**	**4.36**	**2.10**

For notes, see end of the table.

Table A.16. World sugar projections (in raw sugar equivalent) (cont.)

Crop year[a]	CONSUMPTION (kt)		Growth[b] (%)		PER CAPITA (kg)		Growth[b] (%)	
	Average 2007-09 est.	2019	2000-09	2010-19	Average 2007-09 est.	2019	2000-09	2010-19
WORLD	**160 310**	**197 823**	**2.44**	**1.79**	**23.8**	**26.1**	**1.22**	**0.73**
DEVELOPED	**47 433**	**49 708**	**0.16**	**0.37**	**35.3**	**35.8**	**–0.24**	**0.09**
NORTH AMERICA	**10 825**	**11 744**	**0.62**	**0.66**	**31.4**	**30.8**	**–0.37**	**–0.22**
Canada	1 350	1 481	0.56	0.76	40.6	39.7	–0.45	–0.13
United States	9 475	10 263	0.63	0.64	30.4	29.9	–0.35	–0.23
EUROPE	**29 152**	**29 928**	**–0.14**	**0.21**	**40.5**	**41.8**	**–0.20**	**0.26**
EU27	18 616	18 991	0.01	0.19	37.7	38.1	–0.28	0.13
Russian Federation	6 082	5 982	–1.21	–0.22	43.0	44.0	–0.75	0.14
Ukraine	2 327	2 486	0.66	0.49	50.6	57.5	1.40	1.05
OCEANIA DEVELOPED	**1 544**	**1 705**	**2.60**	**0.84**	**61.0**	**60.7**	**1.43**	**–0.10**
Australia	1 311	1 449	2.85	0.86	62.2	61.8	1.67	–0.11
New Zealand	233	256	1.26	0.77	55.0	55.3	0.15	–0.06
OTHER DEVELOPED[c]	**5 912**	**6 331**	**0.23**	**0.46**	**23.4**	**24.2**	**–0.28**	**0.20**
Japan	2 348	2 284	–0.35	–0.24	18.5	18.4	–0.39	0.01
South Africa	1 629	1 795	0.08	0.60	32.8	34.2	–1.15	0.19
DEVELOPING	**112 877**	**148 114**	**3.52**	**2.32**	**20.9**	**23.9**	**2.08**	**1.07**
AFRICA	**13 318**	**17 613**	**2.98**	**2.31**	**14.1**	**14.6**	**0.63**	**0.09**
NORTH AFRICA	**6 301**	**8 111**	**2.58**	**2.12**	**25.7**	**27.2**	**0.66**	**0.36**
Algeria	1 198	1 637	2.69	2.59	34.8	40.8	1.19	1.20
Egypt	2 774	3 404	2.31	1.92	34.0	35.0	0.44	0.34
SUB-SAHARAN AFRICA	**7 016**	**9 502**	**3.36**	**2.47**	**10.0**	**10.4**	**0.86**	**0.10**
LATIN AMERICA and CARIBBEAN	**29 032**	**32 907**	**2.60**	**1.03**	**50.4**	**51.4**	**1.37**	**0.10**
Argentina	1 877	2 159	2.99	1.16	47.1	48.9	2.04	0.32
Brazil	12 857	14 744	3.98	1.04	67.0	70.9	2.79	0.36
Chile	828	995	2.64	1.49	49.3	53.8	1.58	0.63
Mexico	5 237	4 918	0.69	0.25	48.2	41.4	–0.37	–0.55
Uruguay	110	123	–0.66	0.85	32.9	35.4	–0.77	0.49
ASIA and PACIFIC	**70 527**	**97 595**	**4.03**	**2.80**	**18.2**	**22.4**	**2.77**	**1.76**
China[d]	15 593	20 315	7.13	2.00	11.7	14.3	6.46	1.43
India	24 598	37 062	4.16	3.50	20.8	27.4	2.61	2.30
Indonesia	4 789	6 553	3.61	2.75	21.1	26.0	2.35	1.85
Iran, Islamic Republic of	1 845	1 964	–2.18	3.96	25.2	23.7	–3.32	2.85
Korea	1 255	1 624	1.74	2.27	26.1	32.8	1.31	2.06
Malaysia	1 304	1 876	2.82	2.80	48.3	59.3	0.99	1.41
Pakistan	4 625	6 409	3.09	3.02	26.1	28.9	0.88	0.97
Saudi Arabia	808	1 145	3.05	2.97	32.0	36.9	0.69	1.09
Thailand	2 620	3 237	3.82	1.33	38.9	45.5	2.87	0.85
Turkey	2 201	2 810	2.23	1.67	29.8	33.8	0.92	0.63
LEAST DEVELOPED COUNTRIES (LDC)	**6 223**	**8 627**	**4.24**	**2.87**	**8.3**	**9.1**	**1.97**	**0.73**
OECD	**43 488**	**45 736**	**0.53**	**0.50**	**34.8**	**34.8**	**–0.09**	**0.07**
NON-OECD	**116 822**	**152 087**	**3.22**	**2.21**	**21.3**	**24.2**	**1.86**	**1.02**

est.: Estimate.

a) Beginning crop marketing year – see "Glossary of Terms" for definitions.

b) Least-squares growth rate (see glossary).

c) Includes other developed Asian countries: Kazakhstan, Kyrgyzstan, Tajikistan, Turkmenistan, Uzbekistan, Armenia, Azerbaijan and Georgia.

d) Refers to mainland only. The economies of Chinese Taipei, Hong Kong (China) and Macau (China) are included in the other Asia Pacific aggregate.

Source: OECD and FAO Secretariats.

StatLink http://dx.doi.org/10.1787/888932287463

Table A.17. Main policy assumptions for meat markets

		Average 2007-09 est.	2010	2011	2012	2013	2014	2015	2016	2017	2018	2019
ARGENTINA												
Beef export tax	%	15	15	15	15	15	15	15	15	15	15	15
CANADA												
Beef tariff-quota	kt pw	76.4	76.4	76.4	76.4	76.4	76.4	76.4	76.4	76.4	76.4	76.4
In-quota tariff	%	0	0	0	0	0	0	0	0	0	0	0
Out-of-quota tariff	%	26.5	26.5	26.5	26.5	26.5	26.5	26.5	26.5	26.5	26.5	26.5
Poultry meat tariff-quota	kt pw	45.4	45.4	45.4	45.4	45.4	45.4	45.4	45.4	45.4	45.4	45.4
In-quota tariff	%	2.5	2.5	2.5	2.5	2.5	2.5	2.5	2.5	2.5	2.5	2.5
Out-of-quota tariff	%	196.6	196.6	196.6	196.6	196.6	196.6	196.6	196.6	196.6	196.6	196.6
EUROPEAN UNION[a]												
Beef basic price[b, c, d]	EUR/kg dw	2.2	2.2	2.2	2.2	2.2	2.2	2.2	2.2	2.2	2.2	2.2
Beef buy-in price[d]	EUR/kg dw	1.6	1.6	1.6	1.6	1.6	1.6	1.6	1.6	1.6	1.6	1.6
Pig meat basic price[c]	EUR/kg dw	1.5	1.5	1.5	1.5	1.5	1.5	1.5	1.5	1.5	1.5	1.5
Sheep meat basic price	EUR/kg dw	5.0	5.0	5.0	5.0	5.0	5.0	5.0	5.0	5.0	5.0	5.0
Sheep basic rate[e]	EUR/head	21	21	21	21	21	21	21	21	21	21	21
Male bovine premium	EUR/head	0	0	0	0	0	0	0	0	0	0	0
Adult bovine slaughter premium	EUR/head	0	0	0	0	0	0	0	0	0	0	0
Calf slaughter premium	EUR/head	0	0	0	0	0	0	0	0	0	0	0
Suckler cow premium	EUR/head	0	0	0	0	0	0	0	0	0	0	0
Beef tariff-quota	kt pw	216	216	216	216	216	216	216	216	216	216	216
Pig meat tariff-quota	kt pw	167	167	167	167	167	167	167	167	167	167	167
Poultry meat tariff-quota	kt pw	96	96	96	96	96	96	96	96	96	96	96
Sheep meat tariff-quota	kt cwe	285	285	285	285	285	285	285	285	285	285	285
Subsidised export limits[c]												
Beef[f]	kt cwe	990	990	990	990	990	990	990	990	990	990	990
Pig meat[f]	kt cwe	588	588	588	588	588	588	588	588	588	588	588
Poultry meat	kt cwe	431	431	431	431	431	431	431	431	431	431	431
JAPAN[g]												
Beef stabilisation prices												
Upper price	JPY/kg dw	1 010	1 010	1 010	1 010	1 010	1 010	1 010	1 010	1 010	1 010	1 010
Lower price	JPY/kg dw	780	780	780	780	780	780	780	780	780	780	780
Beef tariff	%	38.5	38.5	38.5	38.5	38.5	38.5	38.5	38.5	38.5	38.5	38.5
Pig meat stabilisation prices												
Upper price	JPY/kg dw	480	480	480	480	480	480	480	480	480	480	480
Lower price	JPY/kg dw	365	365	365	365	365	365	365	365	365	365	365
Pig meat import system[h]												
Tariff	%	4.3	4.3	4.3	4.3	4.3	4.3	4.3	4.3	4.3	4.3	4.3
Standard import price	JPY/kg dw	4 099	4 099	4 099	4 099	4 099	4 099	4 099	4 099	4 099	4 099	4 099
Poultry meat tariff	%	7.4	7.4	7.4	7.4	7.4	7.4	7.4	7.4	7.4	7.4	7.4
KOREA												
Beef tariff	%	40	40	40	40	40	40	40	40	40	40	40
Beef mark-up	%	0	0	0	0	0	0	0	0	0	0	0
Pig meat tariff	%	22	22	22	22	22	22	22	22	22	22	22
Poultry meat tariff	%	21	21	21	21	21	21	21	21	21	21	21
MEXICO												
Pig meat tariff	%	45	45	45	45	45	45	45	45	45	45	45
Pig meat NAFTA tariff	%	0	0	0	0	0	0	0	0	0	0	0
Poultry meat tariff-quota	kt pw	41	41	41	41	41	41	41	41	41	41	41
In-quota tariff	%	50	50	50	50	50	50	50	50	50	50	50
Out-of-quota tariff	%	227.7	227.7	227.7	227.7	227.7	227.7	227.7	227.7	227.7	227.7	227.7

For notes, see end of the table.

Table A.17. **Main policy assumptions for meat markets** *(cont.)*

		Average 2007-09 est.	2010	2011	2012	2013	2014	2015	2016	2017	2018	2019
RUSSIAN FEDERATION												
Beef tariff-quota	*kt pw*	474	560	560	560	560	560	560	560	560	560	560
In-quota tariff	%	15	15	15	15	15	15	15	15	15	15	15
Out-of-quota tariff	%	41	40	40	40	40	40	40	40	40	40	40
Pigmeat tariff-quota	*kt pw*	504	490	490	450	450	450	450	450	450	450	450
In-quota tariff	%	15	15	15	15	15	15	15	15	15	15	15
Out-of-quota tariff	%	65	75	75	75	75	75	75	75	75	75	75
Poultry meat tariff-quota	*kt pw*	1 111	780	600	550	550	550	550	550	550	550	550
In-quota tariff	%	25	25	25	25	25	25	25	25	25	25	25
UNITED STATES												
Beef tariff-quota	*kt pw*	697	697	697	697	697	697	697	697	697	697	697
In-quota tariff	%	4.8	4.8	4.8	4.8	4.8	4.8	4.8	4.8	4.8	4.8	4.8
Out-of-quota tariff	%	26.4	26.4	26.4	26.4	26.4	26.4	26.4	26.4	26.4	26.4	26.4
CHINA												
Beef tariff	%	15.5	15.5	15.5	15.5	15.5	15.5	15.5	15.5	15.5	15.5	15.5
Pig meat tariff	%	16	16	16	16	16	16	16	16	16	16	16
Sheep meat tariff	%	15	15	15	15	15	15	15	15	15	15	15
Poultry meat tariff	%	19.1	19.1	19.1	19.1	19.1	19.1	19.1	19.1	19.1	19.1	19.1
INDIA												
Beef tariff	%	100	100	100	100	100	100	100	100	100	100	100
Pig meat tariff	%	100	100	100	100	100	100	100	100	100	100	100
Sheep meat tariff	%	91.9	91.9	91.9	91.9	91.9	91.9	91.9	91.9	91.9	91.9	91.9
Poultry meat tariff	%	87	87	87	87	87	87	87	87	87	87	87
Eggs tariff	%	150	150	150	150	150	150	150	150	150	150	150
SOUTH AFRICA												
Sheep meat tariff-quota	*kt pw*	6	6	6	6	6	6	6	6	6	6	6
In-quota tariff	%	20	20	20	20	20	20	20	20	20	20	20
Out-of-quota tariff	%	96	96	96	96	96	96	96	96	96	96	96

est.: Estimate.

a) EU farmers also benefit from the Single Farm Payment (SFP) Scheme, which provides flat-rate payments independent from current production decisions and market developments. For the new member states, payments are phased in with the assumption of maximum top-ups from national budgets up to 2013 through the Single Area Payment (SAP), and through the (SFP) from 2014. Due to modulation, an increasing share of the total SFP will go to rural development spending rather than directly to farmers.

b) Price for R3 grade male cattle.

c) Year beginning 1 July.

d) Basic price for storage.

e) A supplementary payment of EUR 7 per head is provided for Less Favoured Areas.

f) Includes live trade.

g) Year beginning 1 April.

h) Pig carcass imports. Emergency import procedures triggered from November 1995 to March 1996 from July 1996 to June 1997 from August 2001 to March 2002 from August 2002 to March 2003 and from August 2003 to March 2004.

Source: OECD and FAO Secretariats.

StatLink http://dx.doi.org/10.1787/888932287482

Table A.18. **World meat projections**

Calendar year[a]		Average 2007-09 est.	2010	2011	2012	2013	2014	2015	2016	2017	2018	2019
OECD[b]												
BEEF AND VEAL												
Production	*kt cwe*	27 569	27 094	27 010	26 811	26 836	26 944	27 175	27 423	27 752	27 956	28 204
Consumption	*kt cwe*	27 259	26 895	26 743	26 585	26 637	26 773	27 045	27 310	27 660	27 892	28 151
Ending stocks	*kt cwe*	1 062	1 050	1 084	1 093	1 131	1 179	1 237	1 306	1 374	1 439	1 493
Per capita consumption	*kg rwt*	15	15	15	15	15	15	15	15	15	15	15
Price, European Union	*EUR/t dw*	2 882	2 600	2 441	2 454	2 521	2 542	2 583	2 574	2 562	2 525	2 508
Price, United States[c]	*USD/t dw*	3 165	3 143	3 299	3 403	3 577	3 646	3 675	3 659	3 649	3 624	3 562
Price, Brazil[d]	*USD/t dw*	3 279	3 351	3 138	3 160	3 245	3 266	3 313	3 294	3 273	3 217	3 196
PIG MEAT												
Production	*kt cwe*	39 315	39 238	39 768	39 340	39 569	40 022	40 608	40 854	41 084	41 377	41 729
Consumption	*kt cwe*	37 394	37 232	37 879	37 455	37 691	38 137	38 702	38 941	39 144	39 370	39 666
Ending stocks	*kt cwe*	927	926	913	929	926	928	932	929	925	927	933
Per capita consumption	*kg rwt*	23.3	23.0	23.3	22.9	22.9	23.1	23.3	23.4	23.4	23.5	23.5
Price, Brazil[e]	*USD/t pw*	2 270	2 233	2 399	2 397	2 493	2 397	2 401	2 402	2 318	2 311	2 259
Price, United States[f]	*USD/t dw*	1 384	1 364	1 506	1 554	1 678	1 709	1 695	1 679	1 714	1 718	1 681
POULTRY MEAT												
Production	*kt rtc*	38 870	38 302	38 873	39 476	39 768	40 198	40 612	41 255	41 930	42 470	42 997
Consumption	*kt rtc*	36 741	36 505	37 315	37 964	38 310	38 761	39 220	39 885	40 548	41 121	41 582
Ending stocks	*kt rtc*	1 132	1 089	1 089	1 088	1 089	1 090	1 091	1 091	1 092	1 093	1 095
Per capita consumption	*kg rwt*	25.8	25.4	25.9	26.2	26.3	26.5	26.7	27.0	27.4	27.6	27.9
Price, Brazil[g]	*USD/t pw*	1 607	1 549	1 580	1 602	1 613	1 647	1 682	1 665	1 659	1 638	1 638
Price, United States[h]	*USD/t rtc*	1 066	1 119	1 138	1 161	1 172	1 200	1 227	1 245	1 266	1 284	1 320
SHEEP MEAT												
Production	*kt cwe*	2 893	2 780	2 767	2 762	2 748	2 734	2 728	2 714	2 702	2 695	2 679
Consumption	*kt cwe*	2 389	2 301	2 285	2 271	2 255	2 234	2 221	2 206	2 186	2 176	2 151
Ending stocks	*kt cwe*	81	67	67	67	67	67	67	66	70	76	81
Per capita consumption	*kg rwt*	1.7	1.6	1.6	1.6	1.5	1.5	1.5	1.5	1.5	1.5	1.4
Price, Australia[i]	*AUD/t dw*	3 915	4 560	4 455	4 346	4 232	4 115	3 995	3 873	3 750	3 626	3 511
Price, Australia[j]	*AUD/t dw*	1 522	1 490	1 489	1 490	1 490	1 490	1 489	1 487	1 484	1 482	1 480
Price, New Zealand[k]	*NZD/t dw*	3 801	4 790	4 170	4 210	4 350	4 353	4 415	4 375	4 340	4 304	4 273
TOTAL MEAT												
Per capita consumption	*kg rwt*	66.1	64.9	65.5	65.4	65.5	65.9	66.5	67.0	67.6	68.0	68.4
Non-OECD												
BEEF AND VEAL												
Production	*kt cwe*	37 660	38 147	38 978	39 854	40 592	41 471	42 461	43 351	44 319	45 319	46 435
Consumption	*kt cwe*	36 971	37 359	38 190	39 051	39 751	40 598	41 507	42 392	43 358	44 294	45 422
Per capita consumption	*kg rwt*	4.7	4.6	4.7	4.7	4.8	4.8	4.8	4.9	4.9	5.0	5.1
Ending stocks	*kt cwe*	99	96	108	118	110	96	112	107	81	97	98
PIG MEAT												
Production	*kt cwe*	63 229	69 076	71 415	72 788	74 622	76 679	78 644	80 188	81 841	83 551	84 922
Consumption	*kt cwe*	65 061	70 793	73 085	74 424	76 267	78 322	80 304	81 862	83 545	85 316	86 738
Per capita consumption	*kg rwt*	9.2	9.8	10.0	10.0	10.2	10.3	10.5	10.5	10.6	10.7	10.8
Ending stocks	*kt cwe*	47	51	51	53	54	56	58	59	61	62	64
POULTRY MEAT												
Production	*kt rtc*	52 350	56 726	58 900	60 992	62 934	65 074	66 951	68 797	70 923	72 703	74 852
Consumption	*kt rtc*	54 027	57 540	59 417	61 372	63 156	65 177	66 911	68 632	70 672	72 320	74 463
Per capita consumption	*kg rwt*	8.7	9.0	9.2	9.3	9.5	9.7	9.8	10.0	10.1	10.3	10.4
Ending stocks	*kt rtc*	113	117	122	120	122	123	125	130	135	139	142

For notes, see end of the table.

Table A.18. **World meat projections** *(cont.)*

Calendar year[a]		Average 2007-09 est.	2010	2011	2012	2013	2014	2015	2016	2017	2018	2019
SHEEP MEAT												
Production	*kt cwe*	9 595	9 969	10 243	10 528	10 845	11 096	11 449	11 772	12 060	12 429	12 718
Consumption	*kt cwe*	10 004	10 339	10 616	10 906	11 236	11 475	11 861	12 168	12 465	12 833	13 133
Per capita consumption	*kg rwt*	1.6	1.6	1.6	1.7	1.7	1.7	1.7	1.8	1.8	1.8	1.8
Ending stocks	*kt cwe*	5	5	5	10	4	16	4	7	6	7	7
TOTAL MEAT												
Per capita consumption	*kg rwt*	24.2	25.1	25.5	25.8	26.1	26.5	26.9	27.1	27.5	27.8	28.1

est.: Estimate.

a) Year ending 30 September for New Zealand.

b) Excludes Iceland but includes the EU8 members that are not members of the OECD. Carcass weight to retail weight conversion factors of 0.7 for beef and veal, 0.78 for pig meat and 0.88 for sheep meat. Rtc to retail weight conversion factor 0.88 for poultry meat.

c) Choice steers, 1 100-1 300 lb lw, Nebraska – lw to dw conversion factor 0.63.

d) Price received by producer.

e) Brazil meat of swine export price (HS 0203).

f) Barrows and gilts, No. 1-3, 230-250 lb lw, Iowa/South Minnesota – lw to dw conversion factor 0.74.

g) Brazil meat of poultry export price (HS 0207).

h) Wholesale weighted average broiler price 12 cities.

i) Saleyard price, lamb, 16-20 kg dw.

j) Saleyard price, wethers, < 22 kg dw.

k) Lamb schedule price, all grade average.

Source: OECD and FAO Secretariats.

StatLink http://dx.doi.org/10.1787/888932287501

Table A.19. **Beef and veal projections**

Calendar year[a]	PRODUCTION[b] (kt)		Growth[c] (%)		IMPORTS[d] (kt)		Growth[c] (%)		EXPORTS[d] (kt)		Growth[c] (%)	
	Average 2007-09 est.	2019	2000-09	2010-19	Average 2007-09 est.	2019	2000-09	2010-19	Average 2007-09 est.	2019	2000-09	2010-19
WORLD	**65 228**	**74 639**	**1.29**	**1.51**	**6 623**	**8 168**	**–**	**2.42**	**7 568**	**8 845**	**3.26**	**1.76**
DEVELOPED	**29 854**	**31 300**	**–0.06**	**0.84**	**3 550**	**3 925**	**–**	**1.62**	**3 499**	**3 713**	**–2.81**	**0.71**
NORTH AMERICA	**13 179**	**13 671**	**–0.11**	**0.86**	**1 463**	**1 726**	**–2.22**	**1.66**	**1 243**	**1 664**	**–4.16**	**2.45**
Canada	1 713	1 617	1.48	1.06	206	211	–5.43	–0.92	465	519	–1.99	1.97
United States	11 467	12 054	–0.33	0.83	1 257	1 515	–1.65	2.06	778	1 145	–5.23	2.68
EUROPE	**11 073**	**11 185**	**–0.81**	**0.28**	**1 310**	**1 385**	**6.15**	**2.68**	**211**	**75**	**–18.22**	**–8.41**
EU27	8 081	7 302	–0.57	–0.86	447	643	1.81	5.33	105	32	–24.23	–8.43
Russian Federation	1 751	2 375	–1.33	2.73	741	558	8.87	–0.01	0	0	–	–
Ukraine	486	481	–5.68	3.57	30	57	40.17	1.86	25	0	–27.28	–20.69
OCEANIA DEVELOPED	**2 997**	**3 319**	**0.66**	**1.18**	**9**	**7**	**–**	**0.00**	**2 032**	**1 964**	**1.48**	**–0.05**
Australia	2 358	2 671	0.55	1.40	5	3	13.07	0.00	1 511	1 438	1.52	0.10
New Zealand	639	647	1.07	0.28	4	4	–	0.00	522	526	1.37	–0.45
OTHER DEVELOPED[e]	**2 604**	**3 126**	**3.07**	**2.54**	**769**	**807**	**–3.99**	**–0.06**	**13**	**9**	**–4.11**	**–0.24**
Japan	507	477	0.01	–0.67	667	720	–4.51	0.77	1	1	21.77	–5.44
South Africa	787	773	4.95	2.40	18	76	–14.23	–1.35	7	4	–9.09	0.57
DEVELOPING	**35 375**	**43 339**	**2.54**	**2.03**	**3 073**	**4 243**	**6.53**	**3.21**	**4 069**	**5 133**	**11.58**	**2.57**
AFRICA	**3 974**	**5 476**	**1.31**	**3.12**	**496**	**748**	**9.04**	**6.14**	**57**	**22**	**–0.28**	**–13.03**
NORTH AFRICA	**1 124**	**1 318**	**0.57**	**1.17**	**299**	**489**	**7.92**	**8.90**	**1**	**1**	**11.99**	**–0.52**
Algeria	123	143	0.57	1.06	74	149	24.04	6.79	0	0	21.21	–0.18
Egypt	400	504	–1.31	1.53	180	252	2.63	11.91	1	1	17.39	–0.36
SUB-SAHARAN AFRICA	**2 850**	**4 158**	**1.61**	**3.83**	**197**	**259**	**11.30**	**2.27**	**56**	**20**	**–0.47**	**–13.42**
LATIN AMERICA and CARIBBEAN	**17 688**	**21 056**	**3.32**	**1.80**	**855**	**1 127**	**5.37**	**2.22**	**3 237**	**4 128**	**12.17**	**2.79**
Argentina	3 227	3 240	3.06	0.28	3	5	–8.71	7.70	556	570	10.38	0.31
Brazil	9 159	11 149	3.92	1.91	15	82	–15.15	7.49	1 904	2 556	12.73	3.26
Chile	246	385	2.03	4.55	140	158	2.39	0.41	11	10	44.54	–0.41
Mexico	1 645	1 975	2.17	1.87	318	383	–0.33	1.98	14	18	22.16	2.39
Uruguay	549	817	4.58	4.77	4	2	39.16	–6.03	318	479	7.43	6.03
ASIA and PACIFIC	**13 712**	**16 807**	**1.95**	**1.98**	**1 722**	**2 368**	**6.35**	**2.89**	**775**	**983**	**10.70**	**2.57**
Bangladesh	192	251	1.07	2.61	0	0	16.34	–1.84	0	0	–39.09	1.82
China[f]	6 012	6 728	2.04	1.51	13	33	–8.59	4.88	57	25	2.58	0.00
India	2 917	3 654	0.37	1.83	1	1	63.81	–0.08	629	874	11.91	3.34
Indonesia	413	460	2.86	0.05	80	80	10.46	–0.75	0	0	–20.59	0.18
Iran, Islamic Republic of	372	513	3.87	3.23	115	146	18.64	1.95	0	0	33.04	–0.07
Korea	252	300	0.37	0.68	266	348	–0.99	2.37	0	0	–	–
Malaysia	18	25	9.20	2.97	90	126	–4.01	3.52	4	4	12.24	–3.60
Pakistan	1 169	1 750	3.55	3.58	9	5	85.49	0.64	14	16	33.50	–1.00
Saudi Arabia	21	22	0.55	0.15	127	230	11.70	5.12	9	6	13.10	–5.12
Turkey	443	560	3.51	2.01	2	1	81.72	–0.91	1	2	20.66	1.03
LEAST DEVELOPED COUNTRIES (LDC)	**2 795**	**3 871**	**1.53**	**3.31**	**142**	**81**	**12.08**	**–5.37**	**3**	**3**	**–4.54**	**1.65**
OECD	**27 569**	**28 204**	**0.03**	**0.51**	**3 342**	**4 015**	**–**	**2.05**	**3 409**	**3 691**	**–2.51**	**0.91**
NON-OECD	**37 660**	**46 435**	**2.29**	**2.17**	**3 281**	**4 153**	**–**	**2.79**	**4 160**	**5 155**	**10.48**	**2.40**

For notes, see end of the table.

Table A.19. **Beef and veal projections** *(cont.)*

Calendar year[a]	CONSUMPTION (kt) Average 2007-09 est.	CONSUMPTION (kt) 2019	Growth[c] (%) 2000-09	Growth[c] (%) 2010-19	PER CAPITA (kg) Average 2007-09 est.	PER CAPITA (kg) 2019	Growth[c] (%) 2000-09	Growth[c] (%) 2010-19
WORLD	**64 231**	**73 574**	**1.23**	**1.53**	**9.5**	**9.7**	**0.01**	**0.46**
DEVELOPED	**30 020**	**31 450**	**0.43**	**0.88**	**22.3**	**22.7**	**0.04**	**0.60**
NORTH AMERICA	**13 569**	**13 896**	**0.07**	**0.73**	**39.3**	**36.5**	**–0.92**	**–0.14**
Canada	1 050	1 084	0.58	0.41	22.1	20.6	–0.42	–0.50
United States	12 519	12 812	0.03	0.76	28.1	26.1	–0.96	–0.11
EUROPE	**12 297**	**12 619**	**0.71**	**0.61**	**17.1**	**17.6**	**0.65**	**0.66**
EU27	8 383	7 885	0.60	–0.42	11.9	11.1	0.30	–0.48
Russian Federation	2 605	3 030	1.58	2.08	12.9	15.6	2.04	2.43
Ukraine	491	537	–2.50	3.43	10.7	12.4	–1.76	3.99
OCEANIA DEVELOPED	**784**	**1 049**	**–0.50**	**2.67**	**31.0**	**37.3**	**–1.66**	**1.73**
Australia	668	927	–0.43	2.50	22.2	27.7	–1.60	1.54
New Zealand	116	121	–0.93	4.08	19.2	18.3	–2.03	3.26
OTHER DEVELOPED[e]	**3 370**	**3 886**	**1.16**	**1.86**	**13.3**	**14.9**	**0.65**	**1.59**
Japan	1 167	1 155	–2.91	–0.15	6.4	6.5	–2.96	0.10
South Africa	814	850	4.70	2.00	11.5	11.3	3.47	1.59
DEVELOPING	**34 210**	**42 124**	**1.97**	**2.04**	**6.3**	**6.8**	**0.53**	**0.79**
AFRICA	**4 454**	**6 247**	**1.99**	**3.58**	**4.7**	**5.2**	**–0.36**	**1.37**
NORTH AFRICA	**1 435**	**1 823**	**1.38**	**2.79**	**5.9**	**6.1**	**–0.54**	**1.03**
Algeria	198	293	5.31	3.60	5.8	7.3	3.81	2.20
Egypt	608	787	–0.36	3.84	7.5	8.1	–2.23	2.27
SUB-SAHARAN AFRICA	**3 019**	**4 424**	**2.29**	**3.93**	**4.3**	**4.9**	**–0.21**	**1.56**
LATIN AMERICA and CARIBBEAN	**14 943**	**17 527**	**1.93**	**1.53**	**26.0**	**27.4**	**0.70**	**0.59**
Argentina	2 674	2 675	1.93	0.30	67.1	60.9	0.98	–0.57
Brazil	7 071	8 289	1.75	1.37	25.8	27.9	0.57	0.68
Chile	374	533	1.69	3.23	22.3	28.8	0.63	2.37
Mexico	1 788	2 200	2.40	2.03	11.5	13.0	1.34	1.23
Uruguay	224	332	0.36	3.20	67.0	95.5	0.25	2.85
ASIA and PACIFIC	**14 814**	**18 350**	**1.99**	**2.05**	**3.8**	**4.2**	**0.74**	**1.01**
Bangladesh	193	251	1.07	2.60	1.2	1.4	–0.50	1.38
China[f]	5 968	6 736	2.00	1.53	3.1	3.3	1.34	0.96
India	2 289	2 781	–1.67	1.39	1.4	1.4	–3.22	0.18
Indonesia	558	606	3.75	–0.06	2.5	2.4	2.49	–0.97
Iran, Islamic Republic of	487	659	6.07	2.94	6.6	7.9	4.93	1.83
Korea	500	627	–0.90	1.68	7.3	8.9	–1.34	1.47
Malaysia	113	156	–2.84	3.45	4.2	4.9	–4.67	2.06
Pakistan	1 164	1 739	3.61	3.62	6.6	7.8	1.41	1.57
Saudi Arabia	143	251	9.37	4.83	5.7	8.1	7.01	2.95
Turkey	447	561	3.65	2.00	4.2	4.7	2.34	0.96
LEAST DEVELOPED COUNTRIES (LDC)	**2 872**	**3 887**	**2.15**	**3.02**	**3.8**	**4.1**	**–0.12**	**0.89**
OECD	**27 259**	**28 151**	**0.27**	**0.60**	**15.3**	**15.0**	**–0.35**	**0.17**
NON-OECD	**36 971**	**45 422**	**1.98**	**2.14**	**4.7**	**5.1**	**0.62**	**0.94**

est.: Estimate.

a) Year ending 30 June for Australia and 31 May for New Zealand.

b) Gross indigenous production.

c) Least-squares growth rate (see glossary).

d) Excludes trade of live animals.

e) Includes other developed Asian countries: Kazakhstan, Kyrgyzstan, Tajikistan, Turkmenistan, Uzbekistan, Armenia, Azerbaijan and Georgia.

f) Refers to mainland only. The economies of Chinese Taipei, Hong Kong (China) and Macau (China) are included in the other Asia Pacific aggregate.

Source: OECD and FAO Secretariats.

StatLink http://dx.doi.org/10.1787/888932287520

Table A.20. **Pig meat projections**

Calendar year[a]	PRODUCTION[b] (kt)		Growth[c] (%)		IMPORTS[d] (kt)		Growth[c] (%)		EXPORTS[d] (kt)		Growth[c] (%)	
	Average 2007-09 est.	2019	2000-09	2010-19	Average 2007-09 est.	2019	2000-09	2010-19	Average 2007-09 est.	2019	2000-09	2010-19
WORLD	**102 544**	**126 651**	**1.78**	**1.75**	**5 505**	**6 319**	**7.32**	**1.57**	**5 756**	**6 647**	**7.05**	**1.49**
DEVELOPED	**40 761**	**43 774**	**1.06**	**0.78**	**3 100**	**3 198**	**5.07**	**1.32**	**4 549**	**5 204**	**6.84**	**1.32**
NORTH AMERICA	**12 242**	**13 721**	**2.17**	**1.43**	**578**	**861**	**0.96**	**3.07**	**2 908**	**3 970**	**10.26**	**2.53**
Canada	2 224	2 157	1.63	0.63	178	288	11.43	3.22	1 100	1 182	5.54	0.90
United States	10 018	11 564	2.29	1.58	399	574	–2.00	3.00	1 808	2 788	14.18	3.31
EUROPE	**26 421**	**28 081**	**0.62**	**0.60**	**1 101**	**657**	**11.81**	**–0.44**	**1 580**	**1 156**	**2.60**	**–1.99**
EU27	22 455	23 097	0.48	0.35	38	40	13.84	1.29	1 496	1 009	2.24	–2.95
Russian Federation	1 927	2 620	2.84	1.75	748	248	10.49	0.31	0	0	–	0.69
Ukraine	572	536	–1.99	0.78	144	261	39.87	5.90	0	0	–102.34	–0.15
OCEANIA DEVELOPED	**413**	**381**	**–0.61**	**–0.52**	**250**	**335**	**13.34**	**3.23**	**56**	**76**	**–3.46**	**0.05**
Australia	363	330	–0.84	–0.74	212	280	14.30	3.23	56	76	–3.47	0.06
New Zealand	50	50	1.15	1.01	38	55	9.01	3.26	0	0	2.08	–12.93
OTHER DEVELOPED[e]	**1 686**	**1 591**	**0.68**	**–0.88**	**1 171**	**1 344**	**1.65**	**0.85**	**4**	**2**	**–0.44**	**–2.08**
Japan	1 273	1 130	0.24	–1.56	1 066	1 150	0.67	0.31	1	0	24.11	–8.23
South Africa	149	177	4.31	0.55	18	26	5.64	5.11	3	2	8.56	–0.55
DEVELOPING	**61 783**	**82 877**	**2.26**	**2.29**	**2 405**	**3 121**	**10.99**	**1.85**	**1 207**	**1 443**	**8.11**	**2.10**
AFRICA	**677**	**903**	**0.95**	**3.16**	**129**	**155**	**15.50**	**1.23**	**7**	**5**	**5.51**	**1.74**
NORTH AFRICA	**4**	**7**	**–3.34**	**4.47**	**1**	**2**	**22.81**	**0.09**	**0**	**0**	**17.27**	**–0.01**
Algeria	0	0	0.00	3.82	0	0	21.93	0.17	0	0	–10.25	–0.17
Egypt	2	3	–8.14	3.53	1	1	65.00	0.05	0	0	29.45	–0.05
SUB-SAHARAN AFRICA	**672**	**897**	**0.99**	**3.15**	**127**	**153**	**15.45**	**1.24**	**7**	**5**	**5.31**	**1.77**
LATIN AMERICA and CARIBBEAN	**6 222**	**7 538**	**2.61**	**1.81**	**748**	**722**	**9.21**	**0.33**	**783**	**971**	**13.40**	**1.77**
Argentina	278	383	4.83	3.25	36	35	–4.93	–0.80	4	29	10.46	16.64
Brazil	3 029	3 666	1.73	1.83	9	7	48.16	–1.50	564	706	13.08	1.80
Chile	508	645	7.48	2.19	3	3	14.99	–2.39	135	176	19.80	2.39
Mexico	1 129	1 226	1.29	0.85	561	520	12.34	–0.42	62	46	8.57	–2.15
Uruguay	23	27	0.34	0.78	11	13	3.91	4.71	0	0	57.57	–4.71
ASIA and PACIFIC	**54 885**	**74 436**	**2.24**	**2.33**	**1 529**	**2 244**	**11.54**	**2.43**	**417**	**467**	**2.01**	**2.83**
Bangladesh	0	0	0.00	7.66	0	0	39.80	0.53	0	0	0.00	–0.53
China[f]	45 619	62 300	1.90	2.27	262	401	13.66	8.42	268	351	3.50	4.17
India	500	551	0.52	0.97	1	1	35.73	–1.50	2	3	33.84	1.93
Indonesia	627	855	4.19	2.71	4	3	8.15	–0.12	3	3	8.65	0.11
Iran, Islamic Republic of	0	0	0.00	0.22	1	1	71.63	0.58	0	0	–5.63	–0.58
Korea	942	1 180	–0.50	1.73	445	558	15.72	2.42	12	12	–13.40	0.00
Malaysia	191	234	1.02	2.37	15	0	–1.96	–18.99	3	4	19.75	1.57
Pakistan	0	0	0.00	–1.09	0	0	51.27	2.04	0	0	56.34	–2.04
Saudi Arabia	0	0	0.00	0.00	5	5	10.32	0.00	0	0	35.66	0.00
Turkey	0	0	–72.40	1.44	1	1	44.30	0.28	1	1	22.91	–0.28
LEAST DEVELOPED COUNTRIES (LDC)	**1 044**	**1 425**	**5.71**	**3.15**	**108**	**122**	**19.02**	**0.11**	**0**	**0**	**5.70**	**2.89**
OECD	**39 315**	**41 729**	**1.04**	**0.70**	**2 961**	**3 485**	**5.09**	**1.41**	**4 673**	**5 292**	**6.91**	**1.16**
NON-OECD	**63 229**	**84 922**	**2.24**	**2.29**	**2 544**	**2 833**	**10.47**	**1.78**	**1 083**	**1 355**	**7.85**	**2.84**

For notes, see end of the table.

Table A.20. **Pig meat projections** *(cont.)*

Calendar year[a]	CONSUMPTION (kt)		Growth[c] (%)		PER CAPITA (kg)		Growth[c] (%)	
	Average 2007-09 est.	2019	2000-09	2010-19	Average 2007-09 est.	2019	2000-09	2010-19
WORLD	**102 455**	**126 404**	**1.81**	**1.75**	**15.2**	**16.6**	**0.59**	**0.69**
DEVELOPED	**39 306**	**41 700**	**0.87**	**0.76**	**29.1**	**29.9**	**0.47**	**0.49**
NORTH AMERICA	**9 695**	**10 399**	**0.37**	**1.16**	**27.6**	**26.8**	**–0.63**	**0.30**
Canada	779	837	–1.86	0.64	18.3	17.8	–2.86	–0.27
United States	8 916	9 562	0.58	1.21	22.3	21.7	–0.40	0.34
EUROPE	**26 131**	**27 728**	**0.94**	**0.68**	**36.3**	**38.7**	**0.88**	**0.74**
EU27	20 939	22 057	0.36	0.53	33.1	34.5	0.07	0.46
Russian Federation	2 903	3 067	5.69	1.41	16.0	17.6	6.15	1.76
Ukraine	715	796	1.04	2.21	15.5	18.4	1.78	2.77
OCEANIA DEVELOPED	**607**	**640**	**4.03**	**1.21**	**24.0**	**22.8**	**2.87**	**0.27**
Australia	519	534	4.02	1.04	19.2	17.8	2.85	0.07
New Zealand	88	106	4.07	2.14	16.2	17.8	2.96	1.31
OTHER DEVELOPED[e]	**2 873**	**2 933**	**1.40**	**0.01**	**11.4**	**11.2**	**0.89**	**–0.26**
Japan	2 359	2 280	0.82	–0.50	14.5	14.3	0.78	–0.25
South Africa	164	201	4.48	1.06	2.6	3.0	3.24	0.65
DEVELOPING	**63 148**	**84 704**	**2.42**	**2.27**	**11.7**	**13.7**	**0.99**	**1.02**
AFRICA	**799**	**1 053**	**2.44**	**2.87**	**0.8**	**0.9**	**0.09**	**0.65**
NORTH AFRICA	**5**	**8**	**–0.65**	**3.43**	**0.0**	**0.0**	**–2.57**	**1.67**
Algeria	0	1	9.15	1.32	0.0	0.0	7.65	–0.08
Egypt	2	4	–4.21	2.40	0.0	0.0	–6.08	0.83
SUB-SAHARAN AFRICA	**793**	**1 045**	**2.46**	**2.87**	**1.1**	**1.1**	**–0.04**	**0.50**
LATIN AMERICA and CARIBBEAN	**6 212**	**7 317**	**2.39**	**1.66**	**10.8**	**11.4**	**1.16**	**0.72**
Argentina	309	388	3.13	2.30	7.8	8.8	2.18	1.44
Brazil	2 474	2 966	0.31	1.83	10.1	11.1	–0.88	1.15
Chile	377	472	5.10	2.09	22.4	25.5	4.04	1.24
Mexico	1 654	1 728	3.88	0.53	11.9	11.3	2.82	–0.27
Uruguay	34	40	1.12	1.96	10.1	11.4	1.02	1.61
ASIA and PACIFIC	**56 138**	**76 334**	**2.43**	**2.32**	**14.5**	**17.5**	**1.18**	**1.29**
Bangladesh	0	0	39.80	0.74	0.0	0.0	38.23	–0.48
China[f]	45 614	62 351	1.93	2.29	26.6	34.1	1.27	1.72
India	498	548	0.49	0.96	0.3	0.3	–1.06	–0.24
Indonesia	609	836	4.35	2.78	2.7	3.3	3.09	1.88
Iran, Islamic Republic of	1	1	72.37	0.58	0.0	0.0	71.23	–0.53
Korea	1 395	1 726	4.20	1.97	22.6	27.2	3.76	1.75
Malaysia	203	231	0.64	2.25	7.5	7.3	–1.19	0.86
Pakistan	0	0	47.02	3.05	0.0	0.0	44.81	1.00
Saudi Arabia	5	5	10.34	0.00	0.2	0.2	7.98	–1.88
Turkey	1	1	27.03	1.53	0.0	0.0	25.72	0.49
LEAST DEVELOPED COUNTRIES (LDC)	**1 152**	**1 547**	**6.54**	**2.88**	**1.5**	**1.6**	**4.27**	**0.74**
OECD	**37 394**	**39 666**	**0.75**	**0.71**	**23.3**	**23.5**	**0.14**	**0.28**
NON-OECD	**65 061**	**86 738**	**2.45**	**2.26**	**9.2**	**10.8**	**1.08**	**1.06**

est.: Estimate.

a) Year ending 30 June for Australia and 31 May for New Zealand.
b) Gross indigenous production.
c) Least-squares growth rate (see glossary).
d) Excludes trade of live animals.
e) Includes other developed Asian countries: Kazakhstan, Kyrgyzstan, Tajikistan, Turkmenistan, Uzbekistan, Armenia, Azerbaijan and Georgia.
f) Refers to mainland only. The economies of Chinese Taipei, Hong Kong (China) and Macau (China) are included in the other Asia Pacific aggregate.

Source: OECD and FAO Secretariats.

StatLink http://dx.doi.org/10.1787/888932287539

Table A.21. **Poultry meat projections**

Calendar year[a]	PRODUCTION (kt)		Growth[b] (%)		IMPORTS (kt)		Growth[b] (%)		EXPORTS (kt)		Growth[b] (%)	
	Average 2007-09 est.	2019	2000-09	2010-19	Average 2007-09 est.	2019	2000-09	2010-19	Average 2007-09 est.	2019	2000-09	2010-19
WORLD	**91 220**	**117 849**	**3.24**	**2.36**	**9 635**	**11 375**	**4.3**	**2.1**	**10 132**	**13 175**	**4.62**	**2.69**
DEVELOPED	**38 337**	**43 642**	**2.07**	**1.44**	**3 496**	**3 652**	**2.9**	**1.7**	**4 387**	**4 254**	**1.92**	**0.50**
NORTH AMERICA	**20 109**	**22 651**	**1.67**	**1.54**	**274**	**294**	**7.3**	**0.6**	**3 420**	**3 286**	**3.16**	**0.74**
Canada	1 206	1 366	1.49	1.26	236	253	5.7	0.7	180	190	5.50	1.16
United States	18 903	21 285	1.68	1.55	38	40	26.4	–0.2	3 240	3 096	3.04	0.71
EUROPE	**14 679**	**16 663**	**2.52**	**1.37**	**2 384**	**2 252**	**3.3**	**1.2**	**915**	**924**	**–1.80**	**–0.24**
EU27	11 533	11 840	0.85	0.61	839	1 210	5.4	2.6	874	895	–2.13	–0.14
Russian Federation	1 968	2 923	11.69	3.20	1 106	548	0.2	–2.0	11	8	14.23	0.00
Ukraine	747	1 392	16.78	5.04	186	129	16.1	–1.9	9	7	23.19	–2.34
OCEANIA DEVELOPED	**1 004**	**1 073**	**3.31**	**0.53**	**0**	**0**	**–**	**–**	**40**	**33**	**5.21**	**0.12**
Australia	855	924	3.30	0.64	0	0	–	–	40	33	5.21	0.12
New Zealand	149	150	3.37	–0.12	0	0	–	–	0	0	–	–
OTHER DEVELOPED[c]	**2 546**	**3 254**	**2.23**	**1.43**	**838**	**1 107**	**1.3**	**3.4**	**13**	**11**	**–4.91**	**–0.37**
Japan	1 360	1 463	1.72	0.78	360	234	–5.8	–4.0	2	2	–2.65	0.00
South Africa	1 015	1 509	2.15	1.49	273	539	16.0	7.8	7	6	–9.49	–0.42
DEVELOPING	**52 882**	**74 207**	**4.15**	**2.94**	**6 139**	**7 723**	**5.2**	**2.3**	**5 744**	**8 921**	**7.12**	**3.90**
AFRICA	**2 690**	**3 697**	**1.95**	**3.12**	**621**	**1 008**	**8.4**	**4.4**	**47**	**34**	**6.28**	**–3.88**
NORTH AFRICA	**1 579**	**2 015**	**1.90**	**2.67**	**44**	**80**	**6.7**	**5.1**	**25**	**24**	**28.92**	**–0.21**
Algeria	251	275	0.63	0.93	8	10	7.7	3.1	0	0	21.92	–0.26
Egypt	649	767	0.62	2.09	20	24	13.9	0.2	23	22	39.28	–0.11
SUB-SAHARAN AFRICA	**1 112**	**1 682**	**2.03**	**3.69**	**577**	**928**	**8.6**	**4.4**	**22**	**10**	**–1.92**	**–8.66**
LATIN AMERICA and CARIBBEAN	**20 639**	**29 258**	**6.11**	**2.85**	**1 378**	**1 720**	**9.4**	**2.7**	**3 799**	**6 616**	**14.31**	**4.95**
Argentina	1 375	1 929	7.28	2.56	16	0	–6.2	–	163	371	41.28	6.02
Brazil	11 480	16 712	7.36	2.91	1	1	8.4	0.0	3 484	5 971	13.97	4.91
Chile	615	897	4.49	3.63	32	65	50.5	6.6	79	166	12.36	5.24
Mexico	2 582	2 947	4.00	1.13	607	998	10.7	5.0	1	1	–13.17	–0.20
Uruguay	45	54	–0.92	2.03	1	7	6.6	16.2	4	1	30.46	–16.24
ASIA and PACIFIC	**29 553**	**41 252**	**3.12**	**2.98**	**4 140**	**4 995**	**3.7**	**1.8**	**1 899**	**2 272**	**–0.96**	**1.46**
Bangladesh	118	140	1.01	1.59	2	4	–7.0	1.3	0	0	12.62	–0.43
China[d]	14 956	20 609	1.83	2.83	884	773	2.5	3.2	474	627	–4.22	4.50
India	2 505	3 912	9.28	3.91	0	0	28.0	0.7	5	4	–3.08	–0.83
Indonesia	1 417	1 789	4.79	1.73	10	10	1.2	0.9	0	0	–40.94	–0.42
Iran, Islamic Republic of	1 487	2 381	7.09	4.17	28	33	24.6	3.9	26	18	11.42	–4.63
Korea	467	538	1.35	1.11	114	160	–2.0	1.3	2	3	–1.52	1.83
Malaysia	1 125	1 430	4.35	2.28	52	37	–2.5	–2.0	106	158	–0.01	2.87
Pakistan	552	764	6.62	3.93	7	9	26.7	5.2	5	3	16.75	–3.86
Saudi Arabia	569	737	1.73	2.20	547	1 015	4.5	3.6	40	26	18.80	–3.39
Turkey	1 084	1 465	6.49	2.78	77	65	11.2	–0.5	82	104	25.22	0.47
LEAST DEVELOPED COUNTRIES (LDC)	**1 682**	**2 699**	**6.99**	**4.47**	**566**	**685**	**7.8**	**2.4**	**7**	**3**	**–6.78**	**–13.21**
OECD	**38 870**	**42 997**	**1.75**	**1.26**	**2 357**	**3 078**	**3.9**	**2.2**	**4 501**	**4 491**	**2.23**	**0.68**
NON-OECD	**52 350**	**74 852**	**4.45**	**3.04**	**7 278**	**8 297**	**4.5**	**2.0**	**5 630**	**8 684**	**6.91**	**3.89**

For notes, see end of the table.

Table A.21. **Poultry meat projections** *(cont.)*

Calendar year[a]	CONSUMPTION (kt)		Growth[b] (%)		PER CAPITA (kg)		Growth[b] (%)	
	Average 2007-09 est.	2019	2000-09	2010-19	Average 2007-09 est.	2019	2000-09	2010-19
WORLD	**90 769**	**116 045**	**3.22**	**2.30**	**13.5**	**15.3**	**2.00**	**1.24**
DEVELOPED	**37 467**	**43 039**	**2.18**	**1.57**	**27.9**	**31.0**	**1.78**	**1.29**
NORTH AMERICA	**16 967**	**19 659**	**1.47**	**1.67**	**49.2**	**51.7**	**0.49**	**0.80**
Canada	1 256	1 429	1.69	1.14	33.2	34.2	0.69	0.23
United States	15 711	18 230	1.46	1.71	44.4	46.7	0.47	0.84
EUROPE	**16 161**	**17 990**	**2.93**	**1.44**	**22.4**	**25.1**	**2.87**	**1.49**
EU27	11 507	12 155	1.39	0.85	20.5	21.5	1.10	0.78
Russian Federation	3 067	3 462	6.74	2.21	19.1	22.4	7.20	2.57
Ukraine	924	1 513	16.45	4.28	20.1	35.0	17.19	4.84
OCEANIA DEVELOPED	**964**	**1 040**	**3.24**	**0.56**	**38.1**	**37.0**	**2.08**	**–0.39**
Australia	815	890	3.22	0.67	34.0	33.4	2.04	–0.29
New Zealand	149	150	3.37	–0.12	31.0	28.4	2.26	–0.95
OTHER DEVELOPED[c]	**3 375**	**4 350**	**2.09**	**1.90**	**13.3**	**16.7**	**1.58**	**1.64**
Japan	1 722	1 694	–0.08	0.00	11.9	12.0	–0.12	0.25
South Africa	1 281	2 043	4.29	2.84	22.7	34.3	3.05	2.43
DEVELOPING	**53 302**	**73 006**	**3.99**	**2.76**	**9.9**	**11.8**	**2.56**	**1.51**
AFRICA	**3 264**	**4 671**	**2.89**	**3.46**	**3.5**	**3.9**	**0.54**	**1.25**
NORTH AFRICA	**1 598**	**2 071**	**1.85**	**2.79**	**6.5**	**7.0**	**–0.07**	**1.03**
Algeria	259	285	0.80	1.00	7.5	7.1	–0.70	–0.39
Egypt	646	768	0.54	2.10	7.9	7.9	–1.33	0.53
SUB-SAHARAN AFRICA	**1 666**	**2 600**	**3.97**	**4.04**	**2.4**	**2.9**	**1.47**	**1.67**
LATIN AMERICA and CARIBBEAN	**18 218**	**24 361**	**5.12**	**2.34**	**31.6**	**38.1**	**3.89**	**1.40**
Argentina	1 228	1 558	5.41	1.89	30.8	35.4	4.45	1.03
Brazil	7 996	10 742	5.30	1.93	36.6	45.5	4.11	1.24
Chile	568	796	4.34	3.54	33.8	43.0	3.28	2.68
Mexico	3 187	3 944	5.05	1.99	25.8	29.2	4.00	1.19
Uruguay	42	60	–1.76	3.59	12.7	17.3	–1.87	3.23
ASIA and PACIFIC	**31 820**	**43 975**	**3.49**	**2.92**	**8.2**	**10.1**	**2.24**	**1.88**
Bangladesh	120	143	0.85	1.58	0.8	0.8	–0.72	0.36
China[d]	15 366	20 755	2.08	2.79	10.1	12.8	1.42	2.22
India	2 501	3 908	9.32	3.91	1.9	2.5	7.77	2.71
Indonesia	1 427	1 799	4.78	1.73	6.3	7.1	3.51	0.82
Iran, Islamic Republic of	1 489	2 396	7.07	4.26	20.3	28.9	5.94	3.15
Korea	578	695	0.59	1.14	10.6	12.4	0.16	0.93
Malaysia	1 070	1 310	4.43	2.06	39.6	41.4	2.60	0.67
Pakistan	554	769	6.69	4.00	3.1	3.5	4.48	1.95
Saudi Arabia	1 076	1 725	2.71	3.10	42.7	55.5	0.36	1.22
Turkey	1 070	1 426	6.02	2.79	12.8	15.1	4.71	1.75
LEAST DEVELOPED COUNTRIES (LDC)	**2 242**	**3 380**	**7.26**	**4.05**	**3.0**	**3.6**	**5.00**	**1.92**
OECD	**36 741**	**41 582**	**1.83**	**1.40**	**25.8**	**27.9**	**1.22**	**0.97**
NON-OECD	**54 027**	**74 463**	**4.24**	**2.83**	**8.7**	**10.4**	**2.88**	**1.63**

est.: Estimate.

a) Year ending 30 June for Australia and 31 May for New Zealand.

b) Least-squares growth rate (see glossary).

c) Includes other developed Asian countries: Kazakhstan, Kyrgyzstan, Tajikistan, Turkmenistan, Uzbekistan, Armenia, Azerbaijan and Georgia.

d) Refers to mainland only. The economies of Chinese Taipei, Hong Kong (China) and Macau (China) are included in the other Asia Pacific aggregate.

Source: OECD and FAO Secretariats.

StatLink http://dx.doi.org/10.1787/888932287558

Table A.22. **Sheep meat projections**

Calendar year[a]	PRODUCTION[b] (kt)		Growth[c] (%)		IMPORTS[d] (kt)		Growth[c] (%)		EXPORTS[d] (kt)		Growth[c] (%)	
	Average 2007-09 est.	2019	2000-09	2010-19	Average 2007-09 est.	2019	2000-09	2010-19	Average 2007-09 est.	2019	2000-09	2010-19
WORLD	12 488	15 398	1.73	2.12	872	888	3.38	0.46	981	1 041	1.89	0.69
DEVELOPED	3 356	3 380	0.14	0.55	491	543	3.21	0.80	897	946	1.48	0.83
NORTH AMERICA	104	103	–2.62	0.03	111	119	4.03	1.06	5	5	7.87	0.00
Canada	18	24	1.94	2.13	23	20	3.82	–0.08	0	0	5.16	0.00
United States	86	80	–3.40	–0.53	88	99	4.09	1.30	5	5	8.09	0.00
EUROPE	1 350	1 280	–0.12	–0.56	295	284	4.25	–0.19	11	13	8.77	–1.80
EU27	1 049	895	–0.87	–1.36	269	263	3.77	–0.16	7	8	–	0.00
Russian Federation	194	268	4.99	1.96	13	8	18.56	–2.44	0	0	–	–
Ukraine	16	22	–0.63	3.01	0	0	34.96	1.53	0	0	–10.42	–1.40
OCEANIA DEVELOPED	1 325	1 252	–0.10	0.06	7	8	1.92	0.00	880	928	1.37	0.87
Australia	762	768	–0.34	0.51	0	0	–	–	377	492	0.95	2.80
New Zealand	563	484	0.25	–0.61	7	8	1.92	0.00	503	436	1.72	–0.96
OTHER DEVELOPED[e]	578	744	2.01	3.88	78	132	–0.61	3.07	0	0	18.42	–0.32
Japan	0	0	–	–	40	32	–0.64	–2.38	0	0	–	–
South Africa	142	127	0.41	2.76	38	100	–0.41	5.60	0	0	19.73	–0.17
DEVELOPING	9 131	12 018	2.36	2.60	381	346	3.62	–0.06	84	95	7.29	–0.56
AFRICA	2 139	2 961	2.65	3.48	33	39	9.01	4.78	14	11	2.10	–3.56
NORTH AFRICA	622	760	1.74	2.32	13	20	7.99	5.12	4	0	23.43	–17.97
Algeria	193	233	1.32	2.03	6	10	11.80	4.95	0	0	37.89	–0.30
Egypt	58	73	–3.16	2.34	3	8	11.32	9.64	0	0	–19.45	–0.60
SUB-SAHARAN AFRICA	1 517	2 202	3.05	3.91	19	19	10.49	4.54	9	11	–2.33	–3.32
LATIN AMERICA and CARIBBEAN	367	429	0.76	1.47	56	43	–1.20	–2.76	31	32	8.73	–0.81
Argentina	62	52	0.70	–1.30	0	0	–23.39	0.00	6	6	19.27	0.00
Brazil	78	78	1.46	0.03	8	9	12.58	1.03	1	0	–	–5.94
Chile	16	15	0.08	–0.02	0	0	27.23	9.38	5	2	1.89	–9.38
Mexico	51	70	4.91	2.90	34	29	–3.66	–1.54	0	0	–	0.00
Uruguay	30	36	–5.64	0.86	0	0	57.44	–0.14	19	24	9.44	0.14
ASIA and PACIFIC	6 626	8 628	2.36	2.38	293	264	4.21	–0.13	40	51	9.01	0.53
Bangladesh	209	295	6.27	2.95	0	0	0.83	0.95	0	0	–7.46	–0.58
China[f]	3 479	4 706	3.52	2.76	10	15	0.57	2.96	8	6	2.81	–1.80
India	784	877	1.83	1.07	0	0	32.86	–0.83	15	30	9.04	2.96
Indonesia	122	163	3.37	2.66	1	1	9.27	1.64	0	0	–6.91	–0.82
Iran, Islamic Republic of	500	648	1.77	2.53	0	0	–3.92	0.80	0	0	17.01	–0.80
Korea	2	2	–4.68	0.00	6	9	10.68	4.70	0	0	–35.31	0.00
Malaysia	1	1	3.06	1.77	18	26	3.83	3.62	0	0	–12.47	–0.04
Pakistan	423	615	–1.77	2.84	0	0	34.99	0.96	7	7	18.21	–3.87
Saudi Arabia	30	28	19.89	–0.60	65	78	3.34	1.80	5	4	26.73	–1.80
Turkey	318	313	–1.53	–0.19	0	1	42.29	–0.54	0	0	–24.79	0.51
LEAST DEVELOPED COUNTRIES (LDC)	1 482	2 135	2.89	3.70	9	6	7.95	0.21	1	1	–21.79	–1.79
OECD	2 893	2 679	–0.58	–0.40	477	471	2.74	–0.05	897	943	1.49	0.82
NON-OECD	9 595	12 718	2.51	2.73	396	418	–	1.06	84	98	7.11	–0.50

For notes, see end of the table.

Table A.22. **Sheep meat projections** *(cont.)*

Calendar year[a]	CONSUMPTION (kt)		Growth[c] (%)		PER CAPITA (kg)		Growth[c] (%)	
	Average 2007-09 est.	2019	2000-09	2010-19	Average 2007-09 est.	2019	2000-09	2010-19
WORLD	**12 392**	**15 284**	**1.85**	**2.13**	**1.8**	**2.0**	**0.63**	**1.07**
DEVELOPED	**2 870**	**2 919**	**0.41**	**0.60**	**2.1**	**2.1**	**0.01**	**0.32**
NORTH AMERICA	**204**	**212**	**0.41**	**0.62**	**0.6**	**0.6**	**–0.58**	**–0.26**
Canada	41	43	3.92	1.05	1.1	1.0	2.91	0.14
United States	163	169	–0.36	0.51	0.5	0.4	–1.34	–0.36
EUROPE	**1 625**	**1 538**	**0.46**	**–0.49**	**2.3**	**2.1**	**0.40**	**–0.43**
EU27	1 311	1 150	–0.11	–1.11	2.3	2.0	–0.41	–1.18
Russian Federation	197	261	5.26	1.91	1.2	1.7	5.72	2.26
Ukraine	16	22	–0.36	3.04	0.3	0.5	0.38	3.59
OCEANIA DEVELOPED	**371**	**279**	**–1.88**	**–1.56**	**14.7**	**9.9**	**–3.04**	**–2.50**
Australia	298	227	–0.64	–2.14	12.4	8.5	–1.82	–3.11
New Zealand	73	51	–6.67	1.48	15.2	9.7	–7.78	0.66
OTHER DEVELOPED[e]	**669**	**890**	**1.72**	**3.69**	**2.6**	**3.4**	**1.21**	**3.42**
Japan	40	32	–0.64	–2.38	0.3	0.2	–0.68	–2.13
South Africa	193	240	0.49	3.67	3.4	4.0	–0.75	3.26
DEVELOPING	**9 523**	**12 365**	**2.31**	**2.53**	**1.8**	**2.0**	**0.88**	**1.28**
AFRICA	**2 110**	**2 941**	**2.71**	**3.61**	**2.2**	**2.4**	**0.36**	**1.40**
NORTH AFRICA	**634**	**782**	**1.64**	**2.40**	**2.6**	**2.6**	**–0.28**	**0.64**
Algeria	199	243	1.48	2.14	5.8	6.1	–0.01	0.75
Egypt	63	84	–2.82	2.78	0.8	0.9	–4.69	1.21
SUB-SAHARAN AFRICA	**1 477**	**2 158**	**3.20**	**4.09**	**2.1**	**2.4**	**0.70**	**1.72**
LATIN AMERICA and CARIBBEAN	**394**	**442**	**–0.14**	**1.15**	**0.7**	**0.7**	**–1.37**	**0.22**
Argentina	56	47	–0.61	–1.45	1.4	1.1	–1.57	–2.31
Brazil	85	88	1.58	0.16	0.4	0.4	0.39	–0.53
Chile	11	13	–0.54	2.54	0.7	0.7	–1.61	1.68
Mexico	87	102	0.36	1.32	0.7	0.8	–0.70	0.53
Uruguay	11	11	–15.96	2.71	3.1	3.3	–16.06	2.36
ASIA and PACIFIC	**7 018**	**8 983**	**2.35**	**2.26**	**1.8**	**2.1**	**1.09**	**1.23**
Bangladesh	210	295	6.28	2.95	1.3	1.6	4.70	1.73
China[f]	3 480	4 716	3.51	2.76	2.3	2.9	2.85	2.19
India	765	844	1.69	1.01	0.6	0.5	0.14	–0.19
Indonesia	123	165	3.41	2.66	0.5	0.7	2.15	1.75
Iran, Islamic Republic of	498	645	1.71	2.54	6.8	7.8	0.57	1.43
Korea	8	11	5.66	3.70	0.1	0.2	5.22	3.49
Malaysia	20	27	3.89	3.45	0.7	0.9	2.06	2.06
Pakistan	416	608	–1.93	2.94	2.4	2.7	–4.14	0.89
Saudi Arabia	158	170	0.91	0.74	6.3	5.5	–1.45	–1.14
Turkey	318	312	–1.39	–0.19	3.8	3.3	–2.70	–1.23
LEAST DEVELOPED COUNTRIES (LDC)	**1 450**	**2 100**	**3.16**	**3.78**	**1.9**	**2.2**	**0.89**	**1.64**
OECD	**2 389**	**2 151**	**–0.52**	**–0.73**	**1.7**	**1.4**	**–1.14**	**–1.16**
NON-OECD	**10 004**	**13 133**	**2.48**	**2.68**	**1.6**	**1.8**	**1.11**	**1.48**

est.: Estimate.

a) Year ending 30 June for Australia and 31 May for New Zealand.

b) Gross indigenous production.

c) Least-squares growth rate (see glossary).

d) Excludes trade of live animals.

e) Includes other developed Asian countries: Kazakhstan, Kyrgyzstan, Tajikistan, Turkmenistan, Uzbekistan, Armenia, Azerbaijan and Georgia.

f) Refers to mainland only. The economies of Chinese Taipei, Hong Kong (China) and Macau (China) are included in the other Asia Pacific aggregate.

Source: OECD and FAO Secretariats.

StatLink http://dx.doi.org/10.1787/888932287577

Table A.23. **Main policy assumptions for dairy markets**

Calendar year		Average 2007-09 est.	2010	2011	2012	2013	2014	2015	2016	2017	2018	2019
CANADA												
Milk target price[b]	*CADc/litre*	71	73	75	76	77	78	79	79	80	81	82
Butter support price	*CAD/t*	6 926	7 058	7 164	7 272	7 381	7 492	7 604	7 718	7 834	7 951	8 071
SMP support price	*CAD/t*	6 009	6 353	6 469	6 505	6 639	6 761	6 880	7 024	7 158	7 270	7 377
Cheese tariff-quota	*kt pw*	20	20	20	20	20	20	20	20	20	20	20
In-quota tariff	*%*	1	1	1	1	1	1	1	1	1	1	1
Out-of-quota tariff	*%*	246	246	246	246	246	246	246	246	246	246	246
Subsidised export limits[c]												
Cheese	*kt pw*	9	9	9	9	9	9	9	9	9	9	9
SMP	*kt pw*	45	45	45	45	45	45	45	45	45	45	45
EUROPEAN UNION[d]												
Milk quota[e]	*mt pw*	146	150	151	152	154	154	0	0	0	0	0
Butter intervention price	*EUR/t*	2 462	2 462	2 462	2 462	2 462	2 462	2 462	2 462	2 462	2 462	2 462
SMP intervention price	*EUR/t*	1 731	1 698	1 698	1 698	1 698	1 698	1 698	1 698	1 698	1 698	1 698
Butter tariff-quotas	*kt pw*	90	90	90	90	90	90	90	90	90	90	90
Cheese tariff-quota	*kt pw*	103	103	103	103	103	103	103	103	103	103	103
SMP tariff-quota	*kt pw*	71	71	71	71	71	71	71	71	71	71	71
Subsidised export limits[c]												
Butter	*kt pw*	412	412	412	412	412	412	412	412	412	412	412
Cheese	*kt pw*	332	332	332	332	332	332	332	332	332	332	332
SMP	*kt pw*	323	323	323	323	323	323	323	323	323	323	323
JAPAN												
Direct payments	*JPY/kg*	11	11	11	11	11	11	11	11	11	11	11
Cheese tariff[f]	*%*	31	31	31	31	31	31	31	31	31	31	31
Tariff-quotas												
Butter	*kt pw*	2	2	2	2	2	2	2	2	2	2	2
In-quota tariff	*%*	35	35	35	35	35	35	35	35	35	35	35
Out-of-quota tariff	*%*	733	733	733	733	733	733	733	733	733	733	733
SMP	*kt pw*	116	116	116	116	116	116	116	116	116	116	116
In-quota tariff	*%*	16	16	16	16	16	16	16	16	16	16	16
Out-of-quota tariff	*%*	210	210	210	210	210	210	210	210	210	210	210
WMP	*t pw*	0.3	0.3	0.3	0.3	0.3	0.3	0.3	0.3	0.3	0.3	0.3
In-quota tariff	*%*	24	24	24	24	24	24	24	24	24	24	24
Out-of-quota tariff	*%*	316	316	316	316	316	316	316	316	316	316	316
KOREA												
Tariff-quotas												
Butter	*kt pw*	0.4	0.4	0.4	0.4	0.4	0.4	0.4	0.4	0.4	0.4	0.4
In-quota tariff	*%*	40	40	40	40	40	40	40	40	40	40	40
Out-of-quota tariff	*%*	89	89	89	89	89	89	89	89	89	89	89
SMP	*kt pw*	1.0	1.0	1.0	1.0	1.0	1.0	1.0	1.0	1.0	1.0	1.0
In-quota tariff	*%*	20	20	20	20	20	20	20	20	20	20	20
Out-of-quota tariff	*%*	176	176	176	176	176	176	176	176	176	176	176
WMP	*kt pw*	0.6	0.6	0.6	0.6	0.6	0.6	0.6	0.6	0.6	0.6	0.6
In-quota tariff	*%*	40	40	40	40	40	40	40	40	40	40	40
Out-of-quota tariff	*%*	176	176	176	176	176	176	176	176	176	176	176
MEXICO												
Butter tariff	*%*	0	0	0	0	0	0	0	0	0	0	0
Tariff-quotas												
Cheese	*kt pw*	9	9	9	9	9	9	9	9	9	9	9
In-quota tariff	*%*	50	50	50	50	50	50	50	50	50	50	50
Out-of-quota tariff	*%*	125	125	125	125	125	125	125	125	125	125	125
SMP	*kt pw*	90	90	90	90	90	90	90	90	90	90	90
In-quota tariff	*%*	0	0	0	0	0	0	0	0	0	0	0
Out-of-quota tariff	*%*	125	125	125	125	125	125	125	125	125	125	125
Liconsa social programme	*MXN mn*	500	500	500	500	500	500	500	500	500	500	500

For notes, see end of the table.

Table A.23. **Main policy assumptions for dairy markets** *(cont.)*

Calendar year		Average 2007-09 est.	2010	2011	2012	2013	2014	2015	2016	2017	2018	2019
RUSSIAN FEDERATION												
Butter tariff	%	8	5	5	5	5	5	5	5	5	5	5
Cheese tariff	%	8	10	5	5	5	5	5	5	5	5	5
UNITED STATES[g]												
Milk support price[b]	*USDc/litre*	22	22	22	22	22	22	22	22	22	22	22
Target price[h]	*USDc/litre*	37	37.3	37.3	37.3	37.3	37.3	37.3	37.3	37.3	37.3	37.3
Butter support price	*USD/t*	2 315	2 315	2 315	2 315	2 315	2 315	2 315	2 315	2 315	2 315	2 316
SMP support price	*USD/t*	1 764	1 764	1 764	1 764	1 764	1 764	1 764	1 764	1 764	1 764	1 764
Butter tariff-quota	*kt pw*	13	13	13	13	13	13	13	13	13	13	13
In-quota tariff	%	10	10	10	10	10	10	10	10	10	10	10
Out-of-quota tariff	%	112	112	112	112	112	112	112	112	112	112	112
Cheese tariff-quota	*kt pw*	135	135	135	135	135	135	135	135	135	135	135
In-quota tariff	%	12	12	12	12	12	12	12	12	12	12	12
Out-of-quota tariff	%	87	87	87	87	87	87	87	87	87	87	87
Subsidised export limits[c]												
Butter	*kt pw*	21	21	21	21	21	21	21	21	21	21	21
SMP	*kt pw*	68	68	68	68	68	68	68	68	68	68	68
INDIA												
Milk tariff	%	80	80	80	80	80	80	80	80	80	80	80
Butter tariff	%	40	40	40	40	40	40	40	40	40	40	40
Cheese tariff	%	40	40	40	40	40	40	40	40	40	40	40
Whole milk powder tariff	%	20	20	20	20	20	20	20	20	20	20	20
SOUTH AFRICA												
Milk powder tariff-quota	*kt pw*	4	4	4	4	4	4	4	4	4	4	4
In-quota tariff	%	20	20	20	20	20	20	20	20	20	20	20
Out-of-quota tariff	%	81	81	81	81	81	81	81	81	81	81	81

Note: The source for tariffs and Tariff Rate Quotas (except the Russian Federation) is AMAD (Agricultural Market Access Database). The tariff and TRQ data are based on Most Favoured Nation rates scheduled with the WTO and exclude those under preferential or regional agreements, which may be substantially different. Tariffs are simple averages of several product lines. Specific rates are converted to *ad valorem* rates using world prices in the Outlook. Import quotas are based on global commitments scheduled in the WTO rather than those allocated to preferential partners under regional or other agreements.

est.: Estimate.

a) Year ending 30 June.

b) For manufacturing milk.

c) The effective volume of cheese and SMP subsidised exports will be lower reflecting the binding nature of subsidised export limits in value terms.

d) EU farmers also benefit from the Single Farm Payment (SFP) Scheme, which provides flat-rate payments independent from current production decisions and market developments. For the new member states, payments are phased in with the assumption of maximum top-ups from national budgets up to 2013 through the Single Area Payment (SAP), and through the (SFP) from 2014. Due to modulation, an increasing share of the total SFP will go to rural development spending rather than directly to farmers.

e) Total quota, EU27 starting in 1999.

f) Excludes processed cheese.

g) Year beginning 1 January.

h) The counter-cyclical payment for milk is determined as a percentage difference between the target price and the Boston class I price. The difference is set at 34% in 2007 and 2008 at 45% in 2009-12 and 34% thereafter. The target price is adjusted by 45% of the percentage difference between the National Average Dairy Feed Rations Cost and the target cost of feed rations of 16.20 USD/100 kg between 2009 and 2012 and 20.94 USD/100 kg thereafter.

Source: OECD and FAO Secretariats.

StatLink http://dx.doi.org/10.1787/888932287596

Table A.24. **World dairy projections (butter and cheese)**

Calendar year[a]		Average 2007-09 est.	2010	2011	2012	2013	2014	2015	2016	2017	2018	2019
BUTTER												
OECD[b]												
Production	*kt pw*	3 742	3 688	3 731	3 781	3 802	3 830	3 850	3 877	3 897	3 926	3 944
Consumption	*kt pw*	3 219	3 247	3 246	3 258	3 271	3 282	3 292	3 309	3 323	3 340	3 356
Stock changes	*kt pw*	10	–61	–27	3	6	7	7	11	28	34	34
Non-OECD												
Production	*kt pw*	5 931	6 393	6 623	6 840	7 035	7 248	7 450	7 667	7 901	8 130	8 437
Consumption	*kt pw*	6 464	6 859	7 104	7 328	7 528	7 756	7 967	8 190	8 413	8 648	8 957
WORLD												
Production	*kt pw*	9 673	10 081	10 354	10 622	10 837	11 078	11 300	11 543	11 798	12 056	12 381
Consumption	*kt pw*	9 683	10 106	10 350	10 587	10 799	11 038	11 259	11 498	11 736	11 988	12 312
Stock changes	*kt pw*	0	–54	–27	3	6	7	7	11	28	34	35
Price[c]	*USD/t*	2 979	3 043	2 821	2 716	2 709	2 694	2 742	2 766	2 922	2 919	2 958
CHEESE												
OECD[b]												
Production	*kt pw*	14 946	15 116	15 340	15 504	15 697	15 896	16 144	16 376	16 562	16 798	17 031
Consumption	*kt pw*	14 388	14 615	14 808	15 004	15 201	15 395	15 605	15 829	16 029	16 257	16 489
Stock changes	*kt pw*	0	–44	–22	–16	–9	–4	0	4	7	10	13
Non-OECD												
Production	*kt pw*	4 487	4 679	4 804	5 002	5 197	5 358	5 507	5 658	5 840	6 013	6 193
Consumption	*kt pw*	4 930	5 085	5 213	5 375	5 558	5 720	5 902	6 058	6 222	6 399	6 578
WORLD												
Production	*kt pw*	19 433	19 795	20 144	20 506	20 894	21 254	21 651	22 034	22 402	22 810	23 224
Consumption	*kt pw*	19 319	19 700	20 021	20 379	20 760	21 115	21 507	21 887	22 251	22 657	23 067
Stock changes	*kt pw*	1	–49	–22	–16	–9	–4	0	4	7	10	13
Price[d]	*USD/t*	3 887	3 716	3 272	3 126	3 140	3 251	3 338	3 428	3 542	3 590	3 641

est.: Estimate.

a) Year ending 30 June for Australia and 31 May for New Zealand in OECD aggregate.

b) Excludes Iceland but includes the EU8 members that are not members of the OECD.

c) F.o.b. export price, butter, 82% butterfat, Oceania.

d) F.o.b. export price, cheddar cheese, 39% moisture, Oceania.

Source: OECD and FAO Secretariats.

StatLink http://dx.doi.org/10.1787/888932287615

Table A.25. **World dairy projections (powders and casein)**

Calendar year[a]		Average 2007-09 est.	2010	2011	2012	2013	2014	2015	2016	2017	2018	2019
SKIM MILK POWDER												
OECD[b]												
Production	*kt pw*	2 639	2 503	2 574	2 526	2 515	2 508	2 535	2 548	2 569	2 589	2 592
Consumption	*kt pw*	1 725	1 754	1 804	1 807	1 803	1 805	1 819	1 822	1 798	1 807	1 809
Stock changes	*kt pw*	91	–11	15	–38	–54	–71	–67	–58	–5	–3	–3
Non-OECD												
Production	*kt pw*	791	865	893	918	938	970	995	1 027	1 067	1 097	1 130
Consumption	*kt pw*	1 512	1 591	1 613	1 641	1 669	1 709	1 743	1 777	1 807	1 847	1 880
WORLD												
Production	*kt pw*	3 430	3 368	3 468	3 444	3 453	3 478	3 530	3 575	3 635	3 685	3 722
Consumption	*kt pw*	3 237	3 345	3 418	3 448	3 472	3 514	3 562	3 598	3 605	3 654	3 690
Stock changes	*kt pw*	92	–11	16	–38	–54	–71	–67	–58	–4	–3	–3
Price[c]	*USD/t*	3 309	2 530	2 434	2 417	2 493	2 590	2 653	2 759	2 918	2 943	3 000
WHOLE MILK POWDER												
OECD[b]												
Production	*kt pw*	1 991	2 089	2 086	2 116	2 114	2 117	2 169	2 189	2 183	2 200	2 220
Consumption	*kt pw*	861	847	869	887	892	898	906	914	921	927	933
Non-OECD												
Production	*kt pw*	2 271	2 375	2 491	2 597	2 706	2 825	2 914	3 021	3 144	3 254	3 366
Consumption	*kt pw*	3 343	3 570	3 661	3 779	3 882	3 998	4 130	4 249	4 360	4 481	4 606
WORLD												
Production	*kt pw*	4 262	4 464	4 577	4 712	4 821	4 943	5 083	5 210	5 327	5 454	5 586
Consumption	*kt pw*	4 203	4 417	4 531	4 666	4 774	4 896	5 037	5 163	5 280	5 408	5 539
Price[d]	*USD/t*	3 500	2 808	2 550	2 476	2 588	2 725	2 763	2 832	2 938	2 989	3 042
WHEY POWDER												
Wholesale price, United States[e]	*USD/t*	853	788	710	705	727	751	768	791	827	835	852
CASEIN												
Price[f]	*USD/t*	6 300	5 775	5 529	5 451	5 474	5 689	5 862	6 061	6 268	6 343	6 440

est.: Estimate.

a) Year ending 30 June for Australia and 31 May for New Zealand in OECD aggregate.

b) Excludes Iceland but includes the EU8 members that are not members of the OECD.

c) F.o.b. export price, non-fat dry milk, 1.25% butterfat, Oceania.

d) F.o.b. export price, WMP 26% butterfat, Oceania.

e) Edible dry whey, Wisconsin, plant.

f) Export price, New Zealand.

Source: OECD and FAO Secretariats.

StatLink http://dx.doi.org/10.1787/888932287634

Table A.26. **Butter projections**

Calendar year[a]	PRODUCTION (kt)		Growth[b] (%)		IMPORTS (kt)		Growth[b] (%)		EXPORTS (kt)		Growth[b] (%)	
	Average 2007-09 est.	2019	2000-09	2010-19	Average 2007-09 est.	2019	2000-09	2010-19	Average 2007-09 est.	2019	2000-09	2010-19
WORLD	**9 673**	**12 381**	**3.15**	**2.21**	**822**	**781**	**2.09**	**0.39**	**834**	**800**	**0.35**	**0.38**
DEVELOPED	**4 224**	**4 481**	**0.46**	**0.75**	**377**	**249**	**1.87**	**–1.89**	**742**	**695**	**–0.48**	**0.21**
NORTH AMERICA	**804**	**887**	**2.88**	**1.27**	**23**	**23**	**–2.11**	**0.06**	**49**	**18**	**31.51**	**–5.33**
Canada	85	94	0.68	0.75	7	6	–1.74	0.00	0	0	–25.75	0.00
United States	719	793	3.16	1.33	17	17	–2.15	0.09	49	17	–	–5.35
EUROPE	**2 778**	**2 806**	**0.04**	**0.33**	**279**	**164**	**1.41**	**–2.46**	**242**	**135**	**–2.27**	**–4.40**
EU27	2 131	2 083	–0.47	0.05	72	40	–4.20	–5.16	170	65	–3.60	–6.44
Russian Federation	388	472	4.59	1.69	196	114	5.40	–1.61	3	0	–20.48	0.00
Ukraine	93	92	–6.16	–0.49	1	1	–4.20	3.05	6	0	–20.93	–63.94
OCEANIA DEVELOPED	**526**	**665**	**–0.74**	**1.87**	**15**	**14**	**5.76**	**–0.21**	**449**	**540**	**–0.70**	**2.06**
Australia	136	149	–3.51	2.04	15	14	6.60	–0.21	70	72	–8.70	3.09
New Zealand	389	516	0.40	1.82	0	0	–	–	380	468	1.44	1.91
OTHER DEVELOPED[c]	**116**	**124**	**0.48**	**1.13**	**59**	**48**	**7.74**	**–1.20**	**2**	**2**	**–2.10**	**–2.36**
Japan	79	88	–0.70	–0.01	11	4	16.04	15.46	0	0	–	–
South Africa	10	14	0.93	1.83	2	2	–6.12	3.58	2	1	1.26	–3.44
DEVELOPING	**5 449**	**7 900**	**5.65**	**3.14**	**445**	**532**	**2.47**	**1.62**	**92**	**105**	**9.81**	**1.57**
AFRICA	**233**	**295**	**2.44**	**2.19**	**110**	**155**	**–0.09**	**3.75**	**1**	**1**	**–0.75**	**–0.26**
NORTH AFRICA	**176**	**221**	**3.12**	**2.00**	**93**	**132**	**–0.24**	**3.78**	**0**	**0**	**–0.21**	**–0.36**
Algeria	2	3	2.88	2.90	13	16	4.97	2.45	0	0	32.67	–0.02
Egypt	129	168	3.84	2.12	55	82	0.46	4.12	0	0	–0.35	–0.11
SUB-SAHARAN AFRICA	**57**	**74**	**0.52**	**2.78**	**16**	**23**	**0.77**	**3.61**	**1**	**1**	**–0.48**	**–0.23**
LATIN AMERICA and CARIBBEAN	**228**	**272**	**2.00**	**2.28**	**30**	**39**	**–0.41**	**–0.24**	**42**	**53**	**13.12**	**3.11**
Argentina	49	48	1.93	1.10	0	0	–	–	18	27	24.35	4.33
Brazil	84	110	1.95	2.23	5	6	–12.15	–4.16	3	2	28.38	5.95
Chile	16	15	5.23	1.88	0	0	–15.99	0.94	2	2	32.29	–0.94
Mexico	13	15	–1.73	1.55	6	5	34.39	–1.99	0	1	6.04	8.47
Uruguay	18	21	2.19	2.28	0	0	0.41	–2.32	15	18	8.25	2.32
ASIA and PACIFIC	**4 987**	**7 333**	**6.01**	**3.21**	**305**	**338**	**3.87**	**0.99**	**49**	**51**	**7.85**	**0.21**
Bangladesh	21	33	2.05	4.51	0	0	–28.98	–1.11	0	0	34.07	0.47
China[d]	133	209	6.28	4.75	34	40	8.10	1.57	4	4	43.29	0.00
India	3 842	5 733	7.45	3.06	1	2	–29.97	–2.15	5	5	13.14	2.22
Indonesia	0	0	0.00	1.87	17	23	6.00	2.75	0	0	2.55	–0.03
Iran, Islamic Republic of	155	187	0.64	2.34	42	46	9.84	–0.05	0	0	35.83	0.00
Korea	4	5	–3.46	2.63	4	1	15.59	–11.32	0	0	–	–
Malaysia	0	0	0.00	–22.39	15	19	4.51	2.51	2	2	13.97	–0.03
Pakistan	595	874	2.12	4.13	1	1	24.22	–0.20	0	0	17.05	0.17
Saudi Arabia	5	6	0.05	2.04	35	30	4.70	0.16	2	3	26.44	0.00
Turkey	113	130	–0.25	3.13	7	2	11.87	–5.28	0	0	–18.35	0.18
LEAST DEVELOPED COUNTRIES (LDC)	**144**	**200**	**1.38**	**3.27**	**12**	**16**	**–0.30**	**5.45**	**0**	**0**	**–14.00**	**–0.30**
OECD	**3 742**	**3 944**	**0.18**	**0.71**	**140**	**91**	**–1.16**	**–2.62**	**675**	**629**	**–0.57**	**0.46**
NON-OECD	**5 931**	**8 437**	**5.40**	**2.99**	**682**	**691**	**2.86**	**0.84**	**159**	**171**	**5.17**	**0.09**

For notes, see end of the table.

Table A.26. **Butter projections** *(cont.)*

Calendar year[a]	CONSUMPTION (kt)		Growth[b] (%)		PER CAPITA (kg)		Growth[b] (%)	
	Average 2007-09 est.	2019	2000-09	2010-19	Average 2007-09 est.	2019	2000-09	2010-19
WORLD	**9 683**	**12 312**	**3.39**	**2.14**	**1.4**	**1.6**	**2.17**	**1.08**
DEVELOPED	**3 880**	**3 985**	**0.93**	**0.44**	**2.9**	**2.9**	**0.53**	**0.17**
NORTH AMERICA	**773**	**888**	**2.09**	**1.13**	**2.2**	**2.3**	**1.11**	**0.25**
Canada	90	98	0.45	0.56	2.7	2.7	–0.56	–0.35
United States	683	790	2.32	1.20	2.2	2.3	1.34	0.33
EUROPE	**2 825**	**2 815**	**0.50**	**0.23**	**3.9**	**3.9**	**0.45**	**0.28**
EU27	2 033	2 039	–0.12	–0.08	4.1	4.1	–0.41	–0.15
Russian Federation	591	586	4.66	1.04	4.2	4.3	5.12	1.40
Ukraine	88	93	–3.51	1.22	1.9	2.2	–2.77	1.78
OCEANIA DEVELOPED	**113**	**123**	**2.51**	**0.68**	**4.5**	**4.4**	**1.35**	**–0.26**
Australia	82	91	3.56	0.93	3.9	3.9	2.38	–0.03
New Zealand	31	32	0.00	0.00	7.4	6.8	–1.10	–0.82
OTHER DEVELOPED[c]	**169**	**159**	**1.98**	**0.37**	**0.7**	**0.6**	**1.48**	**0.10**
Japan	87	82	–0.55	0.23	0.7	0.7	–0.59	0.48
South Africa	11	15	–1.15	2.63	0.2	0.3	–2.39	2.22
DEVELOPING	**5 802**	**8 327**	**5.31**	**3.06**	**1.1**	**1.3**	**3.88**	**1.81**
AFRICA	**342**	**449**	**1.58**	**2.71**	**0.4**	**0.4**	**–0.77**	**0.50**
NORTH AFRICA	**269**	**353**	**1.86**	**2.64**	**1.1**	**1.2**	**–0.06**	**0.87**
Algeria	15	19	4.60	2.51	0.4	0.5	3.10	1.12
Egypt	184	250	2.72	2.74	2.3	2.6	0.85	1.16
SUB-SAHARAN AFRICA	**73**	**96**	**0.59**	**3.00**	**0.1**	**0.1**	**–1.91**	**0.64**
LATIN AMERICA and CARIBBEAN	**217**	**257**	**0.22**	**1.71**	**0.4**	**0.4**	**–1.01**	**0.77**
Argentina	32	21	–3.91	–2.01	0.8	0.5	–4.87	–2.88
Brazil	87	114	1.40	1.70	0.5	0.5	0.22	1.01
Chile	14	13	2.04	2.30	0.9	0.7	0.98	1.44
Mexico	18	20	2.41	0.41	0.2	0.2	1.35	–0.39
Uruguay	3	2	–9.94	1.99	0.9	0.7	–10.04	1.64
ASIA and PACIFIC	**5 243**	**7 621**	**5.86**	**3.13**	**1.4**	**1.8**	**4.61**	**2.09**
Bangladesh	21	33	1.04	4.49	0.1	0.2	–0.53	3.27
China[d]	163	245	6.33	4.26	0.1	0.2	5.66	3.69
India	3 838	5 729	7.42	3.06	3.2	4.2	5.87	1.86
Indonesia	17	23	6.02	2.77	0.1	0.1	4.75	1.86
Iran, Islamic Republic of	197	233	1.98	1.81	2.7	2.8	0.85	0.70
Korea	7	6	2.71	–2.13	0.2	0.1	2.27	–2.34
Malaysia	12	16	3.33	2.94	0.5	0.5	1.50	1.55
Pakistan	596	874	2.14	4.12	3.4	3.9	–0.07	2.07
Saudi Arabia	37	34	2.96	0.50	1.5	1.1	0.60	–1.37
Turkey	120	132	0.23	2.93	1.6	1.6	–1.08	1.90
LEAST DEVELOPED COUNTRIES (LDC)	**155**	**216**	**1.42**	**3.42**	**0.2**	**0.2**	**–0.85**	**1.28**
OECD	**3 219**	**3 356**	**0.49**	**0.38**	**2.6**	**2.6**	**–0.13**	**–0.05**
NON-OECD	**6 464**	**8 957**	**5.11**	**2.88**	**1.2**	**1.4**	**3.75**	**1.68**

est.: Estimate.

a) Year ending 30 June for Australia and 31 May for New Zealand.

b) Least-squares growth rate (see glossary).

c) Includes other developed Asian countries: Kazakhstan, Kyrgyzstan, Tajikistan, Turkmenistan, Uzbekistan, Armenia, Azerbaijan and Georgia.

d) Refers to mainland only. The economies of Chinese Taipei, Hong Kong (China) and Macau (China) are included in the other Asia Pacific aggregate.

Source: OECD and FAO Secretariats.

StatLink http://dx.doi.org/10.1787/888932287653

Table A.27. **Cheese projections**

Calendar year[a]	PRODUCTION (kt)		Growth[b] (%)		IMPORTS (kt)		Growth[b] (%)		EXPORTS (kt)		Growth[b] (%)	
	Average 2007-09 est.	2019	2000-09	2010-19	Average 2007-09 est.	2019	2000-09	2010-19	Average 2007-09 est.	2019	2000-09	2010-19
WORLD	**19 433**	**23 224**	**2.38**	**1.77**	**1 707**	**1 937**	**4.15**	**1.39**	**1 818**	**2 080**	**4.63**	**1.29**
DEVELOPED	**15 619**	**18 174**	**2.09**	**1.51**	**1 000**	**1 034**	**3.72**	**0.80**	**1 482**	**1 673**	**2.75**	**1.23**
NORTH AMERICA	**4 877**	**5 933**	**2.41**	**1.94**	**196**	**210**	**–1.80**	**0.31**	**108**	**99**	**6.62**	**–2.04**
Canada	375	424	1.59	1.15	22	22	–0.98	0.00	10	10	–6.87	–0.39
United States	4 502	5 510	2.48	2.00	174	187	–1.90	0.35	98	89	9.20	–2.21
EUROPE	**9 878**	**11 194**	**2.01**	**1.30**	**522**	**530**	**8.03**	**1.10**	**844**	**956**	**3.39**	**1.51**
EU27	8 692	9 537	1.53	1.00	88	94	–6.46	0.73	573	561	0.92	0.18
Russian Federation	460	601	8.17	2.06	341	351	19.23	1.94	25	25	28.47	0.00
Ukraine	274	512	12.45	5.73	14	7	32.82	–5.37	72	141	15.12	6.00
OCEANIA DEVELOPED	**715**	**862**	**0.83**	**1.38**	**65**	**59**	**7.50**	**–0.52**	**522**	**610**	**1.08**	**1.44**
Australia	354	370	–1.11	1.51	65	59	8.34	–0.52	187	147	–2.62	1.74
New Zealand	360	491	2.92	1.28	0	0	–	–	335	463	3.32	1.36
OTHER DEVELOPED[c]	**150**	**185**	**2.76**	**1.69**	**218**	**235**	**0.55**	**0.95**	**6**	**7**	**8.45**	**–1.54**
Japan	44	50	3.44	1.22	195	216	–0.61	1.00	0	0	–	–
South Africa	38	44	0.21	1.39	5	3	7.45	–6.50	1	1	–7.46	1.89
DEVELOPING	**3 814**	**5 050**	**3.65**	**2.78**	**706**	**903**	**4.80**	**2.10**	**337**	**407**	**17.77**	**1.52**
AFRICA	**928**	**1 394**	**3.88**	**3.64**	**96**	**109**	**5.42**	**0.90**	**68**	**76**	**21.18**	**0.03**
NORTH AFRICA	**729**	**1 141**	**4.84**	**3.87**	**80**	**90**	**7.47**	**0.76**	**68**	**76**	**21.33**	**0.03**
Algeria	2	2	2.52	–0.79	23	32	3.28	2.63	0	0	14.57	–0.03
Egypt	690	1 097	4.71	3.94	20	19	6.82	–0.68	19	20	27.25	0.50
SUB-SAHARAN AFRICA	**198**	**253**	**0.84**	**2.68**	**16**	**20**	**–1.94**	**1.56**	**1**	**0**	**9.51**	**–0.77**
LATIN AMERICA and CARIBBEAN	**1 679**	**2 135**	**3.15**	**2.45**	**203**	**262**	**3.09**	**2.27**	**113**	**154**	**8.48**	**3.14**
Argentina	492	638	2.49	3.02	3	0	–4.90	–	43	72	10.44	7.64
Brazil	623	866	4.40	2.68	8	7	–5.56	–0.67	7	22	14.98	10.10
Chile	63	68	1.87	1.60	8	12	10.33	3.10	13	2	22.96	–16.13
Mexico	157	176	2.06	1.21	85	110	4.00	2.11	4	5	26.21	0.03
Uruguay	41	40	6.29	0.18	0	0	–16.12	2.78	29	22	8.42	–2.78
ASIA and PACIFIC	**1 207**	**1 521**	**4.25**	**2.48**	**407**	**531**	**5.56**	**2.28**	**156**	**178**	**29.83**	**0.92**
Bangladesh	1	2	0.00	5.78	0	0	–0.06	0.36	0	0	0.00	–0.36
China[d]	357	553	7.06	4.75	26	38	10.24	3.53	0	0	–	1.57
India	1	1	7.34	1.71	1	1	8.33	–0.47	1	2	31.65	0.47
Indonesia	0	0	0.00	–1.97	15	25	12.70	4.05	0	0	23.36	–0.04
Iran, Islamic Republic of	225	248	0.33	0.72	0	0	20.14	–0.47	8	22	59.70	8.28
Korea	26	31	5.09	1.35	49	58	9.84	1.41	0	0	–	–
Malaysia	0	0	0.00	58.74	10	17	8.74	4.62	0	0	7.28	–4.62
Pakistan	0	0	0.00	21.72	2	2	18.27	–0.06	0	0	63.14	–139.67
Saudi Arabia	135	163	178.40	1.37	70	66	–0.39	–0.28	131	136	44.70	0.28
Turkey	120	118	–0.59	0.71	2	2	–9.42	3.43	2	2	–14.80	–2.48
LEAST DEVELOPED COUNTRIES (LDC)	**254**	**311**	**0.74**	**2.26**	**14**	**14**	**0.63**	**0.36**	**0**	**1**	**11.95**	**4.64**
OECD	**14 946**	**17 031**	**1.75**	**1.32**	**737**	**818**	**–0.04**	**0.84**	**1 293**	**1 346**	**1.43**	**0.45**
NON-OECD	**4 487**	**6 193**	**4.73**	**3.12**	**970**	**1 119**	**8.47**	**1.80**	**525**	**734**	**18.15**	**3.01**

For notes, see end of the table.

Table A.27. **Cheese projections** *(cont.)*

Calendar year[a]	CONSUMPTION (kt)		Growth[b] (%)		PER CAPITA (kg)		Growth[b] (%)	
	Average 2007-09 est.	2019	2000-09	2010-19	Average 2007-09 est.	2019	2000-09	2010-19
WORLD	**19 319**	**23 067**	**2.32**	**1.76**	**2.9**	**3.0**	**1.10**	**0.70**
DEVELOPED	**15 135**	**17 522**	**2.11**	**1.46**	**11.3**	**12.6**	**1.71**	**1.19**
NORTH AMERICA	**4 951**	**6 031**	**2.15**	**1.85**	**14.4**	**15.9**	**1.17**	**0.98**
Canada	384	435	1.84	1.13	11.6	11.8	0.84	0.22
United States	4 567	5 595	2.18	1.91	14.7	16.3	1.20	1.04
EUROPE	**9 561**	**10 769**	**2.13**	**1.27**	**13.3**	**15.0**	**2.07**	**1.32**
EU27	8 212	9 071	1.45	1.05	16.6	18.2	1.16	0.98
Russian Federation	777	927	11.21	2.03	5.5	6.8	11.66	2.38
Ukraine	216	378	12.50	5.27	4.7	8.8	13.24	5.83
OCEANIA DEVELOPED	**261**	**310**	**1.76**	**0.86**	**10.3**	**11.0**	**0.60**	**–0.08**
Australia	233	282	1.99	0.94	11.1	12.0	0.81	–0.02
New Zealand	28	28	0.00	0.05	6.6	6.1	–1.11	–0.78
OTHER DEVELOPED[c]	**362**	**412**	**1.30**	**1.33**	**1.4**	**1.6**	**0.79**	**1.06**
Japan	239	266	0.04	1.04	1.9	2.1	–0.01	1.29
South Africa	42	46	1.38	0.71	0.8	0.9	0.14	0.30
DEVELOPING	**4 183**	**5 546**	**3.10**	**2.76**	**0.8**	**0.9**	**1.66**	**1.51**
AFRICA	**955**	**1 428**	**3.32**	**3.63**	**1.0**	**1.2**	**0.97**	**1.42**
NORTH AFRICA	**742**	**1 156**	**4.23**	**3.89**	**3.0**	**3.9**	**2.31**	**2.13**
Algeria	25	33	3.24	2.42	0.7	0.8	1.74	1.03
Egypt	691	1 095	4.45	3.92	8.5	11.3	2.58	2.35
SUB-SAHARAN AFRICA	**213**	**272**	**0.59**	**2.60**	**0.3**	**0.3**	**–1.91**	**0.23**
LATIN AMERICA and CARIBBEAN	**1 769**	**2 243**	**2.80**	**2.38**	**3.1**	**3.5**	**1.57**	**1.44**
Argentina	450	566	1.61	2.56	11.3	12.9	0.66	1.70
Brazil	625	852	4.16	2.51	3.3	4.1	2.97	1.82
Chile	58	78	0.47	3.01	3.4	4.2	–0.60	2.15
Mexico	237	282	2.46	1.58	2.2	2.4	1.40	0.78
Uruguay	12	19	1.36	4.99	3.4	5.5	1.25	4.63
ASIA and PACIFIC	**1 459**	**1 875**	**3.33**	**2.59**	**0.4**	**0.4**	**2.08**	**1.55**
Bangladesh	1	2	0.02	5.29	0.0	0.0	–1.55	4.06
China[d]	382	591	7.26	4.67	0.3	0.4	6.60	4.10
India	0	0	–40.28	8.30	0.0	0.0	–41.82	7.10
Indonesia	15	25	12.62	4.08	0.1	0.1	11.36	3.17
Iran, Islamic Republic of	218	226	–0.19	0.19	3.0	2.7	–1.33	–0.92
Korea	76	89	7.46	1.39	1.6	1.8	7.02	1.17
Malaysia	10	18	8.76	4.87	0.4	0.6	6.93	3.48
Pakistan	1	2	17.53	0.44	0.0	0.0	15.32	–1.61
Saudi Arabia	74	94	1.54	1.76	2.9	3.0	–0.82	–0.12
Turkey	121	118	–0.42	0.82	1.6	1.4	–1.73	–0.22
LEAST DEVELOPED COUNTRIES (LDC)	**267**	**325**	**0.71**	**2.17**	**0.4**	**0.3**	**–1.55**	**0.03**
OECD	**14 388**	**16 489**	**1.66**	**1.34**	**11.5**	**12.6**	**1.05**	**0.91**
NON-OECD	**4 930**	**6 578**	**4.43**	**2.89**	**0.9**	**1.0**	**3.07**	**1.69**

est.: Estimate.

a) Year ending 30 June for Australia and 31 May for New Zealand.

b) Least-squares growth rate (see glossary).

c) Includes other developed Asian countries: Kazakhstan, Kyrgyzstan, Tajikistan, Turkmenistan, Uzbekistan, Armenia, Azerbaijan and Georgia.

d) Refers to mainland only. The economies of Chinese Taipei, Hong Kong (China) and Macau (China) are included in the other Asia Pacific aggregate.

Source: OECD and FAO Secretariats.

StatLink http://dx.doi.org/10.1787/888932287672

Table A.28. **Skim milk powder projections**

Calendar year[a]	PRODUCTION (kt)		Growth[b] (%)		IMPORTS (kt)		Growth[b] (%)		EXPORTS (kt)		Growth[b] (%)	
	Average 2007-09 est.	2019	2000-09	2010-19	Average	2019	2000-09	2010-19	Average 2007-09 est.	2019	2000-09	2010-19
WORLD	**3 430**	**3 722**	**–0.03**	**1.04**	**1 167**	**1 321**	**–0.33**	**1.13**	**1 244**	**1 317**	**–0.18**	**1.13**
DEVELOPED	**2 850**	**2 916**	**–0.72**	**0.54**	**129**	**74**	**–6.40**	**0.08**	**1 112**	**1 176**	**–0.58**	**1.09**
NORTH AMERICA	**860**	**878**	**1.78**	**0.99**	**3**	**3**	**–7.33**	**0.00**	**302**	**278**	**8.18**	**1.53**
Canada	82	103	–0.37	2.92	3	3	5.96	0.00	12	8	–18.77	–0.52
United States	778	775	2.03	0.76	0	0	–	–	290	269	11.79	1.60
EUROPE	**1 212**	**1 133**	**–2.93**	**–0.65**	**73**	**23**	**–8.22**	**–0.34**	**320**	**282**	**–5.41**	**–2.35**
EU27	913	706	–3.74	–2.15	8	6	–30.28	0.00	199	64	–9.44	–12.04
Russian Federation	81	117	–5.72	1.80	58	6	–0.25	0.93	1	1	–27.24	0.00
Ukraine	116	176	–0.49	4.89	0	0	35.81	–0.33	42	115	–3.94	9.70
OCEANIA DEVELOPED	**593**	**710**	**0.81**	**2.23**	**4**	**4**	**–2.03**	**–1.28**	**480**	**606**	**0.10**	**3.03**
Australia	189	222	–3.17	2.41	4	4	3.61	–1.28	150	193	–4.54	4.05
New Zealand	403	488	3.41	2.15	0	0	–	–	330	413	3.09	2.59
OTHER DEVELOPED[c]	**186**	**195**	**–0.69**	**0.08**	**49**	**45**	**–3.29**	**0.54**	**9**	**11**	**1.80**	**–0.64**
Japan	169	175	–1.10	–0.03	32	29	–6.50	–0.61	0	0	–	–
South Africa	11	15	5.53	1.23	8	6	5.71	8.23	1	1	–0.08	–0.08
DEVELOPING	**579**	**806**	**3.94**	**3.08**	**1 038**	**1 247**	**0.66**	**1.20**	**132**	**140**	**3.39**	**1.47**
AFRICA	**3**	**4**	**–1.01**	**3.60**	**163**	**192**	**–1.33**	**1.28**	**3**	**3**	**–2.35**	**–0.58**
NORTH AFRICA	**0**	**0**	**0.00**	**3.26**	**121**	**135**	**–0.88**	**1.06**	**1**	**1**	**–4.01**	**–0.05**
Algeria	0	0	0.00	–2.53	71	74	–3.49	0.09	0	0	26.17	0.00
Egypt	0	0	0.00	0.46	21	20	–0.33	1.25	0	0	–11.39	–0.01
SUB-SAHARAN AFRICA	**3**	**4**	**–1.02**	**3.60**	**42**	**57**	**–2.46**	**1.81**	**2**	**2**	**–1.18**	**–0.71**
LATIN AMERICA and CARIBBEAN	**235**	**297**	**3.34**	**2.43**	**184**	**197**	**–2.31**	**0.10**	**31**	**34**	**–0.99**	**0.81**
Argentina	26	23	–5.77	0.94	0	0	–	–	13	19	–4.68	1.41
Brazil	134	192	6.23	3.17	5	0	–16.47	–	2	2	–	5.85
Chile	18	17	7.09	–0.11	5	10	–5.71	8.36	1	1	41.34	–8.36
Mexico	32	36	1.51	3.16	135	156	1.29	0.77	0	0	–8.76	0.05
Uruguay	16	16	3.53	–0.30	0	0	–5.96	2.46	11	7	0.51	–2.46
ASIA and PACIFIC	**342**	**505**	**4.44**	**3.48**	**691**	**858**	**2.12**	**1.45**	**98**	**104**	**5.49**	**1.76**
Bangladesh	0	0	0.00	–2.38	14	20	4.12	2.80	0	0	–33.67	–0.03
China[d]	60	108	–1.68	5.07	53	90	11.27	2.49	2	1	–	0.00
India	211	309	5.73	3.28	1	1	19.23	–0.53	30	42	15.26	5.24
Indonesia	0	0	0.00	–3.25	105	142	4.54	2.01	2	2	–23.17	–0.02
Iran, Islamic Republic of	0	0	0.00	0.26	6	9	0.71	0.66	1	1	–6.10	–0.66
Korea	20	30	–3.51	3.10	6	2	5.77	–12.50	0	0	–	–
Malaysia	0	0	0.00	0.00	85	94	4.46	0.33	11	12	13.88	0.00
Pakistan	0	0	0.00	1.36	9	14	30.02	3.23	1	1	56.88	–3.23
Saudi Arabia	0	0	0.00	2.47	39	47	1.11	1.64	17	18	6.09	–0.02
Turkey	0	0	0.00	0.36	2	2	–2.46	0.84	0	0	–13.76	–0.84
LEAST DEVELOPED COUNTRIES (LDC)	**0**	**0**	**0.00**	**19.39**	**54**	**77**	**–2.56**	**2.56**	**1**	**1**	**–11.93**	**–1.02**
OECD	**2 639**	**2 592**	**–0.74**	**0.30**	**196**	**214**	**–4.18**	**0.55**	**995**	**960**	**–0.31**	**0.50**
NON-OECD	**791**	**1 130**	**2.63**	**2.96**	**971**	**1 107**	**0.60**	**1.24**	**250**	**356**	**1.11**	**3.05**

For notes, see end of the table.

Table A.28. **Skim milk powder projections** *(cont.)*

Calendar year[a]	CONSUMPTION (kt)		Growth[b] (%)		PER CAPITA (kg)		Growth[b] (%)	
	Average 2007-09 est.	2019	2000-09	2010-19	Average 2007-09 est.	2019	2000-09	2010-19
WORLD	**3 237**	**3 690**	**–0.06**	**1.03**	**0.4**	**0.4**	**–0.43**	**0.03**
DEVELOPED	**1 754**	**1 774**	**–1.29**	**0.11**	**1.1**	**1.1**	**–0.29**	**–0.16**
NORTH AMERICA	**549**	**603**	**2.42**	**0.48**	**1.5**	**1.4**	**0.68**	**–0.68**
Canada	71	98	8.51	3.14	1.1	1.2	0.37	1.34
United States	478	505	1.68	0.02	1.5	1.5	0.70	–0.85
EUROPE	**903**	**874**	**–3.93**	**0.06**	**1.0**	**0.9**	**–1.65**	**0.16**
EU27	660	648	–4.71	–0.16	0.9	0.9	–1.91	–0.26
Russian Federation	138	123	–3.15	1.57	1.0	0.9	–2.69	1.92
Ukraine	74	61	2.40	–0.78	1.6	1.4	3.14	–0.22
OCEANIA DEVELOPED	**81**	**68**	**1.05**	**–2.38**	**3.2**	**2.4**	**–0.11**	**–3.32**
Australia	46	33	1.91	–4.28	2.2	1.4	0.73	–5.25
New Zealand	35	35	0.00	–0.18	8.3	7.6	–1.11	–1.01
OTHER DEVELOPED[c]	**222**	**229**	**3.51**	**0.20**	**0.8**	**0.8**	**4.99**	**0.63**
Japan	198	204	–1.40	–0.11	1.3	1.6	–1.19	0.88
South Africa	18	20	4.91	2.98	0.4	0.4	3.67	2.57
DEVELOPING	**1 483**	**1 916**	**1.56**	**1.95**	**0.3**	**0.3**	**0.09**	**0.69**
AFRICA	**162**	**193**	**–1.31**	**1.35**	**0.2**	**0.2**	**–3.65**	**–0.87**
NORTH AFRICA	**120**	**134**	**–0.85**	**1.06**	**0.5**	**0.5**	**–2.77**	**–0.70**
Algeria	71	74	–3.49	0.09	2.1	1.9	–4.99	–1.30
Egypt	21	19	0.00	1.27	0.3	0.2	–1.87	–0.30
SUB-SAHARAN AFRICA	**42**	**59**	**–2.42**	**2.04**	**0.1**	**0.1**	**–4.92**	**–0.33**
LATIN AMERICA and CARIBBEAN	**386**	**461**	**0.49**	**1.53**	**0.6**	**0.6**	**–0.98**	**0.51**
Argentina	12	4	–7.26	–0.81	0.3	0.1	–8.21	–1.67
Brazil	136	190	3.16	2.91	0.5	0.7	2.09	2.46
Chile	21	26	2.31	2.93	1.2	1.4	1.25	2.08
Mexico	165	192	1.14	1.28	1.5	1.6	0.08	0.48
Uruguay	6	9	12.63	1.75	1.7	2.5	12.53	1.40
ASIA and PACIFIC	**935**	**1 262**	**2.61**	**2.20**	**0.2**	**0.3**	**1.36**	**1.16**
Bangladesh	14	20	4.15	2.80	0.1	0.1	2.57	1.58
China[d]	111	198	2.84	3.85	0.1	0.1	2.18	3.28
India	182	267	4.68	2.98	0.2	0.2	3.14	1.77
Indonesia	103	140	7.49	2.04	0.5	0.6	6.22	1.14
Iran, Islamic Republic of	5	8	6.17	0.85	0.1	0.1	5.03	–0.26
Korea	25	35	–1.37	1.90	0.5	0.7	–1.81	1.68
Malaysia	74	82	3.74	0.39	2.7	2.6	1.91	–1.00
Pakistan	8	13	29.40	3.88	0.0	0.1	27.19	1.83
Saudi Arabia	22	29	–0.55	2.82	0.9	0.9	–2.90	0.94
Turkey	2	2	–1.16	0.89	0.0	0.0	–2.47	–0.15
LEAST DEVELOPED COUNTRIES (LDC)	**53**	**76**	**–1.97**	**2.68**	**0.1**	**0.1**	**–4.23**	**0.54**
OECD	**1 725**	**1 809**	**–1.38**	**0.19**	**1.2**	**1.2**	**–0.58**	**–0.23**
NON-OECD	**1 512**	**1 880**	**1.67**	**1.90**	**0.3**	**0.3**	**0.28**	**0.69**

est.: Estimate.

a) Year ending 30 June for Australia and 31 May for New Zealand.

b) Least-squares growth rate (see glossary).

c) Includes other developed Asian countries: Kazakhstan, Kyrgyzstan, Tajikistan, Turkmenistan, Uzbekistan, Armenia, Azerbaijan and Georgia.

d) Refers to mainland only. The economies of Chinese Taipei, Hong Kong (China) and Macau (China) are included in the other Asia Pacific aggregate.

Source: OECD and FAO Secretariats.

StatLink http://dx.doi.org/10.1787/888932287691

Table A.29. **Whole milk powder projections**

	PRODUCTION (kt)		Growth[b] (%)		IMPORTS (kt)		Growth[b] (%)		EXPORTS (kt)		Growth[b] (%)	
Calendar year[a]	Average 2007-09 est.	2019	2000-09	2010-19	Average 2007-09 est.	2019	2000-09	2010-19	Average 2007-09 est.	2019	2000-09	2010-19
WORLD	**4 262**	**5 586**	**3.85**	**2.50**	**1 777**	**2 055**	**4.67**	**1.13**	**1 840**	**2 101**	**3.01**	**1.10**
DEVELOPED	**1 894**	**2 109**	**0.87**	**0.67**	**118**	**95**	**11.07**	**–0.08**	**1 261**	**1 417**	**0.82**	**0.54**
NORTH AMERICA	**34**	**21**	**–4.63**	**–4.61**	**44**	**44**	**10.16**	**0.00**	**14**	**11**	**8.35**	**0.00**
Canada	17	15	0.33	–1.74	27	30	5.72	0.00	1	1	–24.38	0.00
United States	18	6	–7.04	–9.71	17	14	23.06	0.00	13	10	–	0.00
EUROPE	**1 010**	**1 047**	**–0.75**	**0.20**	**49**	**30**	**14.43**	**–1.39**	**486**	**449**	**–2.80**	**–0.95**
EU27	845	861	–1.27	–0.05	1	1	–24.13	0.00	435	403	–3.87	–1.02
Russian Federation	71	86	–2.40	1.54	43	23	27.30	–1.74	3	3	2.24	0.00
Ukraine	36	40	11.00	1.57	0	1	38.38	0.04	18	16	15.73	–0.39
OCEANIA DEVELOPED	**807**	**993**	**3.65**	**1.31**	**9**	**7**	**5.63**	**–0.30**	**759**	**950**	**3.89**	**1.33**
Australia	142	166	–4.97	1.63	9	7	8.43	–0.30	97	126	–6.78	1.81
New Zealand	665	827	6.46	1.25	0	0	–	–	662	824	6.47	1.25
OTHER DEVELOPED[c]	**42**	**46**	**–0.63**	**1.27**	**16**	**15**	**9.24**	**3.18**	**2**	**8**	**–17.65**	**1.95**
Japan	13	12	–3.89	–0.27	0	0	–	–	0	0	–	–
South Africa	16	27	3.23	4.83	10	3	24.87	–1.99	2	8	–18.07	1.99
DEVELOPING	**2 368**	**3 477**	**6.82**	**3.77**	**1 659**	**1 960**	**4.31**	**1.19**	**579**	**684**	**9.55**	**2.38**
AFRICA	**6**	**5**	**–4.05**	**–0.83**	**422**	**558**	**6.95**	**1.98**	**11**	**12**	**4.63**	**–0.05**
NORTH AFRICA	**0**	**0**	**0.00**	**2.21**	**189**	**241**	**6.14**	**1.83**	**0**	**0**	**–0.70**	**–0.04**
Algeria	0	0	0.00	–0.82	152	186	4.59	1.36	0	0	43.33	–0.01
Egypt	0	0	0.00	0.59	22	36	16.80	3.83	0	0	–1.24	–0.04
SUB-SAHARAN AFRICA	**6**	**5**	**–4.06**	**–0.83**	**233**	**317**	**7.61**	**2.09**	**11**	**11**	**4.88**	**–0.05**
LATIN AMERICA and CARIBBEAN	**1 213**	**1 701**	**5.03**	**2.72**	**331**	**341**	**–0.08**	**0.15**	**287**	**422**	**7.79**	**3.68**
Argentina	202	277	0.64	1.92	1	0	–0.66	–	136	231	4.16	3.94
Brazil	582	910	8.74	3.65	18	11	–18.90	–4.42	76	134	–	7.23
Chile	77	100	1.80	1.49	0	0	–63.03	12.40	15	4	13.85	–12.40
Mexico	192	211	3.64	1.76	41	39	–0.17	–1.51	10	10	–2.04	0.01
Uruguay	41	38	8.96	0.26	0.0	0.0	18.01	0.35	33	29	9.42	–0.35
ASIA and PACIFIC	**1 149**	**1 771**	**9.08**	**4.90**	**905**	**1 061**	**5.09**	**1.15**	**281**	**250**	**12.40**	**0.61**
Bangladesh	0	0	0.00	–7.96	25	39	1.25	4.00	0	0	0.00	0.52
China[d]	1 117	1 713	9.15	4.87	88	53	4.91	–7.22	46	30	5.88	7.19
India	13	26	128.15	5.70	1	1	19.17	2.69	1	1	61.71	–1.51
Indonesia	0	0	0.00	–1.89	88	130	14.64	3.08	18	19	11.61	–0.03
Iran, Islamic Republic of	1	1	0.43	0.50	3	3	44.98	0.76	0	0	32.99	–0.76
Korea	4	6	–6.42	5.11	1	2	2.10	4.51	0	0	–	–
Malaysia	0	0	0.00	–2.03	81	86	3.69	2.01	26	3	5.10	–0.02
Pakistan	0	0	0.00	1.88	8	13	9.04	3.97	0	0	41.40	–3.97
Saudi Arabia	0	0	0.00	1.51	101	120	11.83	1.79	43	45	45.40	–0.02
Turkey	0	0	0.00	0.24	3	3	–2.79	0.89	0	0	17.15	–0.89
LEAST DEVELOPED COUNTRIES (LDC)	**0**	**0**	**0.00**	**15.58**	**173**	**258**	**9.18**	**2.70**	**7**	**8**	**28.49**	**–0.26**
OECD	**1 991**	**2 220**	**1.00**	**0.73**	**101**	**94**	**2.19**	**–0.58**	**1 236**	**1 381**	**0.62**	**0.48**
NON-OECD	**2 271**	**3 366**	**6.97**	**3.84**	**1 676**	**1 961**	**4.84**	**1.22**	**604**	**720**	**9.75**	**2.41**

For notes, see end of the table.

Table A.29. **Whole milk powder projections** *(cont.)*

Calendar year[a]	CONSUMPTION (kt)		Growth[b] (%)		PER CAPITA (kg)		Growth[b] (%)	
	Average 2007-09 est.	2019	2000-09	2010-19	Average 2007-09 est.	2019	2000-09	2010-19
WORLD	**4 203**	**5 539**	**4.54**	**2.52**	**0.6**	**0.7**	**3.32**	**1.46**
DEVELOPED	**755**	**786**	**2.28**	**0.81**	**0.6**	**0.6**	**1.89**	**0.54**
NORTH AMERICA	**65**	**54**	**0.94**	**–2.05**	**0.2**	**0.1**	**–0.04**	**–2.93**
Canada	43	44	4.89	–0.64	1.3	1.2	3.88	–1.55
United States	22	10	–4.43	–6.78	0.1	0.0	–5.41	–7.65
EUROPE	**577**	**629**	**2.45**	**1.02**	**0.8**	**0.9**	**2.40**	**1.07**
EU27	416	459	2.29	0.90	0.8	0.9	2.00	0.83
Russian Federation	110	106	3.04	0.71	0.8	0.8	3.50	1.06
Ukraine	18	25	7.26	2.98	0.4	0.6	8.01	3.53
OCEANIA DEVELOPED	**57**	**49**	**1.62**	**0.86**	**2.2**	**1.8**	**0.45**	**–0.09**
Australia	53	46	1.77	0.93	2.5	1.9	0.59	–0.04
New Zealand	4	4	0.00	0.01	0.9	0.8	–1.11	–0.82
OTHER DEVELOPED[c]	**56**	**54**	**3.39**	**1.67**	**0.2**	**0.2**	**2.88**	**1.40**
Japan	13	12	–3.88	–0.27	0.1	0.1	–3.92	–0.02
South Africa	23	22	19.25	4.82	0.5	0.4	18.01	4.41
DEVELOPING	**3 448**	**4 753**	**5.07**	**2.83**	**0.6**	**0.8**	**3.64**	**1.58**
AFRICA	**417**	**551**	**6.88**	**1.99**	**0.4**	**0.5**	**4.53**	**–0.22**
NORTH AFRICA	**189**	**240**	**6.16**	**1.83**	**0.8**	**0.8**	**4.24**	**0.07**
Algeria	152	186	4.59	1.36	4.4	4.6	3.10	–0.03
Egypt	22	36	17.20	3.85	0.3	0.4	15.33	2.28
SUB-SAHARAN AFRICA	**228**	**311**	**7.47**	**2.11**	**0.3**	**0.3**	**4.97**	**–0.25**
LATIN AMERICA and CARIBBEAN	**1 258**	**1 620**	**2.70**	**1.90**	**2.2**	**2.5**	**1.47**	**0.97**
Argentina	68	46	–4.79	–4.77	1.7	1.0	–5.75	–5.64
Brazil	524	787	3.62	2.99	2.7	3.8	2.43	2.30
Chile	63	95	–0.92	2.75	3.7	5.2	–1.98	1.90
Mexico	223	240	3.09	1.23	2.1	2.0	2.03	0.43
Uruguay	8	9	6.74	2.40	2.3	2.6	6.64	2.04
ASIA and PACIFIC	**1 773**	**2 582**	**6.58**	**3.65**	**0.5**	**0.6**	**5.33**	**2.61**
Bangladesh	25	39	1.25	4.00	0.2	0.2	–0.33	2.77
China[d]	1 159	1 736	9.08	4.14	0.9	1.2	8.41	3.57
India	13	27	60.60	5.82	0.0	0.0	59.06	4.61
Indonesia	69	111	18.39	3.71	0.3	0.4	17.12	2.80
Iran, Islamic Republic of	4	4	22.29	0.75	0.1	0.0	21.15	–0.36
Korea	5	8	–4.58	4.99	0.1	0.2	–5.02	4.78
Malaysia	54	83	0.37	2.09	2.0	2.6	–1.46	0.70
Pakistan	8	13	8.83	4.17	0.0	0.1	6.62	2.12
Saudi Arabia	58	75	4.63	3.03	2.3	2.4	2.27	1.15
Turkey	2	3	–3.41	0.96	0.0	0.0	–4.72	–0.08
LEAST DEVELOPED COUNTRIES (LDC)	**166**	**250**	**8.74**	**2.82**	**0.2**	**0.3**	**6.47**	**0.68**
OECD	**861**	**933**	**1.82**	**0.96**	**0.7**	**0.7**	**1.21**	**0.53**
NON-OECD	**3 343**	**4 606**	**5.32**	**2.86**	**0.6**	**0.7**	**3.96**	**1.66**

est.: Estimate.

a) Year ending 30 June for Australia and 31 May for New Zealand.

b) Least-squares growth rate (see glossary).

c) Includes other developed Asian countries: Kazakhstan, Kyrgyzstan, Tajikistan, Turkmenistan, Uzbekistan, Armenia, Azerbaijan and Georgia.

d) Refers to mainland only. The economies of Chinese Taipei, Hong Kong (China) and Macau (China) are included in the other Asia Pacific aggregate.

Source: OECD and FAO Secretariats.

StatLink http://dx.doi.org/10.1787/888932287710

Table A.30. **Milk projections**

Calendar year[a]	PRODUCTION (kt)		Growth[b] (%)		INVENTORIES ('000 hd)		Growth[b] (%)		YIELD (t/hd)		Growth[b] (%)	
	Average 2007-09 est.	2019	2000-09	2010-19	Average 2007-09 est.	2019	2000-09	2010-19	Average 2007-09 est.	2019	2000-09	2010-19
WORLD	**677 452**	**848 502**	**2.19**	**2.18**	**590 214**	**680 555**	**1.38**	**1.45**	**1.1**	**1.2**	**0.81**	**0.73**
DEVELOPED	**356 551**	**398 381**	**0.61**	**1.17**	**82 427**	**84 026**	**–0.41**	**0.43**	**4.3**	**4.7**	**1.02**	**0.74**
NORTH AMERICA	**93 799**	**104 680**	**1.53**	**1.16**	**10 202**	**9 677**	**–0.05**	**–0.28**	**9.2**	**10.8**	**1.58**	**1.44**
Canada	8 402	9 298	0.33	0.96	979	896	–1.40	–0.97	8.6	10.4	1.72	1.92
United States	85 396	95 381	1.65	1.18	9 223	8 781	0.10	–0.21	9.3	10.9	1.55	1.39
EUROPE	**208 922**	**224 026**	**–0.01**	**0.76**	**45 395**	**41 417**	**–2.30**	**–0.76**	**4.6**	**5.4**	**2.29**	**1.52**
EU27	147 373	151 065	–0.02	0.35	24 053	21 719	–1.24	–0.82	6.1	7.0	1.22	1.18
Russian Federation	32 314	38 311	–0.30	1.54	9 124	7 188	–4.20	–2.40	3.5	5.3	3.90	3.94
Ukraine	11 699	12 814	–1.77	1.32	4 333	4 103	–4.60	–0.44	2.7	3.1	2.83	1.76
OCEANIA DEVELOPED	**25 532**	**30 429**	**0.72**	**1.55**	**5 911**	**6 228**	**0.59**	**0.46**	**4.3**	**4.9**	**0.13**	**1.09**
Australia	9 689	10 651	–1.89	1.63	1 694	1 714	–3.41	0.96	5.7	6.2	1.52	0.67
New Zealand	15 843	19 778	2.61	1.51	4 217	4 514	2.57	0.27	3.8	4.4	0.05	1.23
OTHER DEVELOPED[c]	**28 298**	**39 246**	**2.38**	**3.58**	**20 919**	**26 704**	**4.22**	**2.91**	**1.4**	**1.5**	**–1.85**	**0.67**
Japan	7 981	7 990	–0.73	0.01	1 025	951	–2.41	–0.38	7.8	8.4	1.69	0.39
South Africa	3 121	3 754	2.42	2.00	970	976	0.04	0.07	3.2	3.8	2.38	1.93
DEVELOPING	**320 901**	**450 121**	**4.18**	**3.16**	**507 788**	**596 529**	**1.69**	**1.60**	**0.6**	**0.8**	**2.49**	**1.56**
AFRICA	**32 720**	**43 225**	**3.04**	**2.70**	**161 965**	**182 841**	**1.50**	**1.28**	**0.2**	**0.2**	**1.55**	**1.42**
NORTH AFRICA	**11 664**	**15 046**	**2.71**	**2.47**	**30 980**	**32 210**	**1.42**	**0.50**	**0.4**	**0.5**	**1.28**	**1.97**
Algeria	2 250	2 807	5.32	2.50	9 040	9 515	1.96	0.87	0.2	0.3	3.36	1.63
Egypt	4 654	6 505	1.93	3.17	6 498	7 123	0.81	0.91	0.7	0.9	1.12	2.26
SUB-SAHARAN AFRICA	**21 056**	**28 180**	**3.23**	**2.82**	**130 985**	**150 631**	**1.52**	**1.46**	**0.2**	**0.2**	**1.72**	**1.36**
LATIN AMERICA and CARIBBEAN	**71 320**	**92 170**	**2.76**	**2.39**	**46 617**	**51 217**	**1.50**	**0.91**	**1.5**	**1.8**	**1.25**	**1.47**
Argentina	9 821	13 826	1.10	3.10	2 149	2 391	–0.31	1.04	4.6	5.8	1.41	2.07
Brazil	27 439	36 300	4.11	2.28	21 747	23 803	2.51	0.61	1.3	1.5	1.60	1.67
Chile	2 533	3 043	2.62	2.06	2 141	2 197	2.03	0.47	1.2	1.4	0.59	1.59
Mexico	10 580	12 127	1.55	1.15	2 341	2 504	0.96	0.51	4.5	4.8	0.59	0.64
Uruguay	1 658	1 896	1.80	1.72	948	930	1.57	0.13	1.7	2.0	0.23	1.58
ASIA and PACIFIC	**216 860**	**314 726**	**4.86**	**3.46**	**299 207**	**362 471**	**1.83**	**1.86**	**0.7**	**0.9**	**3.04**	**1.60**
Bangladesh	3 035	5 003	4.69	4.52	30 815	46 875	5.36	4.02	0.1	0.1	–0.67	0.50
China[d]	36 852	56 974	14.18	4.75	13 558	17 152	10.41	2.56	2.7	3.3	3.76	2.19
India	105 828	147 949	3.59	3.06	107 617	134 150	2.13	2.09	1.0	1.1	1.45	0.97
Indonesia	995	1 326	3.27	3.17	10 626	11 916	3.34	1.44	0.1	0.1	–0.07	1.73
Iran, Islamic Republic of	7 711	9 442	3.53	1.92	35 194	34 512	0.06	–0.16	0.2	0.3	3.47	2.08
Korea	2 162	2 323	–1.20	0.77	255	262	–2.84	0.42	8.5	8.9	1.64	0.35
Malaysia	48	65	3.02	3.37	100	113	1.61	1.61	0.5	0.6	1.41	1.76
Pakistan	35 264	58 599	4.16	4.13	29 178	39 509	3.02	2.65	1.2	1.5	1.14	1.48
Saudi Arabia	1 267	1 582	3.52	1.92	4 091	4 059	0.67	–0.07	0.3	0.4	2.84	1.99
Turkey	12 196	16 465	3.65	2.75	17 140	19 696	–4.52	1.22	0.7	0.8	8.17	1.54
LEAST DEVELOPED COUNTRIES (LDC)	**23 942**	**33 199**	**3.29**	**3.06**	**165 474**	**200 904**	**1.93**	**1.94**	**0.1**	**0.2**	**1.36**	**1.11**
OECD	**307 816**	**333 786**	**0.67**	**0.85**	**63 941**	**64 071**	**–1.76**	**0.09**	**4.8**	**5.2**	**2.43**	**0.76**
NON-OECD	**369 636**	**514 716**	**3.59**	**3.14**	**526 273**	**616 484**	**1.81**	**1.60**	**0.7**	**0.8**	**1.78**	**1.55**

est.: Estimate.

a) Year ending 30 June for Australia and 31 May for New Zealand.

b) Least-squares growth rate (see glossary).

c) Includes other developed Asian countries: Kazakhstan, Kyrgyzstan, Tajikistan, Turkmenistan, Uzbekistan, Armenia, Azerbaijan and Georgia.

d) Refers to mainland only. The economies of Chinese Taipei, Hong Kong (China) and Macau (China) are included in the other Asia Pacific aggregate.

Source: OECD and FAO Secretariats.

StatLink http://dx.doi.org/10.1787/888932287729

Table A.31. Whey powder and casein projections

Calendar year[a]		Average 2007-09 est.	2019	Growth[b] (%) 2000-09	Growth[b] (%) 2010-19
AUSTRALIA					
Net trade, whey	*kt pw*	81.2	89.0	9.72	0.50
Exports, casein	*kt pw*	9.4	7.9	-6.97	1.73
CANADA					
Net trade, whey	*kt pw*	7.6	15.0	4.30	–0.23
EU27					
Net trade, whey	*kt pw*	334.2	367.5	6.53	0.97
Casein, EU15					
Production	*kt pw*	140.3	165.8	–2.75	1.48
Consumption	*kt pw*	116.6	140.1	–4.52	1.69
Net trade	*kt pw*	71.2	30.7	-7.87	6.30
JAPAN					
Net trade, whey	*kt pw*	–52.6	–77.7	0.44	4.72
KOREA					
Net trade, whey	*kt pw*	–37.0	–40.6	–0.37	0.85
MEXICO					
Net trade, whey	*kt pw*	–84.0	–84.9	5.50	–0.07
NEW ZEALAND					
Net trade, whey	*kt pw*	8.4	18.4	24.96	10.22
Exports, casein	*kt pw*	157.6	197.7	4.46	1.45
UNITED STATES					
Whey					
Production	*kt pw*	494.6	497.2	–0.41	0.47
Consumption	*kt pw*	280.9	228.7	–3.64	–2.88
Exports	*kt pw*	211.7	268.4	4.23	4.37
Imports, casein	*kt pw*	68.8	66.9	–0.24	–0.47
ARGENTINA					
Net trade, whey	*kt pw*	28.4	64.2	43.65	4.88
BRAZIL					
Net trade, whey	*kt pw*	–29.9	–41.8	–0.58	3.47
CHINA[c]					
Net trade, whey	*kt pw*	–222.1	–404.8	7.93	5.10
RUSSIAN FEDERATION					
Net trade, whey	*kt pw*	–50.4	–65.2	25.45	2.30

est.: Estimate.

a) Year ending 30 June for Australia and 31 May for New Zealand.

b) Least-squares growth rate (see glossary).

c) Refers to mainland only. The economies of Chinese Taipei, Hong Kong (China) and Macau (China) are included in the other Asia Pacific aggregate.

Source: OECD and FAO Secretariats.

StatLink http://dx.doi.org/10.1787/888932287748

Table A.32. **Main policy assumptions for biofuels markets**

		08/09 est.	09/10	10/11	11/12	12/13	13/14	14/15	15/16	16/17	17/18	18/19	19/20
BRAZIL													
Ethanol													
Import tariffs	%	20.00	20.00	0.00	0.00	0.00	0.00	0.00	0.00	0.00	0.00	0.00	0.00
Incorporation mandate[a]	%	18.26	18.26	18.26	18.26	18.26	18.26	18.26	18.26	18.26	18.26	18.26	18.26
Biodiesel													
Tax concessions[b]	*BRL/hl*	7	7	7	7	7	7	7	7	7	7	7	7
Import tariffs	%	4.60	4.60	4.60	4.60	4.60	4.60	4.60	4.60	4.60	4.60	4.60	4.60
CANADA													
Ethanol													
Tax concessions[b]	*CAD/hl*	17	7	7	6	6	6	6	6	6	6	6	2
Import tariffs	*CAD/hl*	5.00	5.00	5.00	5.00	5.00	5.00	5.00	5.00	5.00	5.00	5.00	5.00
Incorporation mandate[a]	%	1.34	2.89	3.41	3.41	3.41	3.41	3.41	3.41	3.41	3.41	3.41	3.41
Direct support													
Ontario	*CAD/hl*	21.00	21.00	21.00	19.00	18.00	17.00	16.00	15.00	15.00	0.00	0.00	0.00
Prairies	*CAD/hl*	18.94	18.94	18.94	8.00	7.00	6.00	5.00	4.00	4.00	0.00	0.00	0.00
Biodiesel													
Tax concessions[b]	*CAD/hl*	8.00	8.00	8.00	8.00	8.00	8.00	8.00	8.00	8.00	8.00	8.00	8.00
Incorporation mandate[a]	%	0.00	0.40	0.63	1.01	1.61	1.61	1.61	1.61	1.61	1.61	1.61	1.61
Direct support	*CAD/hl*	21.81	21.81	21.81	16.00	14.00	12.00	10.00	8.00	6.00	0.00	0.00	0.00
COLOMBIA													
Ethanol													
Import tariffs	%	9.65	9.65	9.65	9.65	9.65	9.65	9.65	9.65	9.65	9.65	9.65	9.65
Blending target[c, e]	%	5.10	6.80	8.00	9.00	10.00	10.00	10.00	10.00	10.00	10.00	10.00	10.00
Biodiesel													
Blending target[e]	%	0.00	2.00	2.00	2.00	5.00	5.00	5.00	5.00	5.00	5.00	5.00	5.00
EUROPEAN UNION													
Ethanol													
Tax concessions[b]	*EUR/hl*	29.81	29.81	29.81	29.81	29.81	29.81	29.81	29.81	29.81	29.81	29.81	29.81
Import tariffs	*EUR/hl*	19.20	19.20	19.20	19.20	19.20	19.20	19.20	19.20	19.20	19.20	19.20	19.20
Incorporation mandate[a, d, f]	%	1.13	2.17	2.58	3.15	3.71	4.27	4.85	5.45	5.97	6.48	6.90	7.06
Biodiesel													
Tax concessions[b]	*EUR/hl*	34.14	34.14	34.14	34.14	34.14	34.14	34.14	34.14	34.14	34.14	34.14	34.14
Import tariffs	%	6.50	6.50	6.50	6.50	6.50	6.50	6.50	6.50	6.50	6.50	6.50	6.50
Incorporation mandate[a, d, f]	%	1.84	1.98	2.44	3.07	3.71	4.34	4.97	5.60	6.15	6.69	7.15	7.43
INDONESIA													
Ethanol													
Import tariffs	%	26.31	26.31	26.31	26.31	26.31	26.31	26.31	26.31	26.31	26.31	26.31	26.31
Blending target[e]	%	0.00	0.00	0.00	0.00	0.00	0.00	0.00	0.00	0.00	0.00	0.00	0.00
Biodiesel													
Blending target[e]	%	0.00	1.00	1.00	3.00	3.00	3.00	3.00	3.00	7.00	7.00	7.00	7.00
MALAYSIA													
Ethanol													
Import tariffs	%	0.00	0.00	0.00	0.00	0.00	0.00	0.00	0.00	0.00	0.00	0.00	0.00
Blending target[e]	%	0.00	0.00	0.00	0.00	0.00	0.00	0.00	0.00	0.00	0.00	0.00	0.00
Biodiesel													
Blending target[e]	%	0.00	2.00	2.00	2.00	5.00	5.00	5.00	5.00	5.00	5.00	5.00	5.00
PERU													
Ethanol													
Import tariffs	%	0.00	0.00	0.00	0.00	0.00	0.00	0.00	0.00	0.00	0.00	0.00	0.00
Blending target[e]	%	0.00	0.00	7.80	7.80	7.80	7.80	7.80	7.80	7.80	7.80	7.80	7.80
Biodiesel													
Import tariffs	%	0.00	0.00	0.00	0.00	0.00	0.00	0.00	0.00	0.00	0.00	0.00	0.00
Blending target[e]	%	0.00	2.00	2.00	2.00	5.00	5.00	5.00	5.00	5.00	5.00	5.00	5.00

For notes, see end of the table.

Table A.32. **Main policy assumptions for biofuels markets** *(cont.)*

		08/09 est.	09/10	10/11	11/12	12/13	13/14	14/15	15/16	16/17	17/18	18/19	19/20
THAILAND													
Ethanol													
Import tariffs	%	0.00	0.00	0.00	0.00	0.00	0.00	0.00	0.00	0.00	0.00	0.00	0.00
Blending target[e]	%	5.20	5.20	5.20	5.20	5.20	5.20	5.20	5.20	5.20	5.20	5.20	5.20
Biodiesel													
Blending target[e]	%	0.00	2.00	2.00	3.00	5.00	5.00	5.00	5.00	5.00	5.00	5.00	5.00
UNITED STATES													
Renewable Fuel Standard	*mnl*												
Total		34 069	42 018	49 021	52 806	57 538	62 648	68 705	77 601	84 225	90 850	98 420	105 991
of which:													
Maize based ethanol		34 069	39 747	45 425	47 696	49 967	52 239	54 510	56 781	56 781	56 781	56 781	56 781
Cellulosic ethanol		0	0	379	946	1 893	3 785	6 624	11 356	16 088	20 820	26 498	32 176
Biodiesel		0	1 893	2 461	3 028	3 785	3 785	3 785	3 785	3 785	3 785	3 785	3 785
Ethanol													
Import tariffs	*USD/hl*	14.27	14.27	14.27	14.27	14.27	14.27	14.27	14.27	14.27	14.27	14.27	14.27
Import tariffs	%	2.50	2.50	2.50	2.50	2.50	2.50	2.50	2.50	2.50	2.50	2.50	2.50
Blenders tax credit	*USD/hl*	13.47	13.47	11.89	11.89	11.89	11.89	11.89	11.89	11.89	11.89	11.89	11.89
Biodiesel													
Import tariffs	%	4.60	4.60	4.60	4.60	4.60	4.60	4.60	4.60	4.60	4.60	4.60	4.60
Blenders tax credit	*USD/hl*	26.42	26.42	26.42	26.42	26.42	26.42	26.42	26.42	26.42	26.42	26.42	26.42

est.: Estimate.

a) Share in respective fuel type, energy equivalent.

b) Difference between tax rates applying to fossil and biogen fuels.

c) Applies to cities with more than 500 000 inhabitants.

d) Note that for many countries, shares for ethanol and biodiesel are not specified individually in the legislation.

e) Expressed in volume share.

f) Legislated incorporation mandates differ across EU member states. Assumptions for average mandated shares are based on national data received from a limited number of individual countries. Additionally, the EU wide mandate in the context of the Renewable Energy Directive has been taken into account.

Source: OECD and FAO Secretariats.

StatLink http://dx.doi.org/10.1787/888932287767

Table A.33. **Biofuels projections: Ethanol**

	PRODUCTION (MNL)		Growth[a] (%)	DOMESTIC USE (MNL)		Growth[a] (%)	FUEL USE (MNL)		Growth[a] (%)	SHARE IN GAZOLINE TYPE FUEL USE (%)				NET TRADE[b] (MNL)	
										Energy shares		Volume shares			
	Average 2007–09 est.	2019	2010-19	Average 2007–09 est.	2019	2010-19	Average 2007–09 est.	2019	2010-19	Average 2007–09 est.	2019	Average 2007–09 est.	2019	Average 2007–09 est.	2019
NORTH AMERICA															
Canada	1 018	1 891	3.16	1 603	2 609	2.17	1 397	2 403	2.39	2.3	3.4	3.4	5.0	–585	–718
United States	34 888	67 919	4.65	36 919	78 797	5.53	35 273	77 065	5.68	4.3	8.4	6.3	12.1	–2 031	–10 878
WESTERN EUROPE															
EU27	4 890	17 987	11.28	6 336	21 223	10.81	3 907	18 716	13.39	1.8	8.5	2.7	12.2	–1 446	–3 237
OCEANIA DEVELOPED															
Australia	165	409	2.84	165	409	2.84	165	409	2.84	0.6	1.3	0.8	1.9	0	0
OTHER DEVELOPED															
Japan	107	618	18.17	604	1 128	5.60	9	518	36.46	n.a.	n.a.	n.a.	n.a.	–485	–510
South Africa	16	17	0.95	16	17	1.09	0	0	3.31	0.0	0.0	0.0	0.0	0	0
SUB-SAHARAN AFRICA															
Mozambique	21	42	4.93	21	29	3.35	0	9	75.27	0.0	3.4	0.0	5.0	0	13
Tanzania	28	87	13.76	32	56	6.87	0	24	41.52	0.1	3.4	0.1	5.0	–4	31
LATIN AMERICA and CARIBBEAN															
Argentina	319	571	3.02	173	509	4.84	31	268	5.53	0.4	3.4	0.7	5.0	146	62
Brazil	25 308	55 020	7.44	21 182	41 681	6.30	19 747	39 441	6.48	46.1	70.2	56.0	77.9	4 127	13 339
Colombia	343	879	7.13	349	494	1.55	285	429	1.80	4.0	6.9	5.9	10.0	–6	384
Mexico	63	90	2.88	135	270	2.88	0	0	–	0.0	0.0	0.0	0.0	–73	–179
Peru	18	194	14.31	10	124	12.87	0	116	22.88	0.0	5.4	0.0	7.8	0	0
ASIA and PACIFIC															
China	3 917	5 999	3.55	3 725	4 809	3.03	1 970	3 873	7.01	1.8	2.1	2.7	3.1	192	1 190
India	1 949	2 803	4.50	2 026	2 852	3.80	171	895	17.37	0.9	3.4	1.3	5.0	–77	–49
Indonesia	215	648	7.07	162	396	4.95	0	241	57.47	0.0	0.7	0.0	1.0	53	252
Malaysia	65	70	0.61	90	85	0.06	0	0	3.63	0.0	0.0	0.0	0.0	–25	–14
Philippines	132	927	15.21	217	890	7.43	119	793	8.96	1.3	6.9	1.9	10.0	–85	37
Thailand	593	2 207	11.05	510	1 965	11.28	307	1 750	14.06	2.8	14.3	4.1	20.0	84	242
Turkey	54	67	0.32	92	119	2.56	47	72	3.68	0.6	0.7	0.8	1.0	–37	–53
Viet Nam	150	405	10.59	135	387	11.89	0	250	112.78	0.0	3.4	0.0	5.0	15	17
TOTAL	**74 257**	**158 849**	**6.26**	**74 497**	**158 849**	**6.16**	**64 022**	**147 879**	**6.75**	**4.6**	**9.0**	**6.6**	**12.9**	**4 465**	**14 591**

est.: Estimate.

n.a.: Not available.

a) Least-squares growth rate (see glossary).

b) For total net trade exports are shown.

Source: OECD and FAO Secretariats.

StatLink http://dx.doi.org/10.1787/888932287786

Table A.34. **Biofuels projections: Biodiesel**

	PRODUCTION (MNL)		Growth[a] (%)	DOMESTIC USE (MNL)		Growth[a] (%)	SHARE IN DIESEL TYPE FUEL USE (%)				NET TRADE[b] (MNL)	
							Energy shares		Volume shares			
	Average 2007-09 est.	2019	2010-19	Average 2007-09 est.	2019	2010-19	Average 2007-09 est.	2019	Average 2007-09 est.	2019	Average 2007-09 est.	2019
NORTH AMERICA												
Canada	138	457	5.02	137	765	11.48	0.3	1.6	0.4	2.0	0	–307
United States	2 319	3 818	5.27	1 286	3 837	6.80	0.4	1.1	0.5	1.3	1 033	–19
WESTERN EUROPE												
EU27	8 041	20 521	7.68	8 971	24 362	7.97	3.4	8.0	4.2	9.8	–930	–3 841
OCEANIA DEVELOPED												
Australia	515	711	1.17	515	711	1.17	2.3	2.7	2.9	3.3	0	0
OTHER DEVELOPED												
South Africa	48	50	0.96	48	62	3.32	0.0	0.0	0.0	0.0	0	–12
SUB-SAHARAN AFRICA												
Mozambique	45	67	6.10	0	23	82.95	0.0	4.0	0.0	5.0	45	44
Tanzania	44	35	0.78	0	21	134.03	0.0	8.2	0.0	10.0	44	14
LATIN AMERICA and CARIBBEAN												
Argentina	1 286	3 860	6.43	70	903	9.71	0.5	6.0	0.7	7.4	1 216	2 957
Brazil	958	3 057	4.88	958	3 057	4.88	1.7	4.0	2.1	5.0	0	0
Colombia	143	876	9.56	137	438	3.22	1.1	4.0	1.3	5.0	5	437
Peru	140	344	10.06	140	185	3.94	1.1	4.0	1.3	5.0	0	159
ASIA and PACIFIC												
India	130	3 035	34.50	244	3 176	32.05	0.0	6.7	0.0	8.2	–114	–141
Indonesia	102	1 148	18.39	15	1 117	29.67	0.5	5.7	0.7	7.0	87	31
Malaysia	515	972	4.60	50	400	12.60	1.1	4.0	1.3	5.0	465	572
Philippines	102	305	5.91	102	246	3.73	1.1	1.6	1.3	2.0	0	60
Thailand	451	1 585	9.41	424	1 532	8.87	1.1	4.0	1.3	5.0	27	52
Turkey	188	26	0.10	188	36	3.69	0.0	0.0	0.0	0.0	0	–11
Viet Nam	6	306	39.35	0	299	110.93	0.0	4.0	0.0	5.0	6	7
TOTAL	**15 170**	**41 171**	**7.33**	**13 286**	**41 171**	**7.55**	**1.6**	**4.2**	**2.0**	**5.2**	**2 250**	**2 957**

est.: Estimate.

a) Least-squares growth rate (see glossary).

b) Exports for total net trade.

Source: OECD and FAO Secretariats.

StatLink http://dx.doi.org/10.1787/888932287805

ANNEX B

Information on Food Price Changes

Table B.1. **Information on food price changes**

	Total CPI % change		Food CPI % change[a]		Expenditure share of food		Food contribution to total change in CPI[b]	
	2008	2009	2008	2009	2008	2009	2008	2009
OECD								
Australia	4.35	1.82	4.32	3.37	12.8	12.8	0.55	0.43
Austria	3.22	0.51	6.30	0.21	12.2	12.2	0.77	0.03
Belgium	4.49	–0.05	6.07	0.85	17.4	17.4	1.06	0.15
Canada	2.37	0.30	3.95	5.49	11.5	11.5	0.45	0.63
Czech Republic	6.34	1.03	8.12	–3.95	16.3	16.3	1.32	–0.64
Denmark	3.40	1.33	7.57	–0.09	12.5	12.5	0.95	–0.01
Finland	4.06	0.01	8.60	2.02	13.3	13.3	1.14	0.27
France	2.81	0.09	5.07	0.08	14.8	14.8	0.75	0.01
Germany	2.60	0.38	6.07	–1.31	10.4	10.4	0.63	–0.14
Greece	4.15	1.21	5.36	1.91	17.8	17.8	0.96	0.34
Hungary	6.04	4.21	10.40	3.89	18.7	18.7	1.94	0.73
Iceland	12.65	12.04	15.99	17.53	12.1	12.1	1.94	2.12
Ireland	4.05	–4.48	6.54	–3.46	11.7	11.7	0.77	–0.41
Italy	3.35	0.77	5.40	1.83	16.8	16.8	0.91	0.31
Japan	1.38	–1.35	2.96	0.04	19.0	19.0	0.56	0.01
Korea	4.67	2.76	4.97	7.56	14.4	14.4	0.72	1.09
Luxembourg	3.40	0.37	5.37	1.43	13.8	13.8	0.74	0.20
Mexico	5.12	5.30	8.07	8.84	19.1	19.1	1.54	1.69
Netherlands	2.49	1.19	5.63	1.08	11.1	11.1	0.62	0.12
New Zealand	3.96	2.12	8.43	6.07	14.5	14.5	1.22	0.88
Norway	3.77	2.17	4.21	4.17	11.2	11.4	0.47	0.48
Poland	4.16	3.80	5.64	4.66	25.7	25.7	1.45	1.20
Portugal	2.57	–0.83	3.73	–3.44	19.1	19.1	0.71	–0.66
Slovak Republic	4.60	1.62	7.67	–3.23	16.1	16.1	1.23	–0.52
Spain	4.08	–0.29	5.88	–1.05	20.2	20.2	1.19	–0.21
Sweden	3.44	–0.28	6.91	2.89	13.1	13.1	0.91	0.38
Switzerland	2.43	–0.48	3.10	–0.19	11.1	10.9	0.34	–0.02
Turkey	10.44	6.25	12.79	8.02	28.6	28.6	3.66	2.29
United Kingdom	3.61	2.17	9.10	5.42	10.9	11.8	0.99	0.64
United States	3.84	–0.36	6.40	0.47	8.2	8.2	0.52	0.04

For notes, see end of the table.

Table B.1. **Information on food price changes** *(cont.)*

	Total CPI % change		Food CPI % change[a]		Expenditure share of food		Food contribution to total change in CPI[b]	
	2008	2009	2008	2009	2008	2009	2008	2009
Non-OECD								
Brazil[c]	5.68	4.89	13.05	5.79	27.5	27.5	3.59	1.59
Chile	8.72	0.35	17.18	4.38	17.8	17.8	3.06	0.78
China[c]	5.93	–0.95	14.43	0.69	40.3	40.3	5.82	0.28
Estonia	10.36	–0.08	14.19	–4.02	21.7	21.7	3.08	–0.87
Indonesia	10.23	4.39	16.96	6.98	19.6	19.6	3.32	1.37
Israel	4.59	3.33	12.30	1.05	14.3	14.3	1.76	0.15
Russian Federation[c]	14.11	11.65	20.94	11.94	32.8	32.8	6.87	3.92
Slovenia	5.65	0.86	10.06	0.58	17.4	17.4	1.75	0.10
South Africa	10.07	7.23	15.53	9.37	18.3	18.3	2.84	1.71
Senegal	5.76	–1.00	9.60	–2.82	40.3	40.3	3.87	–1.14
Ghana	16.37	19.57	15.01	16.00	44.9	44.9	6.74	7.19
Malawi	8.72	8.41	6.96	7.34	58.1	58.1	4.04	4.26
South Africa	10.07	7.24	15.50	9.58	18.3	18.3	2.83	1.75
Tanzania	10.30	12.10	12.68	17.36	55.9	55.9	7.09	9.70
Niger	11.25	4.53	20.36	11.17	43.6	43.6	8.87	4.86
Rwanda	15.44	10.21	16.35	15.36	35.4	35.4	5.79	5.44

a) CPI food: Food + non alcoholic beverages.
b) Contribution is column 2 × 3/100.
c) Includes restaurant meals.
Sources:
Food CPI
OECD members and Brazil, Chile, China, Estonia, Indonesia, Israel, the Russian Federation, Slovenia and South Africa (*www.oecd.org/document/54/0,3343,en_2649_33715_15569334_1_1_1_1,00.html*).
Developing countries:
Guatemala: Istituto Nacional de Estadistica (*www.ine.gob.gt/index.php/economia*).
Malawi: National Statistical Office of Malawi (*www.nso.malawi.net*); ILO Statistics (Laborsta) (*http://laborsta.ilo.org*).
Pakistan: ILO Statistics (Laborsta) (*http://laborsta.ilo.org*).
Peru: Instituto Nacional de Estadistica e Informatica (*www1.inei.gob.pe*).
Sri Lanka: Department of Census and Statistics (*www.statistics.gov.lk/price/index.htm*).
Senegal: Agence nationale de la statistique et de la démographie (*www.ansd.sn*).
Burkina Faso: Institut national de la statistique et de la démographie (*www.insd.bf*).
Niger: Institut national de la statistique (*www.ins.ne/statistique/download.php?lng=fr*, *www.stat-niger.org/statistique*).
Rwanda: National Institute of Statistics (*www.statistics.gov/rw*).
Ghana: Ghana Statistical Service (*www.statsghana.gov.gh/docfiles/CPI%20Release_pdf/consumer_price_index_numbers.pdf*).
Tanzania: National Bureau of Statistics (*www.nbs.go.tz*).
Share of food in CPI
OECD countries: OECD Statistics Directorate Secretariat.
Indonesia, Israel, Slovenia, South Africa, Chile, Estonia: OECD *Main Economic Indicators*, food price index, meta data, by country.
Brazil: Fundacao Getulio Vargas (*www.fgvdados.fgv.br*).
Chile: Istituto Nacional de Estadisticas, (*www.ine.cl/canales/chile_estadistico/estadisticas_precios/ipc/nuevo_ipc/metodologia_eng/Consumer%20Price%20Index%20-%20ingles.pdf?lang=eng*).
China: National Bureau of Statistics (*www.stats.gov.cn/english/statisticaldata/index.htm*).
The Russian Federation: Federal State Statistics Service Russian Federation (*www.gks.ru/eng*), developing countries from national statistical offices as indicated above.

StatLink http://dx.doi.org/10.1787/888932287824

Glossary of Terms

A-H1N1

This is an influenza virus that had never been identified as a cause of infections in people before the current H1N1 pandemic. Genetic analyses of this virus have shown that it originated from animal influenza viruses and is unrelated to the human seasonal H1N1 viruses that have been in general circulation among people since 1977.

Average Crop Revenue Election (ACRE) programme

A new programme introduced with the 2008 US FCE Act allowing farmers to chose revenue-based protection against yield and market fluctuations.

AMAD

Agricultural Market Access Database. A co-operative effort between Agriculture and Agri-food Canada, EU Commission-Agriculture Directorate-General, FAO, OECD, the World Bank, UNCTAD and the United States Department of Agriculture, Economic Research Service. Data in the database is obtained from countries' schedules and notifications submitted to the WTO.

ASEAN

The Association of Southeast Asian Nations, or ASEAN, was established on 8 August 1967 in Bangkok, Thailand, with the signing of the ASEAN Declaration (Bangkok Declaration) by the Founding Fathers of ASEAN, namely Indonesia, Malaysia, Philippines, Singapore and Thailand. Brunei Darussalam then joined on 8 January 1984, Viet Nam on 28 July 1995, Lao PDR and Myanmar on 23 July 1997, and Cambodia on 30 April 1999, making up what is today the ten member states of ASEAN.

Australia-US Free Trade Agreement (AUSFTA)

A Bilateral Agreement negotiated between the United States and Australia that came into force on 1 January 2005. AUSFTA covers goods, services, investment, financial services, government procurement, standards and technical regulations, telecommunications, competition-related matters, electronic commerce, intellectual property rights, labour and the environment.

Avian influenza

Avian influenza is an infectious disease of birds caused by type A strains of the influenza virus. The disease, which was first identified in Italy more than 100 years ago, occurs worldwide. The quarantining of infected farms, destruction of infected or potentially exposed flocks, and recently inoculation are standard control measures.

Atlantic beef/pigmeat market

Beef/pigmeat trade between countries of the Atlantic Rim.

Baseline

The set of market projections used for the Outlook analysis in this report and as a benchmark for the analysis of the impact of different economic and policy scenarios. A detailed description of the generation of the baseline is provided in the chapter on Methodology in this report.

Biofuels

In the wider sense defined as all solid, fluid or gaseous fuels produced from biomass. More narrowly, the term biofuels comprises those that replace petroleum-based road-transport fuels, *i.e.* bioethanol produced from sugar crops, cereals and other starchy crops that can be used as an additive to, in a blend with or as a replacement of gasoline, and biodiesel produced mostly from vegetable oils, but also from waste oils and animal fats, that can be used in blends with or as a replacement of petroleum-based diesel.

Biomass

Biomass is defined as any plant matter used directly as fuel or converted into other forms before combustion. Included are wood, vegetal waste (including wood waste and crops used for energy production), animal materials/wastes and industrial and urban wastes, used as feedstocks for producing bioproducts.

BRIICs

Refers to the emerging economies of Brazil, the Russian Federation, India, Indonesia and China.

Cereals

Defined as wheat, coarse grains and rice.

CAFTA

CAFTA is a comprehensive trade agreement between Costa Rica, the Dominican Republic, El Salvador, Guatemala, Honduras, Nicaragua, and the United States.

Common Agricultural Policy (CAP)

The European Union's agricultural policy, first defined in Article 39 of the Treaty of Rome signed in 1957.

Coarse grains

Defined as barley, maize, oats, sorghum and other coarse grains in all countries except Australia, where it includes triticale and in the European Union where it includes rye and other mixed grains.

Country of Origin Labelling (COOL)

A provision of the 2008 US Farm Act that requires retailers to inform consumers of the country of origin of different commodities, among them meats.

Conservation Reserve Program (CRP)

A major provision of the United States' Food Security Act of 1985 and extended under the Food and Agriculture Conservation and Trade Act of 1990, the Food and Agriculture Improvement and Reform Act of 1996, and the Farm Security and Rural Investment Act of 2002 is designed to reduce erosion on 40 to 45 million acres (16 to 18 million hectares) of farm land. Under the programme, producers who sign contracts agree to convert erodable crop land to approved conservation uses for ten years. Participating producers receive annual rental payments and cash or payment in kind to share up to 50% of the cost of establishing permanent vegetative cover. The CRP is part of the *Environmental Conservation Acreage Reserve Program*. The 1996 FAIR Act authorised a 36.4 million acre (14.7 million hectares) maximum under CRP, its 1995 level. The maximum area enrolled in the CRP was increased to 39.2 million acres in the 2002 FSRI Act.

Commonwealth of Independent States (CIS)

The heads of twelve sovereign states (except the Baltic states) have signed the Treaty on establishment of the Economic Union, in which they stressed that the Azerbaijan Republic, Republic of Armenia, Republic of Belarus, Republic of Georgia, Republic of Kazakhstan, Kyrgyz Republic, Republic of Moldova, the Russian Federation, Republic of Tajikistan, Turkmenistan, Republic of Uzbekistan and Ukraine on equality basis established the Commonwealth of Independent States.

Common Market Organisation (CMO) for sugar

The common organisation of the sugar market (CMO) in the European Union was established in 1968 to ensure a fair income to community sugar producers and self-supply of the Community market. At present the CMO is governed by Council Regulation (EC) No. 318/2006 (the basic regulation) which establishes a restructuring fund financed by sugar producers to assist the restructuring process needed to render the industry more competitive.

Crop year, coarse grains

Refers to the crop marketing year beginning 1 April for Japan, 1 July for the European Union and New Zealand, 1 August for Canada and 1 October for Australia. The US crop year begins 1 June for barley and oats and 1 September for maize and sorghum.

Crop year, oilseeds

Refers to the crop marketing year beginning 1 April for Japan, 1 July for the European Union and New Zealand, 1 August for Canada and 1 October for Australia. The US crop year begins 1 June for rapeseed, 1 September for soybeans and for sunflower seed.

Crop year, rice

Refers to the crop marketing year beginning 1 April for Japan, Australia, 1 August for the United States, 1 September for the European Union, 1 November for Korea and 1 January for other countries.

Crop year, sugar

A common crop marketing year beginning 1 October and extending to 31 September, used by FO Licht.

Crop year, wheat

Refers to the crop marketing year beginning 1 April for Japan, 1 June for the United States, 1 July for the European Union and New Zealand, 1 August for Canada and 1 October for Australia.

Decoupled payments

Budgetary payments paid to eligible recipients who are not linked to current production of specific commodities or livestock numbers or the use of specific factors of production.

Direct payments

Payments made directly by governments to producers.

Doha Development Agenda

The current round of multilateral trade negotiations in the World Trade Organization that were initiated in November 2001, in Doha, Qatar.

Domestic support

Refers to the annual level of support, expressed in monetary terms, provided to agricultural production. It is one of the three pillars of the *Uruguay Round Agreement on Agriculture* targeted for reduction.

Economic Partnership Agreements (EPAs)

Free trade agreementss currently being negotiated between the EU and the African, Caribbean Pacific (ACP) group of developing countries to replace the Cotonou Agreement which expired in 2007.

Energy Independence and Security Act (EISA), 2007

US legislation passed in December 2007 that is designed to increase US energy security by lessening dependence on imported oil, to improve energy conservation and efficiency, expand the production of renewable fuels, and to make America's air cleaner for future generations.

Ethanol

A biofuel that can be used as a fuel substitute (hydrous ethanol) or a fuel extender (anhydrous ethanol) in mixes with petroleum, and which is produced from agricultural feed-stocks such as sugarcane and maize.

Everything-But-Arms (EBA) initiative

The Everything-But-Arms (EBA) initiative eliminates EU import tariffs and restrictions for numerous goods, including agricultural products, from the least developed countries. The tariff elimination is scheduled in four steps from 2006-07 to 2009-10.

Export credits (with official support)

Government financial support, direct financing, guarantees, insurance or interest rate support provided to foreign buyers to assist in the financing of the purchase of goods from national exporters.

Export restitutions (refunds)

EU export subsidies provided to cover the difference between internal prices and world market prices for particular commodities.

Export subsidies

Subsidies given to traders to cover the difference between internal market prices and world market prices, such as for example the EU *export restitutions*. Export subsidies are now subject to value and volume restrictions under the *Uruguay Round Agreement on Agriculture*.

FCE Act, 2008

Officially known as the Food, Conservation and Energy Act of 2008. This US farm legislation replaces the FSRI Act of 2002 and covers the period 2008-13.

FSRI Act, 2002

Officially known as the Farm Security and Rural Investment Act of 2002. This US farm legislation replaces the FAIR Act of 1996, covering a wide range of commodity programmes and policies for US agriculture for the period 2002-07.

Gur, khandasari or jaggery

Semi-processed sugars (plantation whites) extracted from sugarcane in India.

Health Check

On November 2008 the EU agriculture ministers reached a political agreement on the Health Check of the Common Agricultural Policy. Among a range of measures, the agreement abolishes arable set-aside, increases milk quotas gradually leading up to their abolition in 2015, and converts market intervention into a genuine safety net. Ministers also greed to increase modulation, whereby direct payments to farmes are reduced and the money transferred to the Rural Development Fund.

High Fructose Corn Syrup (HFCS)

Isoglucose sweetener extractd from maize.

Industrial oilseeds

A category of oilseed production in the European Union for industrial use (*i.e.* biofuels).

Intervention purchases

Purchases by the EC Commission of certain commodities to support internal market prices.

Intervention purchase price

Price at which the European Commission will purchase produce to support internal market prices. It usually is below 100% of the intervention price, which is an annually decided policy price.

Intervention stocks

Stocks held by national intervention agencies in the European Union as a result of *intervention* buying of commodities subject to market price support. Intervention stocks may be released onto the internal markets if internal prices exceed intervention prices; otherwise, they may be sold on the world market with the aid of *export restitutions*.

Inulin

Inulin syrups are extracted from chicory through a process commercially developed in the 1980s. They usually contain 83 per cent fructose. Inulin syrup production in the European Union is covered by the sugar regime and subject to a production quota.

Isoglucose

Isoglucose is a starch-based fructose sweetener, produced by the action of glucose isomerase enzyme on dextrose. This isomerisation process can be used to produce glucose/fructose blends containing up to 42% fructose. Application of a further process can raise the fructose content to 55%. Where the fructose content is 42%, isoglucose is equivalent in sweetness to sugar. Isoglucose production in the European Union is covered by the sugar regime and subject to a production quota.

Least squares growth rate

The **least-squares growth rate**, r, is estimated by fitting a linear regression trend line to the logarithmic annual values of the variable in the relevant period, as follows: Ln(xt) = a + r * t.

Loan rate

The commodity price at which the *Commodity Credit Corporation* (CCC) offers *non-recourse loans* to participating farmers. The crops covered by the programme are used as collateral for these loans. The loan rate serves as a floor price, with the effective level lying somewhat above the announced rate, for participating farmers in the sense that they can default on their loan and forfeit their crop to the CCC rather than sell it in the open market at a lower price.

Market access

Governed by provisions of the *Uruguay Round Agreement* on *Agriculture* which refer to concessions contained in the country schedules with respect to bindings and reductions of tariffs and to other minimum import commitments.

Marketing allotments (US sugar programme)

Marketing allotments designate how much sugar can be sold by sugar millers and processors on the US internal market and were established by the 2002 FSRI Act as a way to guarantee the US sugar loan programme operates at no cost to the Federal Government.

Marketing year, protein meals

Refers to the marketing year beginning 1 October.

Marketing year, vegetable oils

Refers to the marketing year beginning 1 October.

Market Price Support (MPS) Payment

Indicator of the annual monetary value of gross transfers from consumers and taxpayers to agricultural producers arising from policy measures creating a gap between domestic market prices and *border prices* of a specific agricultural commodity, measured at the farm gate level. Conditional on the production of a specific commodity, MPS includes the transfer to producers associated with both production for domestic use and exports, and is measured by the price gap applied to current production. The MPS is net of financial contributions from individual producers through producer levies on sales of the specific commodity or penalties for not respecting regulations such as production quotas (*Price levies*), and in the case of livestock production is net of the market price support on domestically produced coarse grains and oilseeds used as animal feed (*Excess feed cost*).

Methyl Tertiary Butyl Ether (MTBE)

A chemical gasoline additive that can be used to boost the octane number and oxygen content of the fuel, but can render contaminated water undrinkable.

Milk quota scheme

A supply control measure to limit the volume of milk produced or supplied. Quantities up to a specified quota amount benefit from full *market price support*. Over-quota volumes may be penalised by a levy (as in the European Union, where the "super levy" is 115% of the target price) or may receive a lower price. Allocations are usually fixed at individual producer level. Other features, including arrangements for quota reallocation, differ according to scheme.

North American Free Trade Agreement (NAFTA)

A trilateral agreement on trade, including agricultural trade, between Canada, Mexico and the United States, phasing out tariffs and revising other trade rules between the three countries over a 15-year period. The agreement was signed in December 1992 and came into effect on 1 January 1994.

Oilseeds

Defined as rapeseed (canola), soybeans, sunflower seed, peanuts and cotton seeds in all countries, except in Japan where it excludes sunflower seed.

Pacific beef/pigmeat market

Beef/pigmeat trade between countries in the Pacific Rim where foot and mouth disease is not endemic.

Payment-In-Kind (PIK)

A programme used in the US to help dispose of public stocks of commodities. Under PIK, government payments in the form of Commodity Credit Corporation (CCC) – owned commodities are given to farmers in return for additional reductions in harvested acreage.

PROCAMPO

A programme of direct support to farmers in Mexico. It provides for direct payments per hectare on a historical basis.

Producer Support Estimate (PSE)

Indicator of the annual monetary value of gross transfers from consumers and taxpayers to agricultural producers, measured at farm gate level, arising from policy measure, regardless of their nature, objectives or impacts on farm production or income. The PSE measure support arising from policies targeted to agriculture relative to a situation without such policies, *i.e.* when producers are subject only to general policies (including economic, social, environmental and tax policies) of the country. The PSE is a gross notion implying that any costs associated with those policies and incurred by individual producers are not deducted. It is also a nominal assistance notion meaning that increased costs associated with import duties on inputs are not deducted. But it is an indicator net of producer contributions to help finance the policy measure (*e.g.* producer levies) providing a given transfer to producers. The PSE includes implicit and explicit payments. The percentage PSE is the ration of the PSE to the value of total gross farm receipts, measured by the value of total production (at farm gate prices), plus budgetary support. The nomenclature and definitions of this indicator replaced the former Producer Subsidy Equivalent in 1999.

Protein meals

Defined as rapeseed meal (canola), soybean meal, sunflower meal, cotton meal, palm kernel meal, peanut meal and coconut meal in all countries, except in Japan where it excludes sunflower meal.

Purchasing Power Parity (PPP)

Purchasing power parities (PPPs) are the rates of currency conversion that eliminate the differences in price levels between countries. The PPPs are given in national currency units per US dollar.

Non-Recourse loan programme

Programme to be implemented under the US FAIR Act of 1996 for butter, non-fat dry milk and cheese after 1999 in which loans must be repaid with interest to processors to assist them in the management of dairy product inventories.

Renewable Energy Directive (RED)

EU directive legislating binding mandates of 20% for the share of renewable energy in all member states' energy mix by the year 2020, with a specific mandate of 10% for the renewable energy share in transport fuels.

Renewable Fuel Standard (RFS and RFS2)

A standard in the United States for the use of renewable fuel use in the transport sector as required by the Energy Independence and Security Act of 2007 (EISA). RFS2 is a revision of the RFS program for 2010 and beyond.

Saccharin

A low calorie, artificial sweetener used as a substitute for sugar mainly in beverage preparations.

Scenario

A model-generated set of market projections based on alternative assumptions than those used in the baseline. Used to provide quantitative information on the impact of changes in assumptions on the Outlook.

Set-aside programme

European Union programme for cereal, oilseed and protein crops that both requires and allows producers to set-aside a portion of their historical base acreage from current production. Mandatory set-aside rates for commercial producers are set at 10% until 2006.

Single Farm Payment

With the 2003 CAP reform, the EU introduced a farm-based payment largely independent of current production decisions and market developments, but based on the level of former payments received by farmers. To facilitate land transfers, entitlements are calculated by dividing the reference amount of payment by the number of eligible hectares (incl. forage area) in the reference year. Farmers receiving the new SFP are obliged to keep their land in good agricultural and environmental condition and have the flexibility to produce any commodity on their land except fruits, vegetables and table potatoes.

SPS Agreement

WTO Agreement on Sanitary and Phyto-sanitary measures, including standards used to protect human, animal or plant life and health.

Support price

Prices fixed by government policy makers in order to determine, directly or indirectly, domestic market or producer prices. All administered price schemes set a minimum guaranteed support price or a target price for the commodity, which is maintained by associated policy measures, such as quantitative restrictions on production and imports; taxes, levies and tariffs on imports; export subsidies; and public stockholding.

Tariff-rate quota (TRQ)

Resulted from the *Uruguay Round Agreement on Agriculture*. Certain countries agreed to provide minimum import opportunities for products previously protected by non-tariff barriers. This import system established a quota and a two-tier tariff regime for affected commodities. Imports within the quota enter at a lower (in-quota) tariff rate while a higher (out-of-quota) tariff rate is used for imports above the concessionary access level.

Uruguay Round Agreement on Agriculture (URAA)

The terms of the URAA are contained in the section entitled the "Agreement on Agriculture" of the Final Act Embodying the Results of the Uruguay Round of Multilateral Trade Negotiations. This text contains commitments in the areas of *market access*, domestic support, and *export subsidies*, and general provisions concerning monitoring and continuation. In addition, each country's schedule is an integral part of its contractual commitment under the URAA. There is a separate agreement entitled the Agreement on the Application of Sanitary and Phyto-sanitary Measures. This agreement seeks establishing a multilateral framework of rules and disciplines to guide the adoption, development and the enforcement of sanitary and phyto-sanitary measures in order to minimise their negative effects on trade. See also *Phyto-sanitary regulations* and *Sanitary regulations*.

Vegetable oils

Defined as rapeseed oil (canola), soybean oil, sunflower seed oil, coconut oil, cotton oil, palm kernel oil, peanuts oil and palm oil, except in Japan where it excludes sunflower seed oil.

Voluntary Quota Restructuring Scheme

Established as part of the reform of the European Union's Common Market Organisation (CMO) for sugar in February 2006 to apply for four years from 1 July 2006. Under the scheme, sugar producers receive a degressive payment for permanently surrendering sugar production quota, in part or in entirety, over the period 2006-07 to 2009-10.

WTO

World Trade Organization created by the Uruguay Round agreement.

OECD PUBLISHING, 2, rue André-Pascal, 75775 PARIS CEDEX 16
PRINTED IN FRANCE
(51 2010 04 1 P) ISBN 978-92-64-08375-2 – No. 57265 2010